Liburnians and Illyrian Lembs

Liburnians and Illyrian Lembs

Iron Age Ships of the Eastern Adriatic

Luka Boršić, Danijel Džino
and Irena Radić Rossi

ARCHAEOPRESS ARCHAEOLOGY

Archaeopress Publishing Ltd
Summertown Pavilion
18-24 Middle Way
Summertown
Oxford OX2 7LG
www.archaeopress.com

ISBN 978-1-78969-915-9
ISBN 978-1-78969-916-6 (e-Pdf)

© Archaeopress, Luka Boršić, Danijel Džino and Irena Radić Rossi 2021

All rights reserved. No part of this book may be reproduced, or transmitted, in any form or by any means, electronic, mechanical, photocopying or otherwise, without the prior written permission of the copyright owners.
This book is available direct from Archaeopress or from our website www.archaeopress.com

Contents

Abbreviations v
    Primary sources v
    Modern literature vi
Acknowledgements vii
Preface ix

1. Introduction 1
    1.1. Research problems and previous scholarship 1
    1.2. Overview of the book 3
    1.3. Terminology 4

2. Geographical context 6

3. Eastern Adriatic populations in the 1st millennium BC 10
    3.1. The Liburni 11
    3.2. Other Iron Age Eastern Adriatic indigenous seafaring groups 16
    3.3. Greek colonising activities in the eastern Adriatic 18
    3.4. Piracy in the eastern Adriatic? 21
    3.5. Conclusion 24

4. Archaeological and iconographic evidence in protohistoric eastern Adriatic 26
    4.1. Underwater finds 26
        4.1.1. Zambratija near Savudrija 27
        4.1.2. Pula 29
        4.1.3. Caska on the island of Pag 32
        4.1.4. Zaton near Nin 38
    4.2. Iconography 42
        4.2.1. Grieves from the Ilijak burial mound on Glasinac 42
        4.2.2. The images of ships from the Daunian Stellae 44
        4.2.3. Situla from Nesactium 45
        4.2.4. Belt buckle from Prozor 46
        4.2.5. Relief from Varvaria (Bribirska glavica) 49
        4.2.6. South Adriatic coinage 50
    4.3. Protohistoric archaeological and iconographical sources for eastern Adriatic ships ... 53

5. Written Sources on Lembs And Liburnians from the 4th c. BC to Late Antiquity ... 59
    5.1. Introduction 59
    5.2. Lemb 61
        5.2.1. Ancient Greek sources 61
        5.2.2. Latin sources 106
    5.3. Liburnian 139
        5.3.1. Ancient Greek sources 139
        5.3.2. Latin sources 148

6. Discussion 173
    6.1. Lemb 173
    6.2. Liburnian 176
    6.3. Etymology 178
    6.4. Overview of usage of the terms lemb and liburnian in ancient sources from the
        4th century BC until Late Antiquity 192
    6.5. Lemb and liburnian: the same ship? 193
    6.6. Conclusion 194

Bibliography 197
    Ancient authors not listed in Chapter 5 197
    Modern sources 197

List of Figures

Figure 1. Distribution of Iron Age Liburnian hillforts (from Batović 1977). 12
Figure 2. The city walls of Varvaria-Bribirska glavica (photo: D. Džino) 13
Figure 3. Aerial photo of Nedinium-Nadin (photo: M. Grgurić). 14
Figure 4. The helmet from the Cape of Jablanac on the island of Cres (from Blečić 2007b, courtesy of the author). 22
Figure 5. The sewn boat of Zambratija (photo: Ph. Groscaux, from Koncani Uhač et al. 2017, courtesy of I. Koncani Uhač. 28
Figure 6. Drawing of the sewn boat of Zambratija (drawing: V. Dumas, from Boetto et al. 2017, courtesy of I. Koncani Uhač. 29
Figure 7. The sewn ships of Pula (photo: T. Brajković, from Boetto et al. 2017, courtesy of I. Koncani Uhač. 30
Figure 8. Drawing of the sewn ships of Pula (from Boetto et al. 2017, courtesy of I. Koncani Uhač. 31
Figure 9. The sewn boat Caska 1 (photo: L. Damelet). 33
Figure 10. Drawing of the sewn boat Caska 1 (drawing: V. Dumas, from Boetto, Radić Rossi 2017). 34
Figure 11. Remains of the sewn boat Caska 3 (photo: T. Seguin). 35
Figure 12. Drawing of the sewn boat Caska 3 (drawing: P. Poveda, from Boetto, Radić Rossi 2017). 36
Figure 13. The sewn boat Caska 4 (photo: L. Roux). 37
Figure 14. Drawing of the sewn boat Caska 4 (drawing: V. Dumas). 37
Figure 15. The sewn boat Zaton 1 during the course of the 1979 research campaign (photo: Z. Brusić). 39
Figure 16. Drawing of the sewn boat Zaton 1 (drawing: Z. Brusić). 39
Figure 17. The sewn boat Zaton 2 during the course of the 1987 research campaign (photo: Z. Brusić). 40
Figure 18. Drawing of the sewn boat Zaton 2 (drawing: Z. Brusić). 41
Figure 19. The sewn boat Zaton 3 during the course of the 2019 research campaign (photo: D. Romanović). 41
Figure 20. Drawings of the grieves from Glasinac/Ilijak (drawing: S. Čerkez, from Benac, Čović 1957). 43
Figure 21. The present state of the grieves from Glasinac/Ilijak (photo: A. Pravidur, courtesy of Zemaljski Muzej Bosne i Hercegovine, Sarajevo). 43
Figure 22. Drawings of the ships on the grieves from Glasinac/Ilijak (drawing: S. Čerkez, from Benac, Čović 1957). 43
Figure 23. The Novillara Stele (courtesy of L. Braccesi). 44
Figure 24. Reconstruction of the situla of Nesactium (from Mihovilić 1996). 46
Figure 25. The ship image on the situla of Nesactium (from Mihovilić 1996). 47
Figure 26. The belt buckle from Prozor (photo: D. Doračić, courtesy of the Archaeological Museum of Zagreb). 48
Figure 27. The belt buckle from Prozor (drawing: K. Rončević, courtesy of the Archaeological Museum of Zagreb). 48
Figure 28. The relief from Varvaria-Bribirska glavica (photo: Z. A. Alajbeg, courtesy of Museum of Croatian Archaeological Monuments). 49

Figure 29. Coins of the Daorsi with images of ships (from Dragičević 2016). 51
Figure 30. Coins of the Daorsi with images of ships (from Kozličić 1993). 52
Figure 31a-b. Coin of king Gentius, with a representation of a ship (photo: Z. A. Alajbeg, courtesy of the Archaeological Museum of Split). 53
Figure 32. Coins from south-Illyrian mints (from Kozličić 1981). 54
Figure 33. The Moken *kabang* (after J. Ivanoff, M. Bountry, http://www.lampipark.org/wp-content/uploads/2017/01/Moken-Sea-gypsies.pdf last accessed 9/7/2020). 55
Figure 34. A Moken man builds a miniature *kabang* (from Hinshiranan 2001). 55
Figure 35. A representation of a liburna from the 16th-century manuscript of *De rebus bellicis*. 58

List of Tables

Table 1: Lemb in Greek and Roman written sources (L. Boršić) 180
Table 2: Liburnian in Greek and Roman written sources (L. Boršić) 188

List of Maps

Map 1. Geography of the Adriatic (D. Džino using Google Earth). 6
Map 2. Distribution of the most important indigenous ethnonyms in the pre-Roman Adriatic and its hinterlands. In white: the ethnonyms not mentioned in the sources related to the Roman conquest (D. Džino using Google Earth). 10
Map 3. The sites related to the East Adriatic Greeks (D. Džino using Google Earth). 19
Map 4. The sites of shipwrecks (pink), iconographic representation of the ships (yellow), and places where the coins with images of ships were minted (white) (D. Džino using Google Earth). 26

Abbreviations

Primary sources

Aesch. *PV*	Aeschilus, *Prometheus Vinctus*
Alciphr.	Alciphron, *Letters*
Amm. Marc.	Ammianus Marcellinus, *Res Gestae*
Anth. Pal.	*Anthologia Palatina*
Ap. Rhod. *Argon.*	Apollonius Rhodius, *Argonautica*
App. *Ill.*	Appian, *Illyrike*
App. *Mith.*	Appian, *Mithridatic wars*
App. *Pun.*	Appian, *Punica*
App. *B Civ.*	Appian, *Bella Civilia*
Arist. *De motu an.*	Aristotle, *De motu animalium*
Ath.	Athenaeus, *The Learned Banqueters*
Caes. *BCiv.*	Caesar, *Bellum Civile*
Cass. Dio.	Cassius Dio, *Historia Romana*
Cl. Mam.	Claudius Mamertinus, *Gratiarum Actio Juliano Augusto*
Dem. *C. Phorm.*	Demosthenes, *Contra Phormionem*
Dem. *Zenoth*	Demosthenes, *Contra Zenothemin*
Diod. Sic.	Diodorus Siculus, *Bibliotheca Historica*
Eutr.	Eutropius, *Breviarium ab urbe condita*
Festus, *Epitoma*	Sextus Pompeius Festus, *Epitoma operis de verborum significatu Verrii Flacci*
Flor.	Florus, *Epitomae de Tito Livio*
Gell. *NA*	Aulus Gellius, *Noctes Atticae*
Hdt.	Herodotus, *Historiae*
Hor. *Epod.*	Horace, *Epodes*
Isid. *Etym.*	Isidore of Seville, *Etymologiae*
Livy	Livy, *Ab urbe condita*
Livy, *Per.*	Livy, *Periochae Ab urbe condita*
Luc.	Lucan, *Pharsalia*
Lycurg. *Leoc.*	Lycurgus, *Contra Leocrates*
Nic. Dam.	Nicolaus Damascenus
Philo Mech.	Philo Mechanicus, *Parasceuastica et poliorcetica*
Philox.	Philoxenus of Alexandria
Plin. *HN*	Pliny the Elder, *Historia Naturalis*
Plut. *Ant.*	Plutarch, *Antony.*
Plut. *Cat. Min.*	Plutarch, *Cato Minor*
Plut. *Pomp.*	Plutarch, *Pompey*
Polyb.	Polybius, *The Histories*
Prop.	Propertius, *Elegies*
Scymn.	*Scymni Chii periegesis*
Scyl.	Pseudo-Skylax's *periplous*
Sisenna, *Hist.*	Lucius Cornelius Sisenna, *Histories* (Fragments)
Solin.	Solinus, *Collectanea rerum memorabilium*
Steph. Byz.	*Stephani Byzantini Ethnicorum*

Stob. *Flor.*	Stobaeus, *Florilegium* (Ἀνθολόγιον)
Strabo	Strabo, *Geography*.
Veg. *Mil.*	Vegetius, *Epitome Re Militaris*
Verg. *Aen.*	Vergil, *Aeneid*
Verg. *G.*	Vergil, *Georgics*
Vell. Pat.	Velleius Paterculus, *Historiae*

Modern literature

AE	*L'Année épigraphique*. Paris
BE	*Bulletin épigraphique*. Paris
BNJ	*Brill's New Jacoby*, ed. I. Worthington. Brill Online
CIL	Corpus Inscriptionem Latinorum
GodCBI	*Godišnjak Centra za balkanološka ispitivanja*. Sarajevo
HAG	*Hrvatski arheološki godišnjak*. Zagreb
HistAntiq	*Histria Antiqua*. Pula
IG	Inscriptione Graeca
IJNA	*The International Journal for Nautical Archaeology*. London
JAZU/HAZU	Jugoslavenska/Hrvatska Akademija znanosti i umjetnosti
LCL	Loeb Classical Library. Cambridge MA, Harvard University Press. Number in the brackets denotes year of publication
LSJ	*Liddell-Scott-Jones Lexicon of Classical Greek*, eds H. G. Liddell, R. Scott, H. S. Jones. Oxford, Oxford University Press
NP	*Brill New Pauly*. Leiden, Brill
P. Cair. Zen.	Cairo Zenon Papyri
P. Oxy.	*The Oxyrhynchus papyri. Part XVI*, ed. B. P. Grenfell *et al.* London, Egypt Exploration Society 1924
PJZ	*Praistorija jugoslavenskih zemalja*, ed. A. Benac. Sarajevo, Svjetlost 1987
PL	*Patrologia Latina*
RFFZd	*Radovi Filozofskoga fakulteta u Zadru*. Zadar
TTH	Translated Texts for Historians. Liverpool, Liverpool University Press. Number in the brackets denotes year of publication
VAHD/VAPD	*Vjesnik za arheologiju i historiju/povijest dalmatinsku*. Split
VAMZ	*Vjesnik Arheološkoga muzeja u Zagrebu*, series 3, Zagreb.

Acknowledgements

This book is the outcome of research conducted within the framework of the 'Archaeology of Adriatic Shipbuilding and Seafaring Project' (IP-09-2014-8211), which was financially supported by the Croatian Science Foundation. The authors would like to thank their institutions: The Institute for Philosophy in Zagreb (L. Boršić), the Department of History and Archaeology at Macquarie University in Sydney (D. Džino), and the Department of Archaeology at the University of Zadar (I. Radić Rossi) for providing research infrastructure and library support. Our gratitude also goes to Rajka Makjanić from Archaeopress, for supporting the publication of this book from its inception. Editing of the English text was carried out voluntarily by Ewan Coopey, a Macquarie University Ancient History postgraduate, and Gerald Brisch from Archaeopress, who did outstanding work and to whom the authors are most grateful. Finally, we want to thank the individuals and institutions who provided us with access to materials, especially Andrijana Pravidur from the Zemaljski Museum in Sarajevo, Tomislav Bilić from the Archaeological Museum in Zagreb, Ida Koncani Uhač from the Archaeological Museum of Istria in Pula and Maja Bonačić Mandinić from the Archaeological Museum in Split.

Preface

Archaeology of Adriatic Shipbuilding and Seafaring Project focused on the technological development of shipbuilding and seafaring in the eastern Adriatic from prehistory to the modern period, considering various categories of available evidence. Until recently, the maritime activities of the Bronze and Iron Age eastern Adriatic population were attested mainly through the material evidence of seafaring, seaborne trade and foreign cultural influences discovered on land sites. Relatively abundant written sources augmented by scarce iconographic evidence provided more direct information on Iron Age seafaring and the maritime enterprises of the local population. In recent times, some promising archaeological discoveries hinted at the prospect of finding new sites that could help reconstruct the development of eastern Adriatic shipbuilding and seafaring within the broader Adriatic and Mediterranean context.

The three authors of various scholarly backgrounds put together their professional experiences and skills in order to provide an overview of what was done in the past, and to complement the current interpretations by systematic examinations of written, iconographic and archaeological evidence on eastern Adriatic shipbuilding tradition. In order to trace the possible origin of the term lemb, often mentioned in relation to south Adriatic (Illyrian) ships, most of the Greek and Latin sources were consulted and contextualised. This demanding task was fulfilled by Luka Boršić, while Danijel Džino and Irena Radić Rossi conducted presented the state-of-the art research in analysis of on archaeological, historical and iconographical sources.

In regard to the ancient ships known as liburnians (*liburnicae* or *liburnae*), the authors are primarily concerned with their relationship to south Adriatic (or 'Illyrian') lembs, which are generally connected in earlier scholarship to the same 'Illyrian' population. Based on our present knowledge of the historical situation and the ethnical diversity of the protagonists of eastern Adriatic seafaring, such conclusions are discussed and significantly revised. The book does not aim to solve the long-lasting discussion on the origin and role of Illyrian lembs and Liburnian liburnians and their shape, but rather to clarify some fundamental notions on the geographical and historical background of the eastern Adriatic in a light of recent research, and to provide the basis for efficient future research.

1. Introduction

1.1. Research problems and previous scholarship

The landscape of the eastern Adriatic coast, until very recently forced its inhabitants to live off the sea. Its rugged coastline is separated from the hinterland by the high chains of the Dinaric mountains, providing scarce resources and severely limiting the degree of connectivity with the hinterland. Thus, the sea remained the only viable source of connectivity – not only with the Italian Adriatic coast across the sea, but also with the rest of the Mediterranean. Orientation on the sea enabled the communities inhabiting the eastern Adriatic coast to participate in and benefit from the Mediterranean networks of exchange and communication.[1] The prosperity of eastern Adriatic communities depended on their shipbuilding and seafaring skills combined with their abilities to control the lines of maritime communications.

Taking all of this into consideration, it is not surprising that the seafaring skills of the communities from this part of the world are noted in ancient, medieval and early modern sources. This study will dissect a small segment of the rich history of eastern Adriatic navigation and shipbuilding, focusing on two types of ancient ships which appear in the written sources connected with this area: south Adriatic ('Illyrian') type of *lembos*, and Liburnian *liburnica* or *liburna*.[2] Both of those ship types have attracted the attention of earlier scholars who gathered existing material and written evidence, attempting to reconstruct their development, appearance and capabilities. The relative abundance of written sources suggests that both ships played significant roles in ancient times, especially *liburnica*, which became the main type of light warship in early Roman imperial fleets and ultimately evolved into a generic name for warships in the Roman Imperial period and Later Antiquity.

The published works can be divided into three general categories: Italian and Croatian archaeology/ancient history, as well as general research on shipbuilding in antiquity. The classical work remains Silvio Panciera's article published in 1956, with the Italian discussions of lemb and liburnian published later also being of note, most notably that of Stefano Medas.[3] Croatian authors have discussed these ships several times, starting from the pioneering work of Bartul (Bare) Poparić, which was continued by Grga Novak and Mladen Nikolanci who addressed this topic tangentially.[4] The topic was revisited several times between the 1970s and 1990s, particularly in the works of Zaninović, Kozličić, Jurišić and Vrsalović. However, in the 21st century it has been rarely addressed.[5] General research on these ship types within the broader context of ancient Mediterranean shipbuilding began at the end of the 19th century by Torr, and was continued later by Casson, Rédde, Morrison and Pitassi. Höckmann examined Illyrian lemb and *liburnica* in two articles, and Bérchez Castaño looked into the period when the *liburnicae* might have been included in the Roman Republican fleets.[6]

---

[1] See e.g. Kirigin *et al.* 2009; Elez 2015: 93-106.
[2] See the section 1.3. below on terminology.
[3] Panciera 1956; Medas 2004; 2016; Anastasi 2003.
[4] Poparić 1899: 1-39; Novak 1962: 12-13, 20 ff.; Nikolanci 1958.
[5] Stipčević 1973; Zaninović 1976; 1988; Jurišić 1983; Vrsalović 2011 [1978]: 142-44; Kozličić 1980/81; 1993; Džino 2003, and most recently Džino, Boršić 2020. Cf. also see the useful overview of Adriatic shipping at the times of the Greek colonisation in Radić Rossi 2010a.
[6] Torr 1895: 16-17, 115-16; Casson 1971: 125-27, 142, 162-63; Rédde 1986: 104-10; Morrison 1995: 72-73; 1996: 203, 248-53, 263-64, 317; Pitassi 2011: 89-90, 106-09, 138-44; 2016: 39, 47; Höckmann 1997; 2000; Bérchez Castaño 2010.

General agreement in the existing scholarship is that Illyrian lemb and the liburnian are the same type of fast ship, initially used for piracy. Illyrian lemb is usually regarded as a general type and liburnian as a regional type or sub-type of lemb, or the late stage of development of that ship type. Earlier Croatian authors and Höckmann stretch the origins of those ships even further, connecting them with the existing visual representations of ships from the Iron Age Adriatic and its hinterland. Such a 'canonic' view was uncritically accepted and rarely challenged in the scientific community, except by Medas, Džino, Radić Rossi and Tiboni. Medas points out that it is difficult to see a clear connection in the iconography of Adriatic early Iron Age ships, while Džino emphasises that the evidence connected with Illyrian *lemb* and *liburnica* comes from different periods, and is related to two different indigenous groups in eastern Adriatic.[7] This idea was further promoted by Radić Rossi, while Tiboni argued that the ships in earlier Adriatic iconography do not present the technical characteristics of indigenous ships but, more likely, rely on the Greek and Etruscan iconographic tradition.[8] It is worth noticing that the earliest English-language discussion touching on eastern Adriatic seafaring, done by Torr in 1895, did not associate lembs and liburnians, most likely because his primary goal was a catalogue of ship types rather than a thorough analysis of seafaring in specific sub-regions.[9]

The problem with the existing evidence is the vagueness of ancient written sources, which were usually produced by writers who were not naval experts and not acquainted with particularities of naval design. Most of these sources lack specific details which would help in recovering more information about the shipbuilding design and origins of these ships. An additional problem is the specificity and changing meanings of the terms *liburnica* and λέμβος, which often depended on the contexts known to the authors and their audience. For example, the term *liburnica* referred to a specific bireme-class of ship in the early Roman imperial fleets, but in later imperial times it becomes the general designation for any light warship. We often do not know for certain if the authors from the second century onwards used this term in a general or specific form, or whether they were referring to the original Liburnian ship or the version of the ship used in the Roman imperial fleets. Similarly, the term *lembos* was a general term, which could refer to ships used in a variety of very different civilian and military purposes. Visual representations also pose interpretative problems. The most important is certainly the question of accuracy, or in other words, how interested were the craftsmen who made these images in creating realistic depictions of these ships. Finally, the ongoing debate regarding the meaning of the ancient terms used to describe different classes of ancient ships, remains an important issue. For example, *liburnica* is usually taken to be a bireme-class warship in accordance with the testimony of ancient sources, as will be discussed later in the book. The scholarship on ancient warships interprets the term 'bireme' as the designation of a ship with two rows of oars, one above another, with one rower per oar. However, this matter is not necessarily solved, for different interpretations of this term are still being suggested.[10]

Although there is relatively abundant written evidence for both of those types of ships, we are still in the dark on crucial questions of their origins, development, shape, and capabilities,

---

[7] Medas 2004: 137-38; 2016: 162-63; Džino 2003. Lewis (2019: 84-86) also identified the lemb and liburnian as different types of ships, but without elaborating on their differences.
[8] Tiboni 2009; 2017; 2018.
[9] Torr 1895: 16-17, 115-16.
[10] Morrison 1996: 262; see Casson 1971: 53-62. The study of Tilley (2007), on the other hand, points out that the prefix which indicates number two in the terms *biremes*/δίκροτα lexicologically indicates the total number of rowers at the rowing bench cross-side, not the number of banks of rowers on one side of the ship.

as well as the connection between the adoption of new shipbuilding technologies and the social development of indigenous communities of the eastern Adriatic in the late Iron Age.[11] While some questions will remain unanswered on account of the lack of relevant sources, we will address (or rather revisit) here the most important problems related to the origins of the Illyrian lemb and Liburnian liburnian: their connection with the existing protohistoric Adriatic traditions of shipbuilding, and their shared relationship.

1.2. Overview of the book

The understanding of geographical and historical context is essential when dealing with ancient shipbuilding and seafaring activities in particular Mediterranean sub-regions. For this reason, Chapter 2 provides more insight into the geographical characteristics and eco-geographical zones of the eastern Adriatic coast where these two types of ships developed. The ethnic and political makeup of this area prior to the Roman conquest must also be taken into account when attempting to understand who built the eastern Adriatic ships, and for what purpose. This, in particular, relates to the protohistoric indigenous groups in the coastal areas of the eastern Adriatic, the Greek colonisation of the central Dalmatian islands from the 4th century BC, and the question of 'endemic' indigenous piracy in these periods. These matters are briefly examined and presented in the Chapter 3.

The next step is the discussion on available archaeological and iconographic sources, which could be directly associated with prehistoric eastern Adriatic seafaring in Chapter 4. Putting aside substantial indirect evidence of intense maritime communication between the eastern and western coast of Adriatic, the archaeological evidence for actual ships is limited to four underwater archaeological sites in areas once populated by the communities known to ancient writers as the Histri and Liburni. These sites revealed the remains of nine boats made of stitched (sewn) planks.[12] One of them (Zambratija) is dated in the end of the 2nd millennium BC, i.e. to the late Bronze Age, while the other eight reflect the surviving prehistoric tradition in the early Roman imperial times. This means that the oldest shipwreck is over half a millennium older than the vessels to which the ancient authors refer when mentioning indigenous eastern Adriatic ships, while the other finds date to the period in which this area was already integrated within the Roman imperial infrastructure. Underwater finds of protohistoric and ancient ships from the north-eastern Adriatic cannot be interpreted as the remains of warships, and, in turn, should not be identified as liburnians or southern Adriatic lembs. These should rather be identified as the *serilia* – cargo ships – mentioned by ancient authors. However, these finds are very important in providing evidence for the existence of distinct shipbuilding traditions maintained in this area: locally in the northeast Adriatic, and inter-regionally on both sides of the northern Adriatic. This understanding of these local shipbuilding traditions is an important facet of the hypothesis that northern and southern Adriatic shipbuilders interacted within two different networks, producing designs of two different types of ship: the liburna and the south Adriatic or 'Illyrian' lemb.

The iconographic evidence is also fairly limited, and spread over time and space. It covers the period from the 7th to the 1st century BC, stretching from the Istrian peninsula in the

---

[11] Some of those questions are tackled in Dzino, Boršić 2020.
[12] We opted for the term 'sewn', although probably not the best choice in English terminology, as it has been widely accepted by scholars in various fields of humanities. See Pomey, Boetto (2019: 6) for clarifications on the argument.

north to present-day northern Albania in the south. The oldest evidence originates from the deeper hinterland of Dalmatia and the western Adriatic coast, but is often associated with the eastern Adriatic seafarers. Although scarce, the visual representations of ships in wider prehistoric Adriatic area attracted the attention of scholars in the past, provoking a range of different opinions discussed below. While some earlier scholars called upon iconographic representations of (typically Roman) warships as evidence of Iron Age lemb and liburnian prototypes, we will demonstrate that there are severe problems with identifying these representations as precursors of the warships mentioned in Greek and Roman sources. However, images of ships upon south Adriatic coinage from the 3rd and 2nd centuries BC seem to be an exception to this rule, providing probably the only significant artistic representations of the ships used by south Adriatic communities in that period: Illyrian lembs.

The limitations of archaeological and iconographic sources have been supplemented by selected quotations from the ancient Greek and Latin texts in Chapter 5. As mentioned previously, the consensus amongst most scholars is that the term *lembos* referred primarily to the south Adriatic 'Illyrian' ships, and that *liburnica* represents a sub-type of lemb developed in later periods. In order to examine different perceptions and contexts in which the Greek and Roman authors used those terms, we decided to collect and comment upon the available written sources, regardless of their specific connection with the eastern Adriatic geographical or historical context. Due to the significant amount of collected texts, they became the core of this publication. Epigraphic evidence from the Roman times is not discussed in more detail. The epigraphic mentions of lemb are very scarce, while the inscriptions mentioning *liburnicae* refer to the Roman liburnians, which were not necessarily of indigenous Liburnian origins, as shown in sections 6.1. and 6.2.

Chapter 6 analyses the written sources to provide detailed overviews of the usage of the terms 'lembos' and 'liburnica', the ships' possible shapes and characteristics, and the etymology of the terms, as well as to present hypotheses of their likely origins and course of development. Finally, the various analyses made within this work have allowed us to form new conclusions, which are presented in sections 6.5. and 6.6.

1.3. Terminology

A few words should be said regarding the terms 'lembos' and 'liburnica' used in this book. There are many different terms used by ancient authors to describe these two ships in both Latin and Greek, as presented in Chapter 5, and these terms have not been used consistently in English scholarship.

In the case of lemb, we suggest the term 'lemb', adapted to English in the same way as other Greek words with the same ending. The term lemb is meant to be an English equivalent of the Greek term λέμβος and the Latin term *lembus*. In scholarly literature it has become customary to use either the Greek or the Latin word, written in italics. This option is not very convenient for the purposes of the present study. Since this text deals with that type of ship, both in Greek and Latin sources, it would be rather unfortunate to choose either the Latin or the Greek version of the word and use it throughout the text. Moreover, since almost all other Greek and Roman ships have their name translated or transposed into English language, there is no need to keep the original name, usually written in italics, since this word appears relatively often in ancient literature. It is not an exotic *hapax legomenon* for which there would be no need of an equivalent in modern languages. With these things in mind, we opted for the English coinage – lemb. It is composed by the same token as many other English words taken from ancient Greek: 'angel' from ἄγγελος, 'abyss' from ἄβυσσος, or even some more recent coinages like 'dinosaur' from δεινός + σαῦρος. We can only hope the word will continue its life in English scholarly literature as to avoid clumsy λέμβος or *lembus*, with their plural forms λέμβοι or *lembi*.

In the case of the other ship, we opted for *liburnica*, used by several important ancient authors like Caesar, Tacitus and Pliny the Elder, or the English version liburnian. The word 'liburnian' occasionally appears in translations and in the secondary literature, sometimes also in the form of *liburna*. This ambivalence between 'liburnian' and 'liburna' reflects that of the Latin original: both are feminine forms of the ethnic noun *Liburnus* or the ethnic adjective *liburnicus*, the latter of which being in the feminine form in relation to *navis* i.e. *navis Liburnica*.

2. Geographical context

Map 1. Geography of the Adriatic (D. Džino using Google Earth).

The complex eco-geographical configuration of the eastern Adriatic coast combined with the archipelago of the islands facing it, can be recognised as a plurality of different 'maritime cultural landscapes'. This concept, first used by Westerdahl and later elaborated by other authors, emphasises the relationship between the nautical environment and the cultural and socio-economic context of its exploitation by populations which inhabit it. In other words, 'maritime coastal landscape' is the result of interaction between human processes and an environment consisting of sea, coast and islands over a longue durée.[13] This relationship between human processes and maritime geo-ecology heavily impacts on the needs of the local population, which transfers onto their specific requirements in shipbuilding design. Such an interaction between humans and the sea could have been an important factor in shaping the development in design of both 'Illyrian' lemb and liburnian, as discussed in Chapter 6. In the same way, the interaction between humans and the sea is effectively illustrated by the prolonged use of sewn plank technology in ancient northern Adriatic shipbuilding, discussed in Chapter 4.

The Adriatic Sea is the deepest gulf of the Mediterranean, which lies between the Apennine and Balkan Peninsula. It is a consistent geographical unit, which since prehistoric times connected people around its shores, and served as communication between central and northern Europe and the rest of the Mediterranean.[14] The sea was named after Adria, the richest emporium

---

[13] Best defined in Westerdahl 2011. See also Flatman 2012; Pungetti 2012, etc.
[14] E.g. Braudel 1972: 125-27.

in the early Iron Ages, and since the very beginnings of Greek seafaring the Adriatic was perceived by the ancients as a gulf.[15] It measures 870 km in length, from the Lagoon of Marano in the northwest to Butrint in modern Albania to the southeast. The maximum width of the Adriatic is 216.7 km, with the average distance between the eastern and western coast being 159.3 km. Up to the line between Pula–Ancona, sea depth never exceeds 50 m, whilst the deepest part, which is in the Southern Adriatic Depression, measures 1233 m.[16] The Adriatic has clear morphological differences along its longitudinal and transversal axis, and is thus divided into three sub-basins. The northern sub-basin spans to the line formed by Giulianova (Italy) and Zadar (Croatia), and is characterised by a shallow depth (about 30 m) and a strong river runoff. The middle Adriatic is a transition zone between the northern and the southern sub-basins, and its conditions are often similar to that of an open sea. This middle zone spans to the Palagruža sill, the line connecting Vieste (Italy) and Split (Croatia). The southern sub-basin extends to the Otranto sill, which divides Adriatic from the Ionian Sea.[17]

The western coast of the Adriatic Sea, in modern-day Italy, is a shallow and sandy area with only a few significant features such as two larger peninsulas: Monte Gargano and Monte Conero near Ancona and the Po valley and delta. The northern arc of the Adriatic, stretching all the way from Venice to Piran (Slovenia), had been used since the early times for communication between the sea and the central European hinterlands through the large mountain passes of the south-eastern Alps. The eastern Adriatic coast is shared at the present by modern countries of Slovenia, Croatia, Bosnia and Herzegovina, Montenegro and Albania. It possesses limited resources thanks to its karstic landscape separated from the hinterland by a high mountain chains of Dinaric Alps. Specific geography shaped its economy through history, directing the coastal population towards the exploitation of marine and agricultural resources of the Adriatic coast and islands, and increased connectivity enabled by maritime links.

A large number of islands, islets, and sea rocks characterise the eastern Adriatic coast north of modern Cavtat – there are 79 islands, 525 islets and 642 sea rocks in total.[18] The western Istrian coast with its settlements, although having an important place in history, never played an important role in seafaring. Further south is the Kvarner gulf, with its small archipelago dominated by the large islands of Cres, Lošinj, Krk and Rab. The present-day area of northern Dalmatia, between Zadar and Šibenik has a large number of islands, the most prominent being Pag, Ugljan, Pašman, Dugi Otok and Kornat. These islands follow the coast parallel with the Dinaric Alps, and gave the origin to the term *Dalmatian coastal type*, which is widely used in oceanographic and geomorphologic terminology.[19] The Ravni Kotari plains in the immediate hinterland of modern Zadar is one of only few regions in the eastern Adriatic to have an abundance of arable land, and is linked to the deep Adriatic hinterland through a system of navigable river valleys, such as those of the Zrmanja or Krka rivers. Due to advantages posed by arable land, the area was densely inhabited in prehistory, antiquity and the medieval periods.[20] The configuration of the coast and presence of numerous islands certainly impacted

---

[15] Braccesi 2010: 62.
[16] Riđanović 2004: 188-89.
[17] Artegiani *et al.* 1996.
[18] Duplančić Leder *et al.* 2004.
[19] Magaš 2013: 178.
[20] Zaro, Čelhar 2018.

on the seafaring skills of local communities, and it is not surprising that it is in this part of the Adriatic Sea that we can trace the development of small and swift ships.

Modern central Dalmatia contains sizeable islands like Brač, Hvar and Vis, and the only significant quantity of arable land is found in the plains between modern Trogir and Split, where the important ancient settlements of Tragurion (Trogir), Salona (Solin) and Spalatum (Split) developed. The coast further south consists of a very narrow stretch of land, limited by the high rising chain of the Biokovo mountain. The mouth of river Neretva, around the ancient city of Narona (the village Vid near Metković) was a swampy area with large quantities of arable land providing excellent communication routes upstream with the hinterland. The mouth of the Neretva is enclosed by the large Pelješac peninsula, which is further connected with the nearby islands of Korčula, Lastovo and Mljet, as well as the Elaphiti islands further south towards modern Dubrovnik. The coast of modern Montenegro is plain, with the exception of Boka Kotorska, a large gulf with some important coastal settlements from prehistoric and ancient times. The Adriatic part of the modern Albanian coast stretches from the mouth of Buna near the Lake of Shkodër in the north, to the Bay of Vlorë in the south. It is ecologically diverse, containing alluvial deposits and marshes which provided beneficial conditions for the development of larger ancient urban structures such as Scodra, Dyrrachium or Apollonia.[21]

Navigation in the Adriatic, especially in the pre-modern period, is seriously affected by its changing weather patterns, particularly its various types of winds. The sailing season in antiquity was limited to the period between March and October, although some long-distance trans-Adriatic sailing could have taken place during the winter months. Thus, it is not surprising that the dangers of sailing in the Adriatic became a *topos* in ancient literature.[22] Serious waves are caused by *bura* (bora, north-easterly wind), *jugo* (sirocco, south-easterly wind), *maestral* (maestrale, north-westerly wind) and *lebić* (garbino, west/south-westerly wind). *Bura* is the strongest wind in all the Adriatic regions. It creates the short and sharp waves raising 'smoke' from finely dispersed particles of seawater which does not allow seafarers to breath and highly reduces the visibility. *Jugo* blows with constant strength, along the coast, and raises long waves. At present, they do not threaten ships, which have enough time to escape in a safe anchorage. The same happens with the *maestral*, which blows from the exactly opposite direction. It blows strongly but briefly, and it usually does not raise the type of waves which endanger navigation. The waves raised by the not so common *lebić* are strong, but not so high. They come at a right angle to many eastern Adriatic ports, penetrating inside port breakwaters with the potential to seriously endanger docked ships and shipping inside port. These winds are also an important factor in local shipbuilding traditions, necessitating the development of fast and sturdy ships which are able to cope with the potential dangers created by changing weather patterns, such as sailing with the side wind for example, which was crucial for trans-Adriatic navigation.[23]

The variety of Mediterranean landscapes, economic resources and seafaring conditions, all of which depended on natural, economic and political factors, resulted in the development of

---

[21] See, for example, Ferriès, Skenderaj 2015.
[22] E.g. Poulain, Racich 2013 (the winds); Brusić 1970; Arnaud 2006; Kozličić, Bratanić 2006; Kirigin *et al.* 2009: 143-50, Kozličić 2012, etc. (ancient navigation in the Adriatic), and Milićević Bradač 2009 (ancient literary *topoi* about sailing in the Adriatic).
[23] Kirigin *et al.* 2009: 143.

various types of ships suitable for specific purposes. As we saw from this very brief overview, the Adriatic Sea unifies the various different micro and macro eco-geographical zones around its shores by providing its inhabitants with the potential for increased connectivity through navigation. However, these different eco-geographical zones around the Adriatic significantly affect local navigation habits and ship design requirements, creating different 'maritime cultural landscapes'. This is particularly visible in the differences between the eco-geography of the southern and northern Adriatic. There is no doubt that these different 'maritime cultural landscapes' played an important role in the development of different local shipbuilding traditions, as well as strategies of selective acceptance of Mediterranean shipbuilding innovations, as we will discuss later.

3. Eastern Adriatic populations in the 1st millennium BC

Map 2. Distribution of the most important indigenous ethnonyms in the pre-Roman Adriatic and its hinterlands. In white: the ethnonyms not mentioned in the sources related to the Roman conquest (D. Džino using Google Earth).

Before discussing indigenous seafaring in historical narratives and the archaeological record, it would be useful to provide a brief overview of protohistoric east Adriatic communities as they existed before their inclusion in the Roman imperial infrastructure of the first centuries BC/AD. As mentioned in the previous chapter, the Adriatic Sea acted as a channel of connectivity for the communities that inhabited its shores, so it is not surprising that a relatively homogenous cultural zone formed there in the late Bronze Age (c. 1000 BC, maybe even earlier) – an Adriatic *koine*. A network of connected communities emerged, linking the Italian and eastern Adriatic coast all the way to central Dalmatia, including the deeper hinterland of the eastern Adriatic coast and the southeast Alps. These common cultural features flourished from the 9th century BC until *c*. 500 BC, when a reorientation of early Iron Age social networks produced increased local differences, resulting in the emergence of distinct material cultures and political institutions around the Adriatic. Some level of trans-regional cultural unity after 500 BC could still be recognised, but in two separate zones: north/central and south Adriatic.[24] It is very likely that this change in the early Iron Age Adriatic social networks around 500 BC was caused by rising influence from Etruscan and Greek maritime penetration into the western Adriatic.[25] In

---

[24] Peroni 1976, cf. Medas 1997: 94-99 and Blečić Kavur 2019/20. Batović (1987: 350-51) recognises this network, but ascribes it to the Liburnian 'thalassocracy' in early Iron Age, which is very problematic concept (see p. 15 below).
[25] Medas 2016: 160.

the period after 500 BC local differences continued to increase, and the indigenous population started to form distinct ethno-political groups, distinguished from one another by ancient sources from the Roman period, such as, for example, Strabo, Pliny the Elder, and Appian, who all, more or less, had an understanding of the ethnic, or at least political reality.[26] The distribution of these groups in these sources needs to be taken with caution, as some of them might have begun forming as political groups prior to the Roman conquest and not existing in earlier periods, whilst the absence of some groups from the later sources does not mean that they completely disappeared.[27]

In the very north of the eastern Adriatic basin the groups known in the sources as the Histri and Liburni formed, and in their hinterland the communities of the Iapodes developed. The central Dalmatian hinterland was dominated by the Delmataean communities, and earlier Greek sources such as the *periploi* of pseudo-Scylax and pseudo-Scymnus identify the coast as being inhabited by the Hyllaei, Bulini, Nesti and Manii – indigenous groups which do not appear in the later written sources describing the Roman conquest.[28] In the southern Adriatic there were a number of smaller communities which ultimately constituted the so-called Illyrian kingdom between the 4th and 2nd centuries BC. Lastly, the deeper hinterland of the eastern Adriatic was inhabited by Pannonian groups which are prominent only in the sources related to the Roman conquest – in particular the Daesitiates, Mezaei, Segestani, or Breuci.[29] In relation to seafaring, several of the most important indigenous groups should be emphasised – the Liburni, Histri, and southern Illyrians – as well as the Greek settlers who established colonies in the central Adriatic in the 4th century BC.

3.1. The Liburni

The Liburni in the last centuries BC inhabited the section of the Ravni kotari around Zadar, the coastline below the Mt Velebit, and the islands of the present-day Gulf of Kvarner and northern Dalmatia (Figure 1). Ancient written sources define Liburnia as a space between the rivers Raša (*fl. Arsia*) and Krka (*fl. Titius*).[30] Archaeology recognises a specific Iron Age material culture in this area, which has distant roots in the Bronze Age. It clearly distinguishes the Liburni from the other indigenous groups to the south, especially from those inhabiting the southeast Adriatic.[31] Indigenous personal names in Liburnia, recorded in the Roman era, bear evident similarity with indigenous personal names of the Histri and Veneti in the northern Adriatic, rather than with their other indigenous neighbours from the hinterland such as the Iapodes and Delmatae.[32] Specific cultural features and the geographic proximity of these indigenous communities very likely resulted with a shared sense of identity, leading modern scholars to recognise Iron Age Liburni as a distinct ethnic group. However, we need to bear in mind that a sense of locality and kin-relationships were much more important for the construction of identity amongst the Iron Age population of Liburnia and other eastern Adriatic groups than

---

[26] Plin. *HN* 3.138-44; Strabo, 7.5; Appian, *Illyrike*, etc.
[27] Dzino 2014a; 2016.
[28] Scyl. 22-24; Scymn. 404-12.
[29] See Džino, Domić Kunić 2013: 62-73, and in English Wilkes 1992: 91-218; Šašel Kos 2005; Dzino 2012; 2014a, etc.
[30] E.g. Plin. *HN* 3.139.
[31] Kurilić 2008: 11.
[32] Katičić 1976: 154-83; Wilkes 1992: 67-87.

Figure 1. Distribution of Iron Age Liburnian hillforts (from Batović 1977).

the shared cultural features which were perceived as markers of 'ethnicity' by ancient Graeco-Roman sources.[33]

These Liburnian communities were distinct from neighbouring indigenous peoples in Dalmatia as their settlements reveal more visible social complexity and continuing connectivity with

---

[33] The criticism of the ethnicisation of Iron Age identities from the western Balkans in earlier scholarship is slowly becoming consistent in scholarly discourse: Džino 2007; 2008; 2012; Kuzmanović, Vranić 2013; Mihajlović 2014; 2019; Dimitrijević 2018.

3. Eastern Adriatic populations in the 1st millennium BC

Figure 2. The city walls of Varvaria-Bribirska glavica (photo: D. Džino).

Italy and Magna Graecia, which peak in the last few centuries BC. Some settlements such as Asseria (Podgrađe near Benkovac), or Varvaria (Bribirska glavica near Skradin) were surrounded by massive walls (Figure 2), and traces of proto-urban organisation can be detected at most of the sites which show continuing habitation into the Roman era, sometimes even later, into Late Antiquity and the medieval period.[34] Other settlements, such as Radovin or

---

[34] On the development of proto-urban Liburnian settlements see Čače 2006 and Chapman, Shiel, Batović 1996: 293-336 for later periods.

Figure 3. Aerial photo of Nedinium-Nadin (photo: M. Grgurić).

Nedinium (Nadin near Benkovac), have similar proto-urban features in the late Iron Age, but lack monumental architecture from Roman times (Figure 3).[35] The Liburnian communities also show regional differences in material culture, which casts doubt on the usual assumption that they represented a monolithic cultural or ethnic group. Some scholars, such as Blečić, see the islands in the Kvarner Gulf as having their own Iron Age social network, distinguishable from the Ravni Kotari communities on the mainland, and closer in material culture to that of the Histrian communities.[36]

The Liburni were well-known to ancient writers who often ascribed to them colonial fantasies about exotic 'Others' – such as those which claim of rule of women or the sharing of sexual partners, or that they practised wife-sharing and raised children together. The Latin grammarian and compiler Gaius Iulius Solinus of the 3rd century AD even ascribed 'Asiatic' origins to the Liburni which is another example of these Graeco-Roman colonial fantasies.[37] They were also known as able and innovative seamen. Along with *liburnica navis*, the Liburni are also connected in the Roman sources, together with the neighbouring Histri, with the development and use of a merchant boat made of sewn planks, described as *serilia*. The use of such vessels in this area is confirmed by underwater finds dateable in the Roman period, when such vessels had already disappeared from the rest of the Mediterranean (see Chapter 4). Although known as able seaman in the sources, more recent analysis shows that the

---

[35] There is comprehensive literature on the Liburni in Iron Age and the Roman period: Wilkes 1969: 159-62, 192-219; 1992: 56-57, 186-88; Lo Schiavo 1970; Čače 1985; 1991; 2006; 2013; Batović 1987; Glogović 1989; Chapman *et al.* 1996; Majnarić Pandžić 1998: 306-18, 349-58; Šašel Kos 2005: 182-88; Blečić 2007a; Kurilić 2008; Batović, Batović 2013; Glogović 2014; Barnett 2016; Miše 2019; Kukoč, Čelhar 2019, etc.
[36] Blečić 2007a; Blečić Kavur 2015.
[37] Scyl. 21; Nic. Dam. *BNJ* 90 F103d (*ap.* Stob. *Flor.* 4.2.25); Solin. 2.51, see Čače 1985: 574-82; Kurilić 2008: 27, 48-51, 59, 77; Džino 2017: 68-72.

population of Liburnia lived primarily off the land. Their diet before the Roman period, for instance, consisted of cereals (wheat, barley, small amounts of millet), and their animals fed upon terrestrial plants rather than marine resources.[38]

While earlier scholarship assumed that the Liburni dominated the Adriatic in the early Iron Age (the so-called *Liburnian thalassocracy*), it is difficult to maintain this opinion today, as archaeology does not provide satisfactory evidence to confirm the concentration of political power in the Liburnian homelands during this period. The numerically modest spread of artefacts in the central Adriatic that show similarity with Liburnian culture is more properly interpreted in the context of the Adriatic *koine* culture, rather than Liburnian political domination in this early period. The term 'Liburni' might have been a general descriptive term for the inhabitants of the Adriatic encountered in early Iron Age contacts by the Greeks. It could have reflected Greek perceptions of the Adriatic *koine* in that period, rather than shared ethnicity of indigenous population. This would explain the reports about the conflicts between the 'Liburni' and early Greek colonists in the southern Adriatic and Ionian Sea in the 8th and 7th centuries BC mentioned by Appian (the Liburni expelled the Taulantii from Epidamnus/Dyrrachium, to be expelled by the Corcyreans in 627 BC) and Strabo (Bacchiad Chersicrates, the Corinthian founder of Corcyra, expelled the Liburni from the island before establishing this colony in 734 BC), so far away from the Liburnian north Adriatic homelands.[39]

There was no central settlement indicating a concentration of political power within the Liburnian homelands. The analysis of the hierarchy of prehistoric settlements indicates that the Liburnian Iron Age society was likely dominated by kinship groups who inhabited enclosed spaces (mostly hillforts), commonly owned the land, and maintained these kinship networks between dispersed settlements.[40] It seems that the Liburni were politically unified only as a loose confederation, with each settlement representing a separate political unit.[41] The mention of the 'Iadasinoi' fighting the Greek colonists in the written and epigraphic sources mentioned below (p. 20) might have been reference to the Liburnian political alliance led by inhabitants of Iader (modern Zadar) in early 4th century BC. However, there is no evidence which allows us to deduce whether this was a pre-existing or ad hoc alliance, or whether the alliance fell apart after the battle. It could be speculated that some of these communities might have become Roman allies in the later 2nd century BC, as there are no explicit reports of the Liburnian conflicts with Rome, but this hypothesis also needs more support in primary evidence. What we know for certain is that the Liburnian communities were included in a Roman system of alliances in roughly the mid 1st century BC at the latest.[42]

Between the 4th and 1st centuries BC, the material record from Liburnian sites shows important changes. Earlier forms of local material culture and burial traditions began to be partially abandoned and replaced with mass-produced Mediterranean goods.[43] A phenomenon which

---

[38] Lightfoot *et al.* 2012: 549-50. The survey analysed stable carbon and nitrogen isotopic analyses of human remains from three Iron Age Liburnian sites (Zadar-Relje, Nadin, and Dragišić – 64 samples) and one Iron Age site unrelated to the Liburni on the island of Korčula (Gumanca – 3 samples).
[39] App. *B Civ.* 2.39; Strabo, 6.2.4; Džino 2014: 52-55; 2017: 68-72, see also Čače 2002; Čače, Kuntić Makvić 2010: 64; Barnett 2016: 71-75.
[40] Chapman *et al.* 1996: 273-92.
[41] Čače 2006; 2013: 15-20; Barnett 2016: 80-84.
[42] Čače 1991; 2013: 24-25; Cerva 1996; Šašel Kos 2005: 323-24.
[43] E.g. Batović, Batović 2013; Glogović 2014; Miše 2017; 2019.

is most evident in the appearance of imported Greek and Hellenistic artefacts used in burial customs, mostly fine Hellenistic wares used for wine-drinking. The quantity of these wares and their association with a large number of burials shows that ownership of imported fine wares was not privilege of the elites, but rather widespread throughout the community. This renders recognising and defining Liburnian elite burials in late Iron Ages difficult, as there are no other distinguishing features appearing in the funeral record, such as, for example, burials with weapons. More evidence and new methodological approaches to material from this period is required in order to better understand late Iron Age Liburnian elites and explain the processes of social change occurring after the 4th century BC.[44] Perhaps we can see a mark of elite status in the wearing of fibulae, especially the Kastav and Nesactium types (which developed under middle La Tène influences and are present in the Histrian culture and north Adriatic), as well as metal plate fibulae and anthropomorphic pendants crafted under Hellenistic influences.[45]

3.2. Other Iron Age Eastern Adriatic indigenous seafaring groups

The Histri, in the very neighbourhood of the Liburni, were somewhat akin to them in matters of language, material culture and personal names as part of a North Adriatic cultural zone, together with the Veneti and Piceni on the Italian coast. They inhabited the Istrian Peninsula, which is a distinct geographical feature separated by the mountain chains from the hinterland. The most important Iron Age sites are the hillfort settlements, which are mostly located in the hinterland upon easily defensible positions placed strategically in the vicinity of coastal positions suitable for ports.[46] It seems that the dominant settlement was Nesactium (Nezakcij) which saw continued habitation in the Roman times, while other significant settlements were: Pula, Picugi-Poreč, Beram, and Monkodonja. The Histrian Iron Age elites had a taste for the Etruscan, Apulian and Greek artefacts and art, which are found in abundance in the elite graves. Contrasting with the material culture of the Liburni, the appearance of weapons is more detectable in Histrian graves throughout Iron Ages,[47] likely the remnants of part of a differently constructed elite funerary image. A leading political institution in the 3rd century BC described in the sources as the 'Histrian kingdom' was an alliance of local communities led by the elite of Nesactium. The indigenous population fought with the Romans in the 3rd and 2nd centuries BC, to be finally crushed in the Second Histrian War of 178-177 BC, although some conflicts might have happened later in the 2nd century BC.[48] As mentioned below (p. 22), the Histri were involved in activities deemed as piratical by ancient writers, which implies the capability to make swift ships able to intercept other cargo ships. The underwater finds of sewn plank ships Pula 1 and Pula 2, discussed in 4.1.2 and dated in the Roman era, imply prolonged use of those ships in Roman Histria, as elsewhere in northern Adriatic.

Another important and distinct group of indigenous communities connected with seamanship and ship construction can be recognised in the south Adriatic. The inhabitants of the south Adriatic and its hinterland belonged to a larger social network, which is described in earlier archaeological literature as the so-called Iron Age 'Glasinac-Mati cultural complex', stretching

---

[44] Barnett 2016: 84-90; Miše 2019.
[45] Blečić Kavur 2009 (fibulae); Brusić 2010 (Liburnian jewelry), cf. Tonc 2012 for a wider context.
[46] Buršić Matijašić 2012.
[47] Batović 1986/87; Gabrovec, Mihovilić 1987; Wilkes 1992: 63-64, 185-86; Majnarić-Pandžić 1998: 251-81; Buršić Matijašić 2008; Cestnik 2009; Mihovilić 2014.
[48] Bandelli 1981; Čače 1989; Matijašić 1991; 2009: 92-94, 99-107; Kuntić Makvić 1997; Starac 1999: 7-10; Cavallaro 2004; Šašel Kos 2005: 271-75, 322, etc.

3. Eastern Adriatic populations in the 1st millennium BC

all the way from the left bank of the Neretva river to the Greek island of Corfu. The shared material culture of these communities also extended deep into the interior of modern-day eastern Herzegovina, south-eastern Bosnia (the Glasinac plateaux), western Serbia, Montenegro, and Albania, all the way to the Lake Ochrid.[49] The written sources, especially in the Greek language, distinguished a number of different local identities, and the most significant political formations arising in the Iron Ages are called the Autariatae and 'Illyrians'. The term 'Illyrians' still occasionally arouses some confusion in scholarly literature due to the fact that it was indiscriminately applied to the wider population of the eastern Adriatic in the imperial period, as far as Histria, Liburnia and the deep hinterland. This was an outcome of earlier Greek ethnographic perceptions, which invented the term 'Illyrians', a term that was then transmitted by Roman imperial geographers.[50] As with other indigenous Iron Age communities on the eastern Adriatic and its hinterland, kin-based relations and cult practices were more important for collective identification in south Adriatic communities than the broader group identities recorded in the ancient sources, such as 'Illyrians' or 'Autariatae'.[51]

The Autariatae were a political alliance of northern communities centred on the Glasinac plateaux, while the term 'Illyrians' was applied by the ancient sources to a number of smaller communities who established the alliance known as the 'Illyrian kingdom' in around the 4th century BC in the southern Adriatic and its hinterland. Written sources indicate political disintegration of the Autariatae after *c.* 300 BC, which is visible in material evidence showing simplification of social complexity in the deep hinterland already starting in the 4th century BC.[52] The 'Illyrian kingdom' reached its peak in the later 3rd and early 2nd century BC when it was ruled by the elites of the Ardiaei from the Boka Kotorska gulf and the Labeatae who lived around the Lake of Scodra. Aside from the Ardiaei and Labeatae, the sources also locate many more smaller groups, such as the Pleraei, Daorsi, Taulantii, Parthini, and Enchelei.[53] The Illyrian kingdom was juggling its position between the Macedonian kingdom and Roman Republic which led to them engaging in three conflicts with Rome in 229, 219 and 168 BC, which all resulted in Roman military victories. The end of Illyrian kingdom came after the Romans defeated its last king, Gentius, in 168 BC in the Third Illyrian War, which was part of wider conflict known as Third Macedonian War. The treaty of Scodra in 167 BC divided the kingdom into three autonomous parts, with tribute-paying obligations to Rome.[54] They were incorporated into the Roman administrative structures by the late 1st century BC.

The vicinity of the Greek world, and its outposts in Corcyra, Apollonia and Epidamnus/Dyrrachium, also strengthened interaction, acculturation, and selective adoption of Mediterranean cultural templates in this area. This can be seen in the gradual urbanisation and rising complexity of indigenous political institutions in the southern Adriatic from *c.*

---

[49] The literature is exhaustive and cannot be listed comprehensively: e.g. Benac, Čović 1957; Prendi 1975; Čović 1987; Parzinger 1991; Govedarica 2002; Vasić 2003; Babić 2004; 2005; Jašarević 2014; Dimitrijević 2018, etc.
[50] Dzino 2014; 2016, see also 6.4. below.
[51] Dimitrijević 2018.
[52] E.g. App. *Ill.* 4, see Čović 1967; Papazoglu 1969: 69-100; Šašel Kos 2005: 121 with fig. 25, 166-82, 188-98.
[53] Čović 1987; Benac 1987: 782-90; Papazoglu 1988; Cabanes 1988; Wilkes 1992: 40-50, 91-99, 117-80. The area of southern Dalmatia, between the Neretva river and the Boka Kotorska gulf shows strong local characteristics, and perhaps should be seen separately – Marijan 2000.
[54] Three Illyrian wars generated immense bibliography, not listed on this occasion. The most important works up to 2013 are listed in Džino, Domić Kunić 2013: 74-96. In English see Šašel Kos 2005: 249-90; Dzino 2010: 44-60. The most recent noticeable works are Bilić Dujmušić, Milivojević 2014; Sampson 2016: 62-89, 185-200; Morton 2017; Bilić Dujmušić 2017, etc.

the 4th century BC, as well as the increased demand for Mediterranean-produced imported goods.[55] Very recent archaeological research from the southern Adriatic, which relies on a number of new underwater finds, confirms the existence of robust trade activities with the Greek world – Corinth and south Italy/Sicily – starting around the 6th/5th century BC. From the 4th/3rd century BC there is a visible increase in shipments of luxury items intended for elite consumption, especially fine wares, jewellery, wine amphorae, and roof tiles.[56] The importation and possession of foreign goods, their presentation in funerary rituals, and the consumption of imported products such as wine become a matter of uttermost importance for these communities, especially after the 5th century BC. It was recently suggested that, aside from trade, warfare (and therefore plunder) was other important means of obtaining imported goods in the later Iron Ages. Material evidence implies that warfare was traditionally connected with social status and had been integrated in socio-economical structures of this area. This is evidenced by the abundance of weapons in funerary and settlement contexts throughout the whole of the Iron Ages, as well as the large concentration of fortified settlements, suggesting the existence of a constant threat from the neighbours and the absence of centralised political institutions.[57]

Knowledge about the southern Adriatic communities prior to the Roman conquest has significantly increased in recent decades thanks to the Polish-Montenegrin joint excavations in Risan, ancient Rhizon in Boka Kotorska. The excavations revealed two elite residential complexes labelled (perhaps a bit eagerly) as 'palatial complexes' by the excavators, the oldest of which is dateable to c. 260 BC, and the youngest from c. a quarter of the century later. These elite residential complexes (as they should perhaps be referred to) show strong Hellenistic influences, and provide evidence for domestic textile production as well as significant quantities of imported wine-amphorae and fine Hellenistic pottery. There is no doubt that these were the residences of members of the upper elite stratum, perhaps even the rulers of the Illyrian kingdom from the Ardiaean dynasty. The evidence for their destruction could be linked with the Roman military intervention in 229 BC.[58]

3.3. Greek colonising activities in the eastern Adriatic

The impact of Greeks on seafaring and naval technologies in the eastern Adriatic cannot be underestimated. However, early contacts with the eastern Adriatic coast are rather obscure, although contacts with the western Adriatic coast are attested in the Archaic period. In the earliest Greek perceptions, the Adriatic was situated at the western 'end of the world'. It was sometimes named *The Sea of Kronos* and *The Gulf of Rhea*, considered as the place where the divine couple was exiled after their son Zeus overthrew them.[59] The Adriatic was well incorporated in different Graeco-Roman mythological schemes, especially in relation to two heroes from the Trojan mythological cycle. One version of the myth, preserved in Vergil and Livy, mentions the sailing of the Trojan hero Antenor along the eastern Adriatic coast. To him is attributed the foundation of Adria and Spina, and his resting place was imagined by ancients to be nearby Patavia – present-day Padua. The name of the hero Diomedes is also connected with several

---

[55] Ceka 1985; Cabanes 1988: 207-33; Wilkes 1992: 126-36; Galaty 2002; Siewert 2004.
[56] Royal 2012: 435-38; cf. Lindhagen 2016: 238.
[57] Dimitrijević 2016; 2018, esp. 10-18.
[58] Dyczek 2009; 2017a; 2017b.
[59] Ap. Rhod. *Argon.* 4.327, 509 (Sea of Kronos); Aesch. *PV* 837 (Gulf of Rhea), Katičić 1995: 64-66.

3. Eastern Adriatic populations in the 1st millennium BC

Map 3. The sites related to the East Adriatic Greeks (D. Džino using Google Earth).

places around the Adriatic, in particular the islands of Palagruža and the sanctuary of Diomedes on Cape Ploče, which is mentioned below. The Palagruža islands, known also as The Islands of Diomedes in the past, served for centuries as logistical supports for Greek long-distance sailing in the central Adriatic, providing the seafarers with a necessary stop and linking the eastern and western Adriatic coast, as well as southern and northern Adriatic areas.

Such mythological perception significantly changed through time, with increasing Greek activities in the area. The Greek advancement towards the Adriatic was manifested through the colonization of Corcyra (present day Corfu), a strategically positioned island controlling the Straights of Otranto. As said earlier (p. 15 above), the colony of Corcyra was established in 734 BC, by the Corinthians, and it opened Greek access to southern Italy and deeper into the Adriatic Sea. Sometime in the 7th century BC the Euboeans founded Oricum in the Gulf of Valona/Vlorë, where there was a previously established Eritrean emporium. In 625 BC the citizens of Corcyra and Corinth founded the colony of Epidamnus (Roman Dyrrachium, present-day Durrës), while Corinth on its own, around 600-590 BC, south of Epidamnus established the settlement of Apollonia near today's village of Pojan, northeast of Valona/Vlorë.[60]

Most of the attractive areas around the Adriatic were already occupied by the indigenous population, so it seems that the first Greek settlers along the Italian (western) Adriatic coast focused more on trade, and settling in combination with the local communities.[61] According to Herodotus, the first Greek seafarers that explored the Adriatic Sea were the Phocaeans from

---
[60] Cabanes 2008: 165-73. See also Milićević Bradač 2010: 51; Čače, Kuntić-Makvić 2010: 64.
[61] Čače, Kuntć Makvić 2010: 64.

Asia Minor, who apparently reached the Adriatic with their *penteconters*. This, however, is not confirmed in the material evidence.[62] The Adriatic route was also used by the Athenians, who from the 6th century BC intensified exchange within this wider region. 'To sail in the Adriatic' was the Athenian expression for extremely dangerous and demanding sailing conditions, economically worthy of effort. Athenians even decided to found a colony in the Adriatic in 325/324 BC, although the outcome of that enterprise remains unknown and was probably unsuccessful.[63]

It seems as though the eastern Adriatic in the early Iron Age was a zone of interest for the Corcyreans, whose influence reached the central Dalmatian islands. The position of the settlement of Heraclea in the Adriatic remains unknown, although coins that this colony minted exist in the archaeological evidence, especially from the island of Hvar. The Greek colonisation of the 'upper' part of the eastern Adriatic is better known from the 4th century BC, much after the main colonising wave, when it occurs in the context of the Syracusan imperial ambitions under Dionysius the Elder. The two most significant settlements in the central Adriatic were established at this time: the Syracusan settlement of Issa (the town of Vis on the island of Vis) and the Parian colony, Pharos (Stari Grad on the island of Hvar).[64] The indications that the Cnidians founded a colony on the island of Korčula are not confirmed by the limited material evidence, perhaps because the colony was destroyed early by the indigenous population or because the information from the written sources derives from some kind of local 'invented traditions'. The Issaean attempt to establish settlements on this island, on its eastern part, Lumbarda, in the early 3rd century very likely failed in the long run, but current ongoing excavations at this site should provide more evidence.[65]

There is only one written source that could be related to Adriatic shipping before 3rd century BC, and that is the depiction of the Battle of Pharos by Diodorus Siculus in 385/384 BC. Diodorus mentions 'small boats' used for the transport of indigenous troops from the mainland to the island of Hvar, where they were invited by the locals to help expel the Parian settlers. These boats were no match for the Syracusan war fleet, sent to help the Parian settlers, which easily dispersed them.[66] Who were the 'Illyrians' from the mainland mentioned by Diodorus, who used this term as a label for the indigenous population, it is not possible to say with certainty.[67] Some scholars connect this battle with the Greek inscription from the *tropaeum* found on the island of Hvar, which celebrates a Parian victory over the 'Iadasinoi'. The Iadasinoi was very likely the ethnonym describing the Liburnian settlement of Iader, although it was also suggested that this term describes the indigenous community from the surroundings of Salona. However, it is very likely that the battle of Hvar described by Diodorus and the inscription from the *tropaeum* are not directly connected.[68]

---

[62] Hdt. 1.163; Čače, Kuntić Makvić 2010: 65, with bibliography.
[63] IG 2² 1629; Ferone 2004: 335; Cabanes 2008: 178-79; Čače, Kuntić Makvić 2010: 66.
[64] Cambi et al. 2002; Zaninović 2004: 1-20; Kirigin 2006a; 2009; Cabanes 2008: 175-78; Poklečki Stošić 2010. Early Corcyrean influences: Čače 2002: 92-97.
[65] Black Corcyra (Korčula) as Cnidian colony: Scymn. 1.421; Strabo, 7.5.5; Plin. HN 3.152. On archaeology of Korčula see Kirigin 2010, and recently Radić, Borzić 2017.
[66] Diod. Sic. 15.13-14.
[67] Čače 1994: 44-52; Stylianou 1998: 193-96; Jeličić Radonić 2005; Kirigin 2006b: 64-67; Cabanes 2008: 176-78.
[68] BE 1953, 147-48 No.122, see Suić 1975; Čače 1993/94; 1998: 72-81; Kuntić-Makvić, Marohnić 2010: 75 No. A2; Cambi 2012; 2013: 9-10; Jeličić Radonić, Katić 2015: 15-17, etc.

The diminishing influence of Syracuse resulted with the Issaeans becoming the dominant political and economic force in the central Adriatic. They established settlements on the mainland, such as Tragurion (Trogir) and probably Epetion (Stobreč), and controlled the important emporiums of Salona and Narona. Issa led a political alliance (commonwealth, league, *sionoikia*) which included some indigenous communities such as the Hyllei who lived in the area between Trogir and river Krka, but also the inhabitants of the islands of Brač and Šolta.[69] This Issaean commonwealth also established and controlled the Sanctuary of Diomedes (*Promunturium Diomedis*) on Cape Ploče between Trogir and Šibenik, as one of the sanctuaries in the 'Adriatic network', in time, also taking trade routes towards the northern Adriatic.[70] The second most important Greek foundation in the Adriatic was the Parian settlement of Pharos near Stari Grad on the island of Hvar.[71] Recently, in light of archaeological excavations, it was suggested that Pharos was founded on an earlier indigenous settlement which was razed to ground together with other indigenous settlements on the island. Nevertheless, such interpretation of the evidence was criticised for using inadequate methodological approaches, so the problem of establishing a beginning for the Greek settlement at Hvar in the material record remains open.[72]

The Greek settlements in the Adriatic seem to be producing significant quantities of Hellenistic pottery such as Gnathia or Hellenistic Relief Ware intended for export to eastern Adriatic indigenous communities. Equipment for wine-drinking, such as *kantharoi* and *skyphoi*, is the most common form of such wares, however, transportation amphorae are also common. This is most evident at Issa, Pharos and probably the Hellenistic port of Resnik, close to Trogir.[73] Another important product made by the Adriatic Greeks was wine, produced on the islands and possibly in the Neretva valley around the Hellenistic trading post, and future Roman city, of Narona. Some of this wine was exported to indigenous communities, but it was also intended for wider Mediterranean markets, as shown by the finds of Pharian amphorae.[74] It seems that the Greek colonists integrated relatively well with the local population, especially in Issa. This is best illustrated by the burial customs in both of the Issaean necropoleis (Vlaška Njiva and Martvilo), which reveal that Issaean identities were complex, and strongly influenced by the pre-Greek indigenous population.[75]

3.4. Piracy in the eastern Adriatic?

There is no doubt that the indigenous population in the eastern Adriatic engaged in naval warfare before the last two or three centuries BC. Raiding enemy coastal settlements and intercepting and engaging with enemy ships on the sea must have been part of life for all communities living off the Adriatic Sea in prehistory and protohistory. Unfortunately, we know very little about it. Some iconographic representations of naval warfare discussed in

---

[69] Zaninović 2004: 20-32; Kirigin 2009: 25-30. See Čače, Šešelj 2005: 168-69 and Kirigin 2009: 29-30 on the extent of the Issaean political control. The Hylleans were portrayed in positive way in older Greek literary tradition, which could be reflection of their alliance with the Corcyreans and later Issaeans; Kirigin, Čače 1998: 74; Čače 2002: 89-97. Generally Epetion is considered a colony of Issa, but Maršić (1996/97) considers it, with good arguments, to be indigenous settlement allied with the Issaeans.
[70] Bilić Dujmušić 2004: 140; Kirigin 2004: 148-49; Čače, Šešelj 2005: 167-69.
[71] Kirigin 2006b; Jeličić Radonić, Katić 2015.
[72] Jeličić Radonić, Katić 2015: 12-13, 33-43, criticised by Kirigin 2018a and Kirigin, Barbarić 2019.
[73] Čargo, Miše 2010; Miše 2015.
[74] Kirigin *et al.* 2006; Lindhagen 2009; Kirigin 2018b. See also Lindhagen 2016 for Narona.
[75] Cambi *et al.* 1980; Kirigin, Marin 1985; Kirigin 1985; Čargo 2010; Ugarković 2019.

section 4.2 below, could have represented real events, but could also be a construction of reality intended to satisfy needs of the people who commissioned the production of such images. Other finds evidencing naval warfare are almost non-existent. Recent finds of 'Illyrian' helmets of the IIIA-1 type, found at a depth of 15-16 m near Cape Jablanac on the island of Cres, might be a witness of such event (Figure 4). These helmets were used between ca. 550 and 400 BC as a defensive weapon, but also as an important status-symbol in the wider Balkan peninsula.[76] Whether this particular helmet ended up underwater as a consequence of ritual activities, naval warfare, or something else, will probably remain unknown.

The question of piracy in the Adriatic is important for the present topic because ancient writers commonly attributed piracy to the local population in this part of the world. Both ship types discussed here, the liburnian and south Adriatic lemb, were described in the sources as being originally types of pirate ships.[77] While some of the more general assessments attributing piracy

Figure 4. The helmet from the Cape of Jablanac on the island of Cres (from Blečić 2007b, courtesy of the author).

to the eastern Adriatic inhabitants were clearly a traditional or conventional literary theme or *topos*,[78] there are a few sources that mention piracy of the indigenous communities in specific cases. These mentions include Livy's description of the expedition of the Spartan leader Cleonymus in 302 BC, the causes of First Illyrian War, and the Histrian-Illyrian piracy leading to the Second Illyrian war. Piracy was also mentioned in the context of Octavian's operations in the eastern Adriatic and its hinterland in 35 BC. Appian mentions that Liburnian piracy in the 1st century led to the confiscation of their ships by Octavian. He also mentions, in the same context, that the piracy conducted by a population from the island of Korčula led to their severe punishment and enslavement by the Romans.[79] It is also interesting to note Strabo's claim that the Romans punished the Ardiaei for their piratical activities and resettled them to the eastern Adriatic to till the soil.[80] The Roman action is reminiscent of the resettlement of the Cilician pirates by Pompey the Great in 67 BC, an act considered as a long-term Roman strategy for solving the causes of sea-raiding activities.[81]

---

[76] Blečić 2007b.
[77] E.g. App. *Ill.* 3.7.
[78] Dzino 2006: 124-25.
[79] Livy, 10.2.4-7 (Cleonymus); Polyb. 2.4-5, 2.8 (beginning of First Illyrian war); Cass. Dio, 12.4.9; App. *Ill.* 8.23 (Histrian-Illyrian piracy); App. *Ill.* 16.47 (35 BC).
[80] Strabo, 7.5.6. This might have been a consequence of the raid of the Ardiaei and Pleraei on 'Roman Illyria' in 135 BC, mentioned in App. *Ill.* 10 and Livy, *Per.* 56.
[81] Plut. *Pomp.* 28; Vell. Pat. 2.32.4; Flor. 1.41.14; Cass. Dio, 36.57.5; Livy, *Per.* 99; App. *Mith.* 14, see Tröster 2009: 23-33.

The increased quantity of ancient shipwreck-finds belonging to merchant cargo ships in the southern Adriatic, as noticed by Royal, is not evidence for Illyrian piracy for two reasons. First, the most important aim of the pirates would be to preserve the structural integrity of the merchant ship which was captured in order to keep the cargo intact. Consequently, it is clear that piracy could not be recorded by underwater finds because robbed ships would be taken to port rather than sunk with cargo. The other reason is that the number of newly found shipwrecks peaks in the period between the 1st century BC and 1st century AD, when rampant piracy is not recorded by the written sources.[82] While the dating of these ships certainly excludes the possibility of connecting them with the south Illyrian piratical activities in the 3rd and 2nd centuries BC, it is not always clear what the ancient pirates did with captured ships. Lewis recently pointed out that the ancient pirates would generally prefer captives over cargo as they were easier to transport than bulky cargo, although it is clear that in some instances the pirates would indeed capture a whole ship or even have auxiliary boats for carrying captured cargo.[83]

It is not necessary to go here into the debate about what defines piracy – this term gets easily confused depending on the observer's standpoint. Trade and piracy were parts of the same system of exchange, the only difference being the negative perception, and the lack of legitimacy for those who were labelled as pirates. It is also clear that, in most cases involving the ancient Mediterranean world, piracy was initiated and coordinated by the elites rather than by socially excluded groups and outcasts.[84] Piracy in the ancient and medieval world could be seen as a mode of production, an income supplement, or even a means of capital accumulation necessary for turning pirates into tradesmen.[85] The appearance of ancient Adriatic piracy in c. the 4th or 3rd century BC as a social process, in many ways reminds one of the appearance of medieval piracy in the same area in the 9th century AD. The rise of piracy in both of these historical periods was the consequence of intense social transformations amongst the local eastern Adriatic communities, as well as the rapid expansion of contacts with the wider Mediterranean world and (in the case of the 9th-century eastern Adriatic) central Europe.[86] Therefore, piracy was an unavoidable part of life in the ancient world, and it is easy to refute the notions of 'endemic' piracy in the Adriatic, which was present in earlier scholarship and most of the recent works dealing with the topic agree on this matter.[87]

There is no reason to doubt the existence of piracy in the eastern Adriatic reported by the Graeco-Roman written sources, although its extent could have certainly been over exaggerated.[88] However, explanations for its appearance should be sought in social transformation and change, rather than it being seen as part of a local economic pattern or 'endemic' habit of the local population. Intensified cultural exchange with the Mediterranean in the late Iron Age Adriatic resulted in the complexification of the social relationships within the existing

---

[82] Royal 2012: 440-41.
[83] Lewis 2019: 99-100.
[84] De Souza 1999 Gabrielsen 2001; 2003; 2013; Jasper, Kolditz 2013: 15-21; Beek 2015 and Wendt 2016.
[85] Horden, Purcell 2000: 156-58.
[86] Borri 2017, esp. 21-23 on piracy in early medieval eastern Adriatic as a social process.
[87] Dell 1967; Fuscagni, Marcaccini 2002; 2004; Čače 2005; Prusac Lindhagen 2009. In recent times, for example Ferone (2004) and Medas (2016) argued differently.
[88] Royal (2012) argues that the enforcement of tariffs or trade agreements by south Illyrian agents may have been viewed at times as piracy from Rome's perspective, which is perhaps not the best explanation as sources do not indicate anything of the sort.

indigenous communities. Local elites, especially those at the eastern coast of the Adriatic that either controlled the cultural exchange routes with the Mediterranean world, or were positioned closer to the points of direct contact, incorporated new symbols within existing traditions, finding new ways to construct and express their identities. The social structure of these groups transformed rapidly, giving birth to new elite groups. These new elites tried to establish themselves within their communities and fought for control of the importation and possession of prestige goods, either through trade, mercenary service, or through organised raids that most certainly included acts of piracy.[89]

3.5. Conclusion

This short overview of Iron Age protohistoric communities on the eastern Adriatic is by no means complete, but we have no useful evidence available for some groups belonging to the Central Dalmatian archaeological culture, such as, for example, the Hyllaei, or the inhabitants of the east Adriatic coast along the Biokovo mountain-chains. Powerful indigenous political alliances of the Delmatae and Iapodes, rising in importance in 2nd and 1st centuries BC, did not have naval capabilities which would be noticeable in the sources describing their extensive conflicts with Rome. The absence of naval capabilities amongst the Delmatae and Iapodes is believable because the centres of their political power were placed in the hinterland – modern-day Lika (Iapodes) and Dalmatian Zagora/southwestern Bosnia/western Herzegovina (Delmatae).

From the archaeological and written evidence, it is clear that the indigenous groups extensively engaged in maritime activities were connected with different parts of the east Adriatic coast. In the north, there were the Histri and Liburni, and in the south the so-called south Illyrian communities. They belonged to different social networks, which might indicate cultural and even ethnic differences – the Liburni and Histri were related to the Veneti and perhaps Iapodes, whereas the southern Illyrians were part of a wider network stretching from modern-day south-eastern Bosnia and western Serbia, all the way to Lake of Ochrid, between modern Albania and Northern Macedonia. These indigenous groups also controlled (or were intensely connected to) the different parts of the Adriatic coast discussed in Chapter 2 – the northern/central (Histri and Liburni) and southern Adriatic sub-basin (southern Illyrians). The noticeable geo-ecological differences between these sub-basins and their different 'maritime cultural landscapes' impacted the sailing techniques of the local prehistorical populations and, most likely, their ship-designs.

While different in ethno-cultural matters and their patterns of inclusion in wider social networks, north and south Adriatic indigenous societies from around the 5th century started to experience significant social transformations. Archaeologically confirmed changes in material culture, which started to prioritise mass-produced imports from c. the 5th/4th century BC, reflected substantial changes in the ways the Liburnian communities and their elites defined themselves. Very similar contemporary processes occurred in the southern Adriatic and Istria as well. Increased communal and individual needs for imported artefacts must have forced east Adriatic indigenous communities to look into more efficient ways to obtain them and significantly improve their naval capacities, regardless of whether they were used for trade,

---

[89] Džino 2012; Džino, Domić Kunić 2013: 80-82, cf. Džino, Domić Kunić 2018: 82-83.

piracy or both of these naval enterprises. Thus, the appearance of piracy in the eastern Adriatic must be understood as the result of social transformations amongst local communities, which arose because of intensified contact with the 'global' Mediterranean world, rather than some endemic habit of the local population.

4. Archaeological and iconographic evidence in protohistoric eastern Adriatic

Map 4. The sites of shipwrecks (pink), iconographic representation of the ships (yellow), and places where the coins with images of ships were minted (white) (D. Džino using Google Earth).

This chapter revisits the available archaeological and iconographic evidence for ships in the protohistoric eastern Adriatic, in search for possible ancestors of the south Adriatic lemb and the liburnian, characteristics of those ship types and their mutual relationship. Two particular types of evidence appear; underwater finds of ships, and visual representation of ships dateable to the pre-Roman period, including the images of ships upon south Adriatic coinage. The number of underwater finds is modest, but it significantly increased in the last decades, while visual representations of the ships are widely known and discussed in the literature, with the exception of the relief from Varvaria (Bribirska glavica), which was never published properly in the past.

4.1. Underwater finds

As already pointed out earlier, although we do not lack the evidence of maritime contacts of the Adriatic populations, the prehistoric shipwreck sites are extremely rare. A group of Bronze Age vessels, recovered by fishermen in the waters around the island of Korčula,[90] indicate the possible presence of a shipwreck site similar to the one discovered in the waters of the Ionian island of Cephallonia,[91] but its exact position remains unknown. The outstanding discovery of

---
[90] Radić Rossi 2011: 119-20.
[91] Dellaporta 2011: 21; Knapp, Demesticha 2017: 73-74.

4. Archaeological and iconographic evidence in protohistoric Eastern Adriatic

the Zambratija boat in 2008, dated to the last centuries of the 2nd millennium BC, showed the potential of new discoveries, as it revealed the remains the oldest Mediterranean boat made by sewn planks.[92] A recently published summary on the Mediterranean sewn boat tradition of shipbuilding gathered 64 finds of such ships, dating from the late Bronze Age to Late Antiquity or the Middle Ages. It is important to notice that 29 of these boats, most dateable to early imperial times, were found in the wider north Adriatic area, nine of which are ascribed to the 'north Adriatic' or 'Istro-Liburnian' tradition, which are discussed below (p. 57).[93] With our current state of knowledge, it is possible to assume that the oldest boats of the entire eastern Adriatic Late Bronze and Early Iron Age population were produced by the same stitching technique.

This section will provide a brief overview of eastern Adriatic underwater finds of these ships, which testify to the long tradition of sewn plank boats, and its persistence in the north Adriatic, including the Histrian and Liburnian area, after its inclusion in the Roman imperial architecture. The archaeological finds are attested by the ancient authors, who mention the Histrian and Liburnian rush-rope boats (*serilia*), mentioned on p. 14 and 56-57. The finds of nine sewn plank boats were attested from a total of four sites, which are discussed here in geographical order from north to south.[94]

4.1.1. Zambratija near Savudrija

The Bay of Zambratija near Savudrija is a complex archaeological site at the north-western tip of the Istrian peninsula, which provided the land and underwater archaeological evidence stretching from the Eneolithic all the way to the Middle Ages. On the small peninsula of Kaštel in the area of Sipar, at the southern edge of the bay, a Late Roman settlement covered the remains of a prehistoric site, presumably from early Iron Age.[95] The position of the site at the extreme point of a long and narrow isthmus, that is slowly disappearing under the sea level, recalls the situation on several similar sites in northern Dalmatia, dated to the middle and late Bronze Age.[96]

The shipwreck was discovered in 2008 in the northwest part of the bay, and was subsequently surveyed and excavated until 2013.[97] The radiocarbon dating suggested a date between the last quarter of the 12th and last quarter of the 10th century BC,[98] which is contemporary with the earliest phase of the Histrian archaeological culture.[99] Judging from the form of its 'keel', the vessel turned out to be a unique Mediterranean example of a transitional form from dugout to keeled plank boat (Figure 5). It was preserved in the muddy sediment at a depth of 2.2-2.5 m, with a length of 6.7 m and 1.6 m wide. Its suggested reconstructed length slightly exceeds 9 m. The element that functioned as the keel was made of elm. It was 6 cm wide and 21-23 cm thick at the extremity of the ship, 39 cm wide and 3 cm thick at the mainframe. On the western

---

[92] Koncani Uhač 2009.
[93] Pomey, Boetto 2019. The 65th ship that presents evidence of stitching together with mortise-and-tenons, as recently briefly reported, seems to be the ship found at Mazotos, Cyprus https://www.ucy.ac.cy/marelab/en/research/mazotos-shipwreck, last accessed 26/6/2020).
[94] See also Pomey, Boetto 2019: 8-12.
[95] Mihovilić 1995: 55-57.
[96] E.g. the islet of Ričul in the Pašman Channel, Čelhar *et al.* 2017.
[97] Koncani Uhač 2009; Koncani Uhač *et al.* 2017; Pomey, Boetto 2019: 8.
[98] Koncani Uhač, Uhač 2012: 534; Pomey, Boetto 2019: 8.
[99] Mihovilić 2014: 117-57.

Figure 5. The sewn boat of Zambratija (photo: Ph. Groscaux, from Koncani Uhač *et al.* 2017, courtesy of I. Koncani Uhač.

side of the boat, the planking was preserved up to the second strake, while on the eastern side the fifth preserved strake was identified as a sheer-strake (Figure 6). The strakes were made of elm, 13.5 to 24.6 cm wide. The second and the third strake had ends pointing to the north, i.e. towards the extremity of the ship, and the fourth strake was composed of two planks, joined by an oblique scarf that maintained the north-south direction of the laying of planks. The fifth strake, or the sheer-strake, was L-shaped, oriented towards the interior of the hull.[100]

The strakes were joined together by simple vegetal stitches (/// pattern), in a longitudinal sewing system. The seams between the planks were sealed by wadding made of marine plants of the *Cymodoceaceae* family, covered by laths 3-4 cm wide and made of fir, which conserved the imprints of the vegetal string that continuously passed through the diagonally-cut stitch holes. Two or three small pegs per stitch hole, driven from inside the hull at a distance of 6 cm, fixed the string in place. The three simple floor-timbers in the trapezoidal section, with a narrow foot with rectangular limber holes, and a broader, rounded upper side, were lashed to the planking. They were made of alder and wild pear. The imprints of three other frames were also noted on the hull. The interior of the boat was covered by a thick layer of pitch.[101] Considering the date, and the transitional form of the boat from dugout to keeled plank type, it is possible to assume that these sewn plank vessels developed during the transitional period from the late Bronze to the early Iron Ages, at the time when the distinct Iron Age Histrian

---

[100] Koncani Uhač *et al.* 2017: 36-38; Pomey, Boetto 2019: 8. The term 'gunwale' is used here for sheer strake.
[101] Koncani Uhač *et al.* 2017: 40-42; Pomey, Boetto 2019: 8-9.

4. Archaeological and Iconographic Evidence in Protohistoric Eastern Adriatic

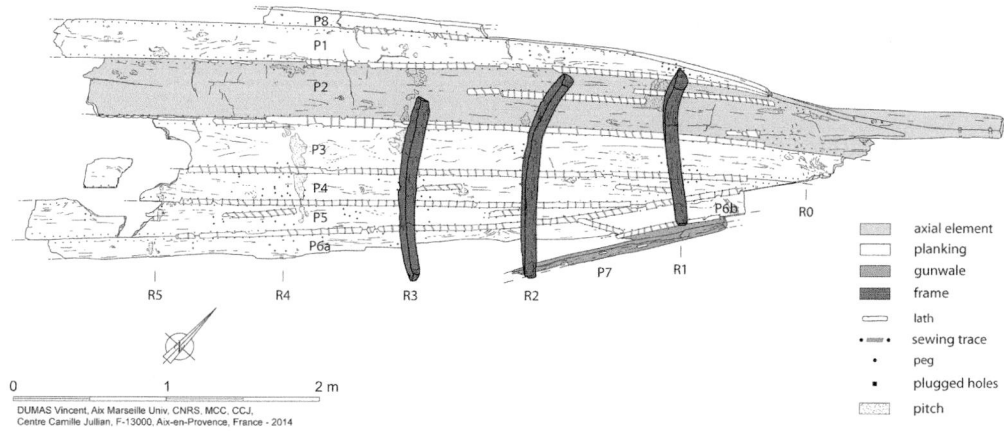

Figure 6. Drawing of the sewn boat of Zambratija (drawing: V. Dumas, from Boetto *et al.* 2017, courtesy of I. Koncani Uhač.

archaeological culture started to develop on the Istrian Peninsula. The boat, at the moment, remains in situ.

4.1.2. Pula

Pula is situated at the south-western end of the Istrian Peninsula, at the bottom of a deep and sheltered bay. At the beginning of the 1st millennium BC, the local population established a fortified settlement at the hill called Kaštel, situated in the present-day historical part of the city.[102] During the lifetime of Julius Caesar, *Colonia Pietas Iulia Pola* was established at the site and it soon developed into the significant and prosperous imperial city of Pola.[103] The main prehistoric settlement in the vicinity of Pula was Nesactium,[104] in which a *situla* with the scene of *naumachia*, dated around 500 BC, was found (see 4.2.3. below). In 2013 the remains of two sewn plank ships, of an estimated length of 10 and 15 m respectively, were rescued in the area of the ancient port of Pola (Figure 7). Because of the rescue nature of the excavation, only the extant remains inside the trench were accessible. Thus, only the central part of Pula 1 and one extremity of Pula 2 were available for research (Figure 8). Stratigraphic and radiocarbon dating suggests a date from the 1st to the first half of the 3rd century AD.[105]

Pula 1 was excavated at a length of 8.1 m, and a width of 4.1 m. It consisted of the keel, rectangular in sections, with six 17.5-21 cm sided and 17 cm moulded strakes on each side, along with 14 floor-timbers. However, it was not possible to determine the position of the bow and stern. Imprints of four additional floor-timbers were also recorded. The planks that formed the strakes were 2.4 cm thick, whilst the thickness of the garboard strakes reached

---

[102] Mihovilić 2014: 71-75.
[103] Matijašić 1996. For the establishment of the Roman colony, see Fraschetti 1983.
[104] Buršić Matijašić 2008: 153-55; Mihovilić 2014: 61-69.
[105] Boetto *et al.* 2014: 2017.

Figure 7. The sewn ships of Pula
(photo: T. Brajković, from Boetto *et al.* 2017, courtesy of I. Koncani Uhač).

3.5 cm. The maximum width of the strakes was 40-49 cm. The floor-timbers were, on average, 9.2 cm sided and 11.7 cm moulded, and the space between them measured on average 35 cm. These timbers had a series of recesses from the bottom side, placed in order to avoid excess pressure on the seams, and the others at the bottom of the ship, rectangular in section, acted as limber holes. They were tree-nailed to the rest of the hull, and only a single floor-timber, labelled R19, was fixed to the keel by an iron bolt. The floor-timbers R13 and R19 bore evidence of rectangular recesses likely meant for hoisting the mast-step. The keel of the ship was made of frames of deciduous oak, evergreen oak, elm, and walnut, as well as planking of elm, and pegs of fir and spruce for fixing the vegetal strings.[106]

Pula 2 was excavated at a length of 6.1 m, and was preserved at a width of 2.1 m. It had a keel, rectangular in section, which terminated with a scarf which probably originally functioned as hosting for the knee or post. The keel/post was locked in place by a horizontal key and a vertically positioned treenail and iron nail. There were eight strakes on the northern side, and five on the southern side of the ship, with no details that would indicate the position of the bow and stern. The planks that formed the strakes were 1.5-2 cm thick, and reached a width of 16 cm. The wale, discovered amongst 20 scattered pieces, was about 5.8 cm thick, and about 6.7 cm wide. At the edges of two planks, two edge fasteners were noted, 0.5 x 0.6 cm thick and 7 cm long. The traces of 11 frames were also detected, 40 cm from each other. The 20 displaced

---

[106] Boetto *et al.* 2017: 195-96.

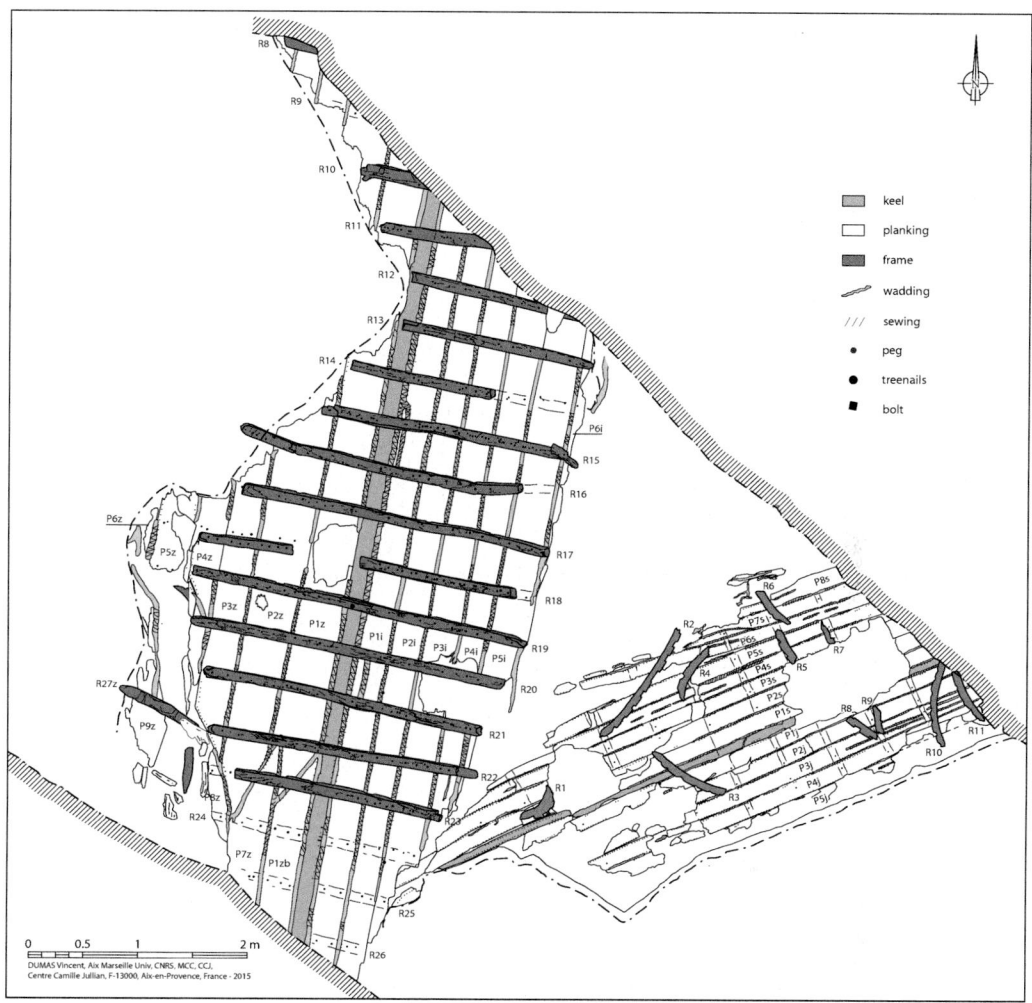

Figure 8. Drawing of the sewn ships of Pula (from Boetto *et al.* 2017, courtesy of I. Koncani Uhač).

frames consisted of floor-timbers, futtocks, and one cant frame, rectangular in section, with an average measurement of 4.8 cm sided and 7.5 cm moulded. The frames were crenelated from the bottom side and fixed to planking with treenails and copper nails. The keel of the ship was made of oak, the knee or post of evergreen oak, the displaced frames of deciduous oak, and the planking of beech.[107]

The stitching on the two ships is slightly different. On Pula 2 it follows the simple pattern (/// pattern), and on Pula 1 the pattern was described as 'overedge stiches with a clamping turn' (/I/I/I pattern). The second pattern is considered stronger, and suitable for the bigger of the two ships.[108] The stitches pass through diagonal stitch holes, over which a wadding

[107] Boetto *et al.* 2017: 196.
[108] Pomey, Boetto 2019: 12.

pad was used for sealing the seams. The holes are also blocked by little pegs, separated by an average distance of 4.6 cm (Pula 1), and 2.4 cm in the case of Pula 2. The interiors of both ships were heavily pitched.[109] The ship named Pula 1 has been reconstructed as a 15 m-long sailing ship. Pula 2 was probably a coastal craft, propelled by oars and sail, and its length has been estimated as 10 m.[110] Both vessels were recovered and sent to the company Arc Nucleart for PEG (Polyethylene Glycol) treatment.

4.1.3. Caska on the island of Pag

The Bay of Caska is situated at the northernmost end of the Gulf of Pag, on the north Dalmatian island of the same name. The island of Pag was mentioned by Pliny the Elder as Cissa or *Cissa portunata*, with an indigenous hillfort acting as a centre of economic and political power, probably situated on the hill of Košljun.[111] The settlement controlled the fertile plains of Novalja and the three adjacent bays of Novalja, Stara Novalja and Caska, along with the whole Gulf of Pag. It seems that the present-day port of Novalja served as important port on the eastern Adriatic seafaring route in the Roman times. The evidence from the Bay of Caska attests to the presence of a large economic estate, probably owned by the senatorial family of Calpurnii Pisones at the end of the Roman Republic and during early Roman Empire.[112]

The systematic coastal research project in the Bay of Caska, which started in 2005, revealed many interesting components of the Roman settlement complex at the site, including the remains of various submerged coastal structures made of rocks, wooden elements (logs, poles, beams, planks, etc.), and scuttled ships. The remains of the sewn boat Caska 1, measuring 8 m in length, and 1.66 m in width, were discovered in 2007, and excavated in 2009 and 2010 (Figures 9, 10).[113] The boat was filled with smaller rocks and mortar, and scuttled as part of linear structures of unknown function. These structures were made of wooden poles, probably once supporting walking surfaces made of planks. The boat was composed of a keel, six strakes on each side, and seven floor-timbers, all found *in situ*. Due to the absence of typical elements, for instance, just a small part of one post is extant, it was not possible to determine between the bow and stern of the boat. The keel was preserved at a length of 6.55 m. It was 5.7 cm sided and 6 cm moulded, and quadrangular in section, with filleted interior edges. At both sides of the keel, there were scarfs which aimed to keep the stem and stern-post secured by a horizontal key and a vertical treenail. The strakes were made of 1.5-2 cm thick planks (pressed by the filling of rocks), which reached a maximum width of 16 cm. The garboard strakes were slightly thicker, measuring 3.5 cm. Seven oblique scarfs connected the planks on the western side of the boat, while on the eastern side, no scarfs were noted. A part of a wale was found out of place on the eastern side of the hull. It was 4.1 cm thick, and 6 cm wide. The planks were stitched together by vegetal string, using simple stitches (/// pattern) which passed through diagonal openings perpendicular to the plank edges. The string was blocked by small pegs, placed at a distance of 2.4 cm between them. The seams were covered by a wadding pad of vegetal origin, and the hull was heavily pitched from the inside and outside. One planking repair was noted on the eastern strake P12, between the floor-timbers labelled F7 and F9.

---

[109] Boetto *et al.* 2017: 195-96.
[110] Pomey Boetto 2019: 10.
[111] Plin. *HN* 3.140.
[112] Kurilić 2011: 69-70.
[113] Čelhar 2008; Radić Rossi 2007; Boetto, Radić Rossi 2017: 280-83.

4. Archaeological and iconographic evidence in protohistoric Eastern Adriatic

Figure 9. The sewn boat Caska 1 (photo: L. Damelet).

As already noted, seven floor-timbers remained *in situ*, and the traces of other seven were identified on the inner surface of the hull. They were 4.5-5 cm sided, and 6.5 cm moulded, and the average distance between them measured 39.5 cm. The floor-timbers were crenelated from the bottom side, and the openings in the lower part of the boat, rectangular in section, figured as limber holes. They were not attached to the keel, rather, they were fastened to the planking by tapered treenails driven from outside the hull and measuring 0.9 to 1.6 cm in diameter. Three pieces of frames found out of their original position bore elaborated ends with grooves on each side, probably meant to host the washboards above the sheer strake. The only entirely preserved piece measured 86.5 cm in length. The keel, a small remaining part of the post, and part of the wale were made of evergreen oak, the planks were made of beech, and the floor-timbers of deciduous oak. The treenails were made of olive tree and evergreen oak, and the pegs mostly of fir.[114]

The ship Caska 2, had a preserved length of 13 m, and its planking was built by applying the mortise-and-tenon joining technique. It was found in 2012 supporting the south-western end of a large pier made of rocks, and was then excavated in 2013-2015.[115] Before scuttling, the bottom of the ship was reinforced by several layers of flat wooden elements, in order to host the filling of rocks. Among other finds, another big portion of a sewn plank vessel, Caska 3, was found, measuring 3.25 x 0.83m (Figures 11, 12). Caska 3 was composed of seven strakes, with

---

[114] Boetto, Radić Rossi 2017: 280-283.
[115] Radić Rossi, Boetto 2011; Boetto, Radić Rossi 2017.

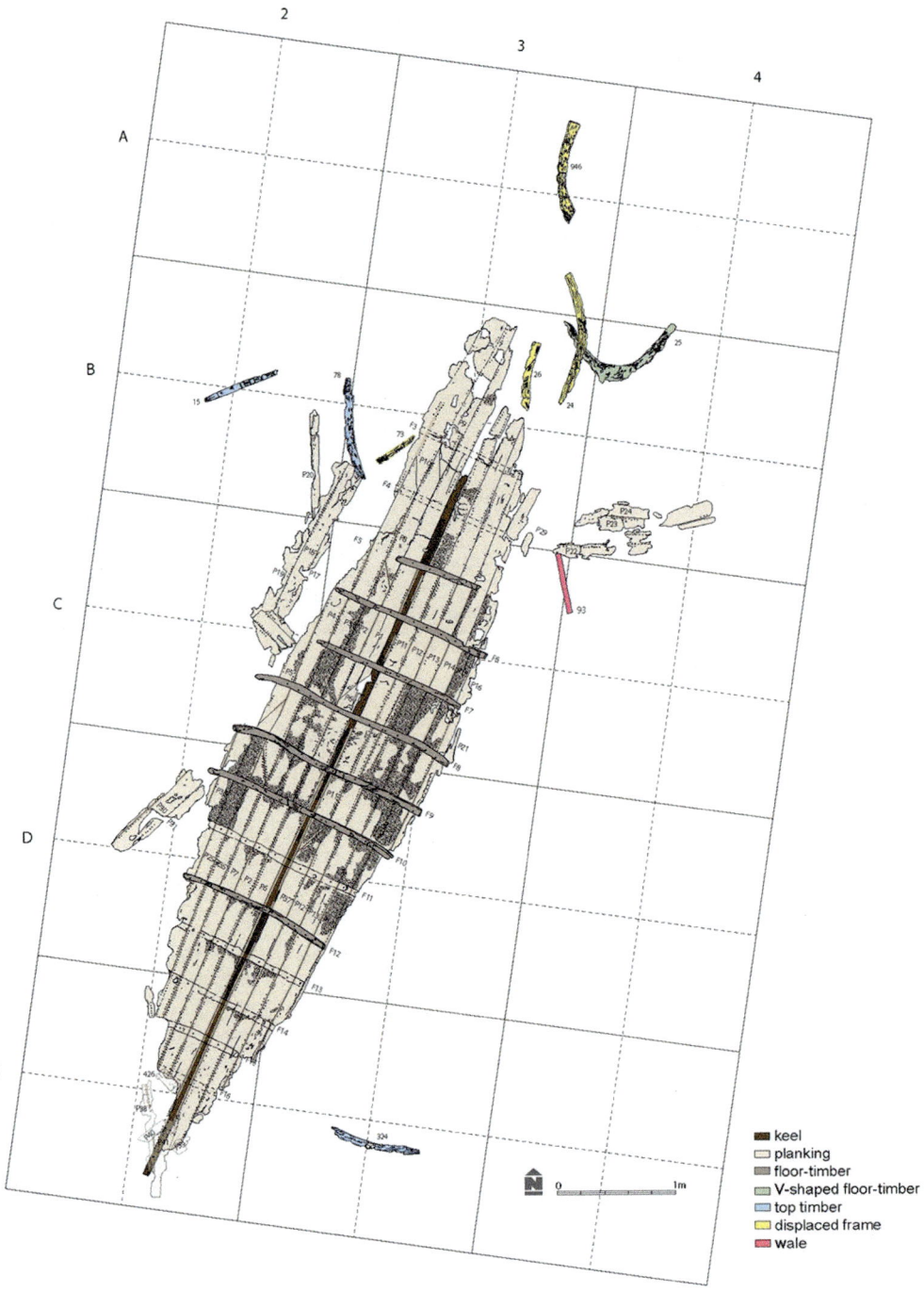

Figure 10. Drawing of the sewn boat Caska 1 (drawing: V. Dumas, from Boetto, Radić Rossi 2017).

Figure 11. Remains of the sewn boat Caska 3 (photo: T. Seguin).

no preserved keel or frames. Three of the preserved planks were stealers (P8, P9 and P10a). The maximum width of the planks was 20 cm, and the maximum thickness 1.5 cm (pressed by the filling of rocks). They were stitched together by simple stitching (/// pattern), similar to Caska 1, and blocked with pegs which were spaced on average every 3.8 cm. The wadding pad made the seams watertight, and the whole structure was covered by a thick layer of pitch. In the pitch layer there were the imprints of six frames, which were 6.2 cm sided, and 39 cm from each other. The planking was made of beech.[116]

One more sewn plank boat known as Caska 4 was discovered in 2016, as part of the rectangular structure consisting of wooden caissons made of logs, and filled with rocks (Figures 13, 14). The boat was obviously scuttled to complete the south-eastern end of the structure, which seems to have been built with the scope of protecting something that existed on land, and needed to be protected from the destructive effects of the south-eastern wind. Although in quite a bad state, it was preserved at a length of about 7 m, and a width of about 3 m, with ten strakes on the northern side, eleven strakes on the southern side, and fragments of just three floor-timbers *in situ*. As was the case with the Caska 1 and Caska 3, the position of the bow and stern was impossible to determine.

The keel of the ship was preserved at a length of 5.7 m, measuring 4.5-5.5 cm sided, and 7.5-11.4 cm moulded. It was connected by scarfs to eastern and western transitional timbers. The scarf to the west was secured by one vertically driven treenail, and the scarf to the east was secured

[116] Boetto, Radić Rossi 2017: 285-286.

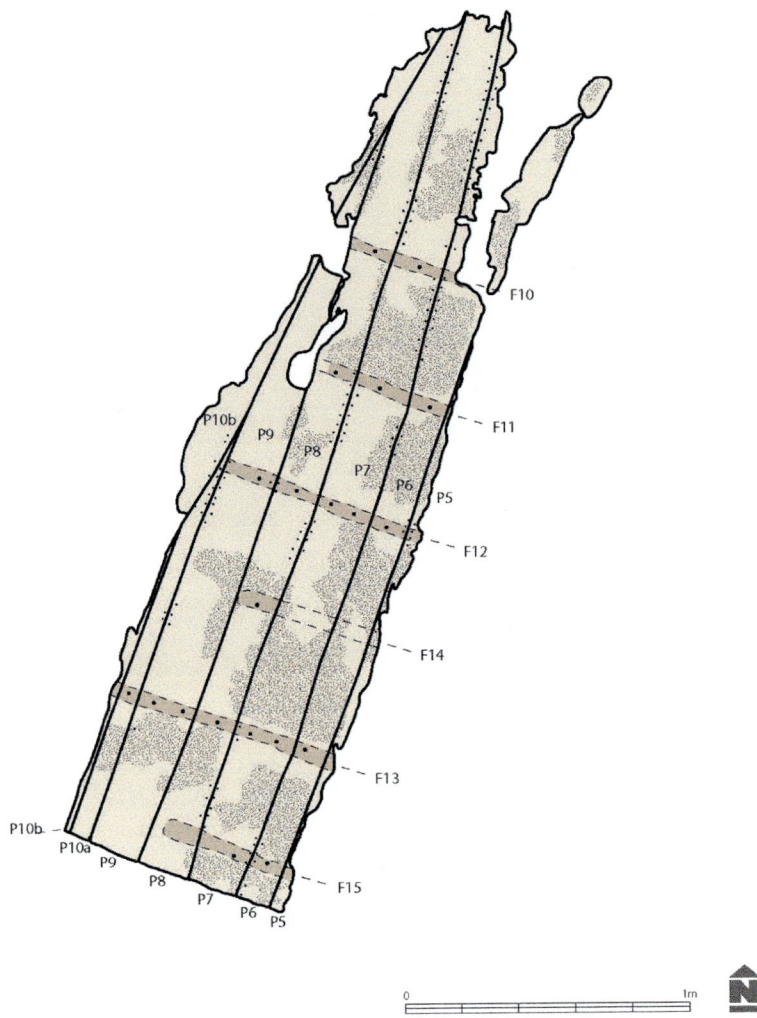

Figure 12. Drawing of the sewn boat Caska 3 (drawing: P. Poveda, from Boetto, Radić Rossi 2017).

by a horizontal key and two vertically driven treenails. The maximum width of the strakes was 16 cm, and the average thickness 1.65 cm. Just one strake to the south (P10) was thicker, measuring 2.9 cm. The planks were stitched together by simple stitching (/// pattern), similar to Caska 1 and Caska 3, and fixed into holes by small pegs, 0.7 cm in diameter, and spaced on average 2.87 cm apart. The stitching passed over the wadding pad, and the interior of the boat was covered by thick layer of pitch. One edge fastener, 4.4 cm long and 8 x 5 cm thick, was spotted in a detached fragment of plank. There were three floor-timbers and four other frame elements, rectangular in section, 4-5 cm sided, 7.2-10 cm moulded, and crenelated on the bottom sides. The floor-timbers were connected to the futtocks by hook scarfs, fixed by two or three vertically driven treenails. The frames were not treenailed but lashed to the planking, which certainly points to the older shipbuilding traditions. The keel of the boat was made of

Figure 13. The sewn boat Caska 4 (photo: L. Roux).

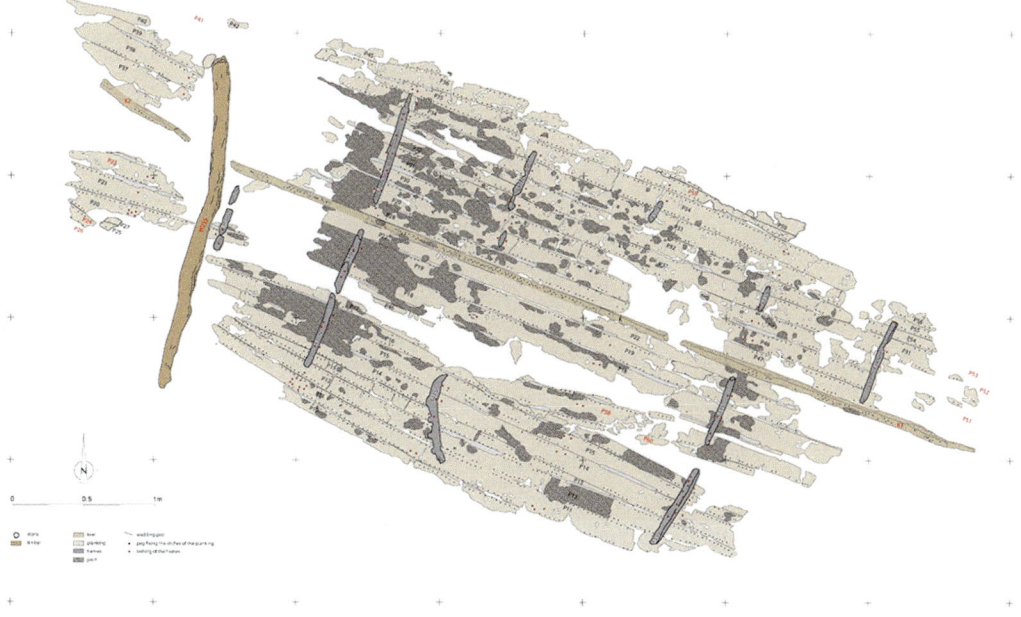

Figure 14. Drawing of the sewn boat Caska 4 (drawing: V. Dumas).

evergreen oak, the transitional timber to the west of maple, and the transitional timber to the east of deciduous oak. The planking was made of beech, the frames of deciduous and evergreen oak, whilst the pegs were made of fir, spruce and larch.[117]

The AMS radiocarbon dating of Caska 1 provided a dating between AD 42 and 107, meaning that the boat was probably scuttled sometime in the 1st or 2nd century.[118] The dendrochronological dating of the other finds is still in progress, but the archaeological context suggests that all the submerged coastal structures belong, more or less, to the same period, perhaps exploiting boats of various ages. The issues with the archaeological remains from the Bay of Caska are numerous, and will be presented on another occasion, but the presence of three sewn vessels may be ascribed to the older indigenous (Liburnian) shipbuilding traditions, which continued to exist in the Roman era. Caska 1 and Caska 4 remained *in situ*, whilst Caska 3 was deposited underwater, after proper documentation.

4.1.4. Zaton near Nin

The port of Zaton is situated on the mainland coast, between the island of Pag, and the city of Zadar. It is a vast and shallow sandy bay where the well-known tourist village of Zaton is located; the remains of Roman-era architecture are visible under the sea and on the coast. Judging from underwater archaeological finds, during the first centuries of the Roman Empire it acted as an active outer port of Roman Aenona (present-day Nin), a Roman town which succeeded a wealthy Iron Age settlement, situated on the small island hidden in the inner lagoon of the Gulf of Nin. A layer of local Iron Age pottery, documented in shallow waters in the locality of Kremenjača,[119] confirms the use of the port of Zaton by local Liburnian communities for their seafaring activities and seaborne trade, activities which are evidenced in the archaeological assemblages found in and around Nin.[120] However, it seems that the port was used most intensively between the early 1st and late 3rd century AD, which is evidenced by the abundant finds of pottery and glass from that period.[121]

In 1966, the first remains of the sewn boat Zaton 1 emerged in the immediate vicinity of the large submerged breakwater at the earlier mentioned locality of Kremenjača.[122] The boat was excavated and recovered in 1979.[123] It measured 6.5 m in length, and 3.5 m in width, with a keel, ten strakes on each side, six floor-timbers *in situ*, and five additional imprints of other floor-timbers (Figures 15, 16). The planks that formed the strakes were from 13 to 17 cm wide. They were stitched together by simple stitches (/// pattern), passing across the wadding pad that sealed the seams. They were fixed by small pegs, placed 2-2.5 cm from each other. The interior of the boat was covered by pitch. The frames, 4-7 cm moulded, and placed at a distance of 40-50 cm, were crenelated on the bottom sides, with bigger trapezoidal holes at the bottom of the

---

[117] Radić Rossi, Boetto, forthcoming.
[118] Radić Rossi 2010; 2011; Radić Rossi, Boetto 2010; 2011; Boetto, Radić Rossi 2017: 286.
[119] Brusić 1973.
[120] Brusić, Domijan 1985: 67-69; Brusić 2002.
[121] Gluščević 1986a; 1986b.
[122] Brusić 1969: 207-09. The boat was introduced to the wider scholarly community as Nin 1 (Brusić, Domijan 1985), which caused confusion with the two mediaeval boats found in Nin (Brusić 1978; Radić Rossi, Brusić 2014). Over the last decade, it was replaced by the term Nin – Zaton 1 (Radić Rossi 2011: 110), which helped the international scholarly community to accept and introduce the name in its correct form, Pomey, Boetto 2019: 10.
[123] Brusić, Domijan 1985: 69-71, Brusić 1995: 40-41.

4. Archaeological and iconographic evidence in protohistoric eastern Adriatic

Figure 15. The sewn boat Zaton 1 during the course of the 1979 research campaign (photo: Z. Brusić).

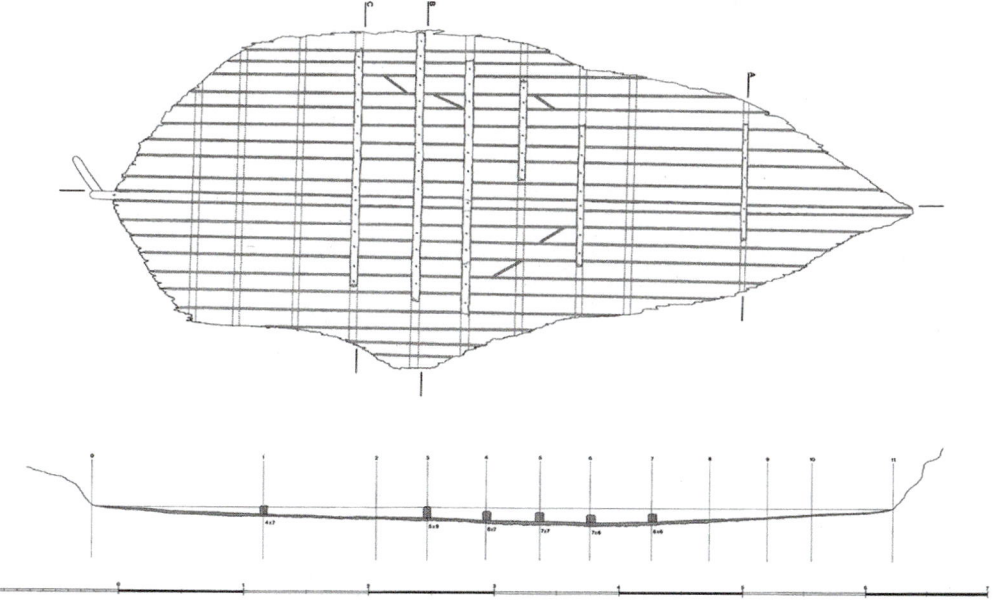

Figure 16. Drawing of the sewn boat Zaton 1 (drawing: Z. Brusić).

ship serving as limber holes. They were fixed to the planking by vertically driven treenails.[124] After conservation, pieces of the boat were deposited in the store rooms of the Archaeological Museum in Zadar, but over time they were misplaced, and at present there is no possibility of further proof or adding information.

The second sewn boat, Zaton 2, was discovered in 1982, and was excavated and recovered in 1987.[125] It was preserved at a length of 8.5 m, and a width of 2 m. The boat consisted of a keel, planking, 26 floor-timbers *in situ*, six limber boards, and two timbers found towards the very end of the preserved part of the hull, with recesses for hosting stanchions of an unknown function, originally interpreted as double mast-step (Figures 17, 18).[126] The planks were 13 to 17 cm wide. The original length of the vessel was estimated as 10 m.[127] After the recovery, the pieces of the boat were deposited in basins with fresh water. Recently they were PEG treated, and are still waiting to be reassembled. The third boat, Zaton 3, was discovered and partly excavated in 2002 and 2003.[128] It remained *in situ*, and the systematic excavation was performed in 2019, with the results still to be published (Figure 19).[129] The 6.6 m long, and 2.7 m wide boat remains were unearthed, consisting of a keel, fifteen strakes on the starboard side and six on the port side, ten frame stations *in situ*, a long mast-step with two recesses, and several limber boards.

Figure 17. The sewn boat Zaton 2 during the course of the 1987 research campaign (photo: Z. Brusić).

The frames of Zaton 2 and Zaton 3 are rectangular in section, crenelated, and tree-nailed to planking. The sewing is made by continuous passage of vegetal string (*///* pattern) over a wadding pad which seals the seams, and the only difference is in the small holes for hosting the string, which, in the case of Zaton 3, are smaller and of different shape.[130] The interiors of the hulls were covered by pitch. The xylological analysis of wooden material of Zaton 2 showed that the frames were made of oak, keel of sycamore, and planks mostly of beech, but also fir

---

[124] Brusić, Domijan 1985: 71-77; Brusić 1995: 40-42.
[125] Brusić, Domijan 1985: 77; Gluščević 1987; Brusić 1995: 41.
[126] Brusić, Domijan 1985: 77, 81; Brusić 1995: 42.
[127] Brusić 1995: 47.
[128] Gluščević 2002; 2004; 2008.
[129] The operation was directed by Dušanka Romanović from the Archaeological Museum in Zadar and Irena Radić Rossi from the University of Zadar.
[130] Gluščević 2004: 48-49.

4. Archaeological and iconographic evidence in protohistoric eastern Adriatic

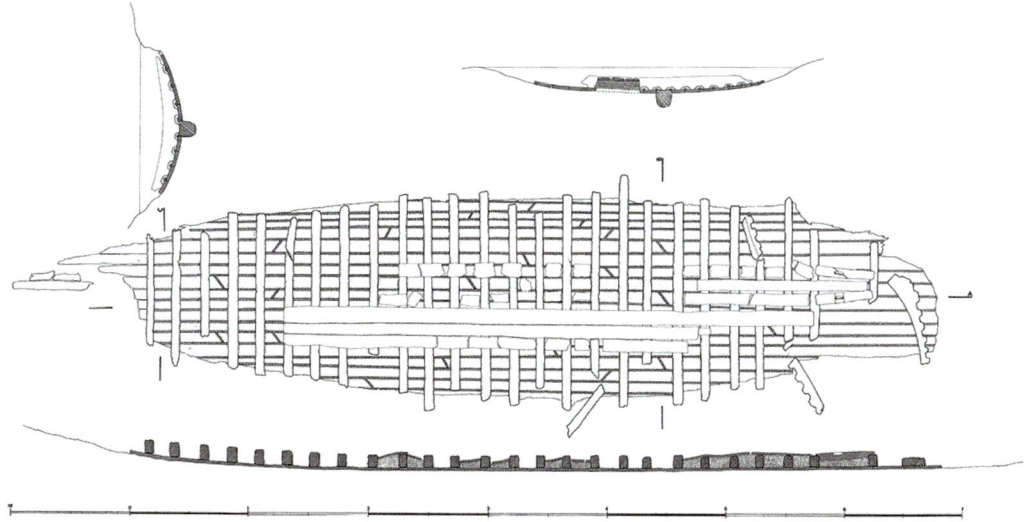

Figure 18. Drawing of the sewn boat Zaton 2 (drawing: Z. Brusić).

Figure 19. The sewn boat Zaton 3 during the course of the 2019 research campaign (photo: D. Romanović).

and sycamore.[131] The radiocarbon dating of two samples of Zaton 2 gave the result of 326±142 BC.[132] As the oldest dated currently known boat, with the exception of Zambratija, Zaton 2 was therefore likely used by the late Iron Age population in Liburnia.[133] The date of the scuttling of one ship was provided by a coin of the emperor Vespasian, found directly over the ship Nin 1, dating the event to the second half of the 1st century AD.[134]

4.2. Iconography

The current state of iconographic evidence for pre-Roman eastern Adriatic ships is rather modest. It consists of several visual representations on functional objects and stone monuments, and a number of images of ships on south Adriatic coinage from the 3rd and 2nd centuries BC. This small number of visual representations of ships in the eastern Adriatic during the Iron Age is probably the consequence of general features of local indigenous visual taste, which was more inclined to geometric decorations rather than figurative. In this section, the iconographic evidence is presented in chronological order. Due to the obvious limitations of such sources, which have already been addressed at length by a range of scholarly works, we do not intend to explore the specific features of the depicted ships, but to instead present their general features and their possible connections (or lack of them) with the ships mentioned in the Graeco-Roman written sources listed in Chapter 5 and discussed in Chapter 6.

4.2.1. Grieves from the Ilijak burial mound on Glasinac

The oldest images of ships come from two bronze grieves, now kept in the National (Zemaljski) Museum of Bosnia and Herzegovina in Sarajevo (Figures 20, 21). The grieves were discovered in 1893 as part of elite grave-assemblage in Barrow III (grave III/9) on the site of Ilijak on the Glasinac plateaux in eastern Bosnia, far from the sea.[135] They were made of breast armour, which was originally produced around the 8th century BC. Close parallels with this breast armour were detected in the valley of the Mati river in northern Albania and are also dated to the 8th century BC. The deposition of the grieves into a grave-assemblage from the Ilijak tumulus is dated around the 7th century BC.[136] The origins of the armour are assumed to be either from the workshop in northern Albania, or travelling craftsman coming from this area.[137] There are all together four images of ships, one on the upper and one on the lower parts of both grieves (Figure 22). These images show highly stylised long ships in iconography reminiscent of the Aegean Geometric style, with raised masts and sails. They have large bows with zoomorphic figureheads facing outwards, with stylised triangular representations of rowers, and a figure of a helmsman on one of the ships. The ships have visible extensions on the lowest part of the bow, usually interpreted as a ram or cutwater. Taking into account the debate on the appearance of rams in the Mediterranean, implying an earlier date for the

---

[131] Lipschitz, Gluščević 2015: 159-160.
[132] Brusić 1995: 43, cf. Srdoč *et al.* 1984: 451 (no. Z1041 130±120 BC).
[133] Brusić, Domijan 1985: 83.
[134] Brusić, Domijan 1985: 81-82; Brusić 1995: 43.
[135] Fiala 1895: 11-12, fig. 23-24. See re-assessment of Fiala's finds and redrawn images in Benac, Čović 1957: 36-38, T.16,2-3, and more detailed drawing of the T.16,2 in Čović 1976: 276, fig. 154.
[136] Čović 1976: 235-36, 275-77; 1984: 19-20: 1987: 592, cf. Kilian 1973 and Tiboni 2018. The dating of the deposition goes from ca. 700 BC (Benac, Čović 1957: 31, 36; Govedarica 2002: 318, 322) to after 650 BC (Vasić 2010: 110 n.5, cf. Čović 1987: 601).
[137] Čović 1984: 20, Tiboni 2018 (the area of Mati in northern Albania), cf. Marijan 2000: 155, who ascribes them to central Italian workshops.

4. Archaeological and iconographic evidence in protohistoric eastern Adriatic

Figure 20. Drawings of the grieves from Glasinac/Ilijak (drawing: H. Volfart, from Benac, Čović 1957).

Figure 21. The present state of the grieves from Glasinac/Ilijak (photo: A. Pravidur, courtesy of Zemaljski Muzej Bosne i Hercegovine, Sarajevo).

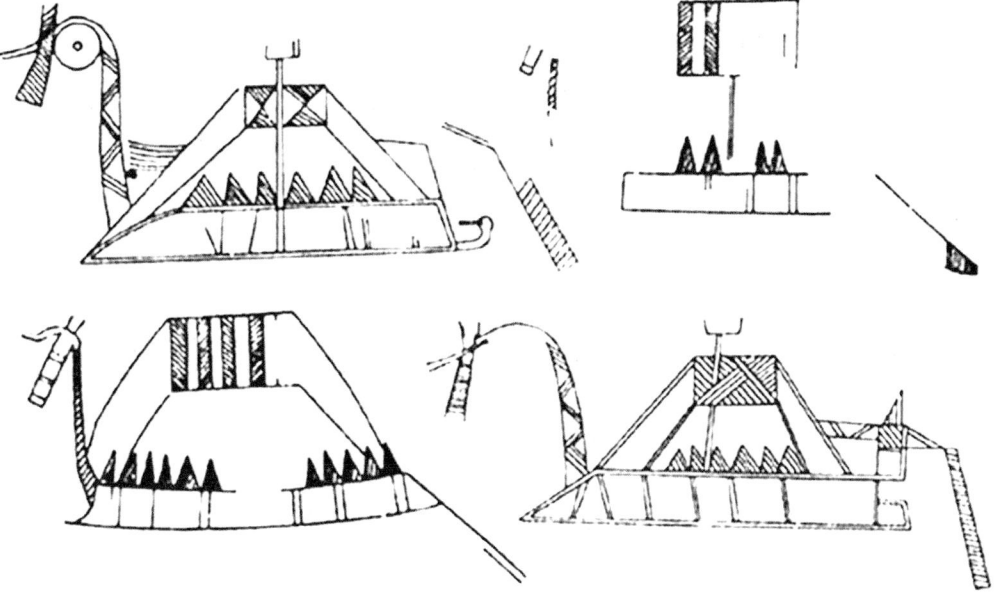

Figure 22. Drawings of the ships on the grieves from Glasinac/Ilijak (drawing: S. Kudra, from Čović 1976)

Figure 23. The Novillara Stele (courtesy of L. Braccesi).

appearance of cutwaters,[138] and the length of the long protruding bow ornament (*antiprosopon*), the second solution seems more probable.[139] There are small differences between the four ships in the shape of their zoomorphic figureheads, the thickness of protruding bows, the size of their cutwaters, and the shape of the sterns, characterised by differently represented additions mounted on the hull.[140] The steering devices are still visible on the original finds, although we have to rely on drawings to interpret them as steering oars. Medas explained the ships as the consequence of inter-Adriatic connections, relating them to the Novillara ships (see below), particularly with regard to their centrally positioned rudders.[141]

4.2.2. *The images of ships from the Daunian Stellae*

Seemingly slightly younger than the Glasinac grieves' images are the ships shown on the Novilara sandstone tablet from the western Adriatic coast, kept in the Museo Oliveriano di Pesaro. The tablet is dated *c.* 7th/6th century BC, with the engravings of three stylised ships (Figure 23). The largest of the ships is shown in an upper visual register with a raised mast with a square sail, whilst the two smaller ships are without masts and sails, and are clearly fighting each other in the lower register. The ships look very similar in design, with characteristic long animal-shaped bow ornaments facing outwards, similar to the ships from Glasinac. The bows

---

[138] See recently Murray *et al.* 2017.
[139] Similar to the Novilara ships, as pointed out by, for example, Basch (1987: 405) and Medas (2016: 148).
[140] The ships on the better preserved grieve (Benac, Čović 1957: T.16,2; Čović 1976: 276, fig. 154) have larger bows and visible cutwaters, while the ships on the other grieve have barely visible cutwaters and much thinner bows. There are also differences in shape of steering oar between ships in the upper part on two grieves.
[141] Medas 2016: 153-55.

have small extensions in their lowest part, representing probably a cutwater, which also recalls the Glasinac images. One possible explanation is that the stela showed the scene of battle in the lower register, and the celebration of the winning crew in the upper.[142] The Novilara images have been interpreted as a representation of north Adriatic naval traditions, either as examples of 'original' local traditions, or as the result of ongoing interactions between locals and Aegean shipbuilding traditions.[143] However, we know almost nothing about the original context of the find, as this stela was discovered on a different site from the other Novilara tablets, embedded in the pavement of the Roman villa near San Nicola di Valvamente in Pesaro, the Roman town of Pisaurum.[144] It was probably originally a funerary stela. The recent analysis of the Novilara ships by Tiboni points out that these visual representations show the ships in a Greek visual manner, increasing the level of uncertainty in regards to provenance, and rendering the idea that these ships are the product of local shipbuilding traditions very problematic.[145]

More recently, another Daunian stela with the image of a ship was found in the township of Cattolica in Italy during construction works in 2007. It is dated to the later 6th century BC.[146] On one side of the stela there are fighting scenes between two horsemen and a swamp shipping scene. On the other side there are damaged graffiti representations of two ship hulls parallel to one another. The larger one has its rowers shown, and the smaller one is depicted without rowers. Below the ships, sea creatures are visible – likely dolphins or large fish. Recent iconographic analysis by Medas establishes that the images of the ships were made in two different successive chronological phases with several adjustments, probably by the same craftsman. The general iconography of the Cattolica ships is consistent with other representations of ships on Daunian stelae. The later image of the larger ship with rowers, with its elongated bow and its distinct stern shape, fits with the iconography of other 7th-/6th-century BC images from the Adriatic area, such as those from Novillara and Glasinac. The author concludes that the image represents the peculiarities of Adriatic naval construction in the early Iron Ages, as dictated by the unique environmental conditions of the central-northern Adriatic basin.[147]

4.2.3. Situla from Nesactium

In 1981, the excavations at Nesactium, the leading settlement of ancient Histri in Istria, uncovered a grave vault with rich finds under the temples built on the Roman forum (*capitolium*). The grave vault comprised of several Iron Age cremation burials, belonging almost certainly to the elites of the settlement.[148] One of the damaged bronze *situlae* found in the grave vault, dated to *c*. 500 BC, was decorated with different images showing scenes of ploughing, and hunting, alongside a scene of a ship with rowers and warriors engaged in a sea-battle, and a poorly preserved scene of a possibly related land battle (Figures 24, 25). The sea-battle scene could

---

[142] Bracessi 2000: 33; cf. Tiboni 2009: 400-01.
[143] Bonino 1975. The Croatian scholarship, e.g. Jurišić 1983: 6-9; Zaninović 1988: 46; Kozličić 1993: 22-28 (with earlier bibliography) assumes that some of those ships from the Novilara tablet are liburnians, relying on older works of Novak and Nikolanci. However, the Italian scholars assess those ships differently – mostly as local Picene ships, or as a depiction of a battle between the local ships and the Greek vessel, see e.g. Cobau 1994: 31; Medas 1997: 122; 2016: 146-52; Bracessi 2000.
[144] Montebelli 2007.
[145] Tiboni 2009.
[146] Nava 2019.
[147] Medas 2019.
[148] Mihovilić 1996.

have been an idealised representation of a real event in which the owner of the *situla* participated.¹⁴⁹ The image of the bow of the ship is not preserved, but the stern of the ship is extant, and has a rounded hull with an inverted spar. Large figures of warriors with helmet-types known from Histrian (but also neighbouring) sites are on the deck, while only the heads of the rowers holding disproportionally large oars are shown. Mihovilić compared the ship from the Nesactium *situla* with the Etruscan cargo ship of Paglieri's 'rounded type', and recognised it as a representation of a battleship, or *serilia lembos (sic!)*, from the 7th or 6th century BC. Such an interpretation should be taken with caution, however, because the lemb as a type of ship appears in the sources only in 4th-century Aegean contexts (see p. 174 below), and even in this later period it has never been associated with the Histri.¹⁵⁰ The rounded hull of the ship from Nesactium looks very different from those depicted on the Glasinac

Figure 24. Reconstruction of the *situla* of Nesactium (from Mihovilić 1996).

grieves and the stela of Novilara, and it is very unlikely that these were images of the same ship types.¹⁵¹ Tiboni actually argued that the ship represented on the *situla* showed clear Etruscan influences, i.e. it was created by an artisan well aware of the Etruscan style. Although locally produced, the *situla* conserves a strong link with Padanian Etruria. Stylistically, it belongs to the objects grouped under the term of 'Situla Art' (*Arte delle situle*), which were produced in the southeast Alps in the 6th and 5th centuries BC.¹⁵²

4.2.4. Belt buckle from Prozor

The belt buckle from Prozor in Lika (central Croatia), dated *c.* 3rd/2nd century BC, depicts in its lower part supersized figures of two warriors fighting while standing on the bows of two ships (Figures 26, 27). The finds from the site of Prozor are associated with Iron Age Iapodean culture and include both locally made and imported artefacts. The circumstances of the find are unclear, probably coming from a grave assemblage, and an original drawing published in 1889 omits some details from the image.¹⁵³ The ship on the right has a visible inward-facing

---

¹⁴⁹ Mihovilić 1992; 1996: 19, 45-48, 58 (No. 66); 2004. Whether the scene of the sea-battle represented a real event is certainly open to discussion.
¹⁵⁰ Mihovilić 1992; 1996: 19; 45-47; 2004. Cf. different interpretations in Höckmann 1997: 192-93; 2000: 136-37 and Kozličić 1993: 28-29.
¹⁵¹ The idea of similar ship types is promoted by Medas (2016: 156), based on the presumed central steering device.
¹⁵² Tiboni 2017. On 'Situla Art', see concisely Frey 2011.
¹⁵³ Ljubić 1889: T.25. The photography of the belt buckle is reproduced in Kukoč 2009: fig. 322 and more accurate drawings in Jurišić 1983: 10, or Rendić-Miočević 1989: T.44,1. Jurišić (1983: 10) sees the images as ships, shown disproportionally in comparison with human figures, while Kozličić (1983: 116-17) thinks that images are proportional

4. Archaeological and iconographic evidence in protohistoric Eastern Adriatic

Figure 25. The ship image on the *situla* of Nesactium (from Mihovilić 1996).

bow ornament, proportionally much smaller than those shown in the images from Novilara and Glasinac, recalling more the bows of ships from southern Adriatic coinage discussed in 4.2.6. Even if it is not clearly visible on the drawing, the ship on the left probably had the same inward-facing bow ornament, ruined by the deterioration processes endured by the artefact. There are also visible extensions on the lower side of the bows, which could be depictions of the same cutwaters as on the above-mentioned coins. It is not possible to determine whether the differences between the warriors, with different head-protection equipment, and their boats should be attributed to the intention of the artisan, or to the state of the buckle's preservation.

The other motifs on this and other belt buckles of the same type from Prozor show similarities with belt buckles from the sites of Gostilj in the southern Adriatic and Ošanići in eastern Herzegovina. As such, it is likely that this Prozor belt buckle, with its scene of warriors standing on ships, was either imported from the southern Adriatic or was strongly influenced by these

---

and that these were small boats.

Figure 26. The belt buckle from Prozor (photo: D. Doračić, courtesy of the Archaeological Museum of Zagreb).

Figure 27. The belt buckle from Prozor (drawing: K. Rončević, courtesy of the Archaeological Museum of Zagreb).

sites' workshops.[154] The connection with southern Adriatic workshops and the dating of this belt buckle provide good grounds for cautiously recognising the possibility that these images might have depicted very stylised representations of south Adriatic lembs – see 4.2.6. and p. 173 below.

---

[154] Rendić-Miočević 1989: 556; Kukoč 1998, and Balen-Letunić 1996: 25.

4. Archaeological and Iconographic Evidence in Protohistoric Eastern Adriatic

4.2.5. Relief from Varvaria (Bribirska glavica)

There is only one known visual representation of a ship found in the Liburnian area. This damaged and thus far only superficially published stone relief from the settlement of Varvaria (Bribirska glavica) was discovered in 1986 as a spolia placed very low down in the wall of the Roman or medieval house at the site. It is currently kept in the on-site museum collection. More details are unfortunately unknown as only the photo of the original finding position exists and no other record from excavation diaries is known. From the photo, it is possible to locate the original finding place in the sub-locality of Tjeme, which was the centre of the ancient and medieval settlement on Bribirska glavica. The relief is difficult to date precisely, especially considering that it is one of only two stone reliefs from pre-Roman times discovered in Liburnia, together with an unpublished relief depicting a sexual act, also discovered in Varvaria. The relief with the ship shows a slightly different visual taste to other 1st- and 2nd-century AD figural representations discovered at this site, which are much more in tune with the Roman imperial visual templates. Thus, it seems, for the time being, justified to date the relief to pre-Roman times, perhaps within the last three centuries BC.[155] This is by no a means certain dating, and the relief will need proper stylistic analysis in the future.

The relief shows a scene of a phallus-like ship penetrating the vulva of a female figure (Figure 28). This was probably a symbolic representation of a ship entering a port, which is indicated by one hand of the woman holding a column which looks like berth for tying up ships.

Figure 28. The relief from Varvaria-Bribirska glavica (photo: Z. A. Alajbeg, courtesy of Museum of Croatian Archaeological Monuments).

---

[155] Dated to the 1st century AD by N. Uroda, in the exhibition catalogue Brstilo Rešetar, Gotić 2019: 108 no. 3.19, but no arguments for dating were presented.

Milošević and Krnčević assume that the woman is symbolically representing a neighbouring port of Scardona, with the small spherical object behind her back representing the settlement of Varvaria.[156] Part of a raised sail tied to the bow is visible, as well as three very stylised sailors on the deck – two noticeably smaller ones might have been rowers. There are no oars or oarlocks visible (there is an arrow-like ornament below one of the rowers – perhaps a raised oar?), the bow has no protruding ornaments, and the hull is decorated with the zig-zag and other ornaments characteristic for Liburnian art. No ram or any other extension is visible on the bow, which is shown in an unusually looking circular shape.[157] Damage to the relief makes it very difficult to make out much more about the origins of this image and the meaning of the scene. This image certainly had local meanings which are unknown to us and represented a symbolic ship which could have been the product of the imagination of the artist, or inspired by locally used vessels. Although the relief was discovered in a secondary, or even tertiary position, the location of its finding place in the central area of the settlement might cautiously indicate that its original context was public, rather than funerary. Another reason for such a hypothetical opinion is that amongst the Liburni, stone stelae as funerary markers are unknown before Roman times.

4.2.6. South Adriatic coinage

The images of boats of peculiar shape are shown on the coins of south Adriatic communities, politically connected with the Illyrian kingdom such as the Daorsi (ΔΑΟΡΣΩΝ), the Labeatae (ΛΑΒΙΑΤΑΝ), and the cities of Scodra (ΣΚΟΔΡΙΝΩΝ) and Lissus (ΛΙΣΣΙΤΑΝ).[158] These are indeed very likely to be the images of the ships described in the sources as the south Adriatic (Illyrian) lemb from the later 3rd and 2nd centuries BC, discussed in 6.1 below (Figures 29-32). Judging by the images on these coins, this type of ship might have had a recognisable shape.[159] The bow ornaments were facing inwards, not outwards like the early Iron Age images of ships from Novilara and Glasinac discussed earlier in the chapter,[160] and the visible projections were present on both sides in the form of a prolongation of the keel towards bow and stern.

It is difficult to recognise the rams in these extensions for three valid reasons. First, for efficient use of rams, sturdier ships than Illyrian lembs were needed. Second, the form of the bow with a bow ornament protruding further out than the bow prolongation of the keel, would not be suitable for ramming, and third, there is no evidence in the written sources that Illyrian lembs originally had rams attached. Improvements, such as lembs as biremes, and the attachment of rams to lembs for the battle of Chios (see p. 175 below) should be ascribed to Macedonian shipbuilders, not to those from the southern Adriatic. Therefore, we should see these extensions as cutwaters, which extended the waterline and increased stability and speed of the ship.[161] This feature in ship design is observable in some Bronze Age Aegean ships, represented by the clay model from Mochlos on Crete. The extensions of the keel in both directions on the

---

[156] Milošević, Krnčević 2017: 34-35.
[157] Reproduced but not discussed in Brajković 2008: 67, Pic.14. See also Milošević, Krnčević 2017: 34-35 fig.18 where the ship is described as 'sewn *liburnica*', and Uroda in Brstilo Rešetar, Gotić 2019: 108 no. 3.19.
[158] Brunšmid 1898; Ceka 1972; Islami 1972; Rendić-Miočević 1989: 283-94, 307-20; Dragičević 2016: 112-114.
[159] Kozličić 1981: 164-75. Thorough analysis of Kozličić also indicates that the images of ships on coins display ships of similar type, but with differences in ratio between length and width. This, in his opinion, implies that these images of the ships show both – the lemb as a merchant ship and as a warship or pirate ship.
[160] Höckmann (1997: 194; 2000: 138; cf. Medas 2004: 132-33) sees these as stern rams and Kozličić (1981: 167) describes these extensions as depictions of waves caused by movement of the ship.
[161] Džino 2003: 25, cf. Kozličić 1981: 167; Radić Rossi 2010: 94.

Figure 29. Coins of the Daorsi with images of ships (from Dragičević 2016).

Mochlos model testify to the existence of such a feature at the end of the 3rd millennium BC, and was interpreted as the transitional form from dugout to planked boat.[162] Considering the shape of the 'keel' of the Zambratija boat (above, 4.1.1), we should acknowledge the possibility of looking to the archaeological evidence for such a presumption. If so, perhaps the south Adriatic shipbuilding techniques changed over time, finding the solutions to maintain the features that proved to be efficient for local needs The bow and stern extensions would, first of all, extend the waterline, but also allow beaching of ships in both directions.[163] Moreover, the already expressed idea about the ships being designed to easily change sailing direction,[164] could be a reasonable justification for such a shipbuilding choice in the complex south Adriatic

---

[162] Basch 1983: 396-97; 1987: 132-33.
[163] Basch 1983: 396-97.
[164] Raban 1984. Raban's interpretation refers to the Minoan ships represented on the famous Miniature freeze from Akrotiri. Although it did not attract much attention, it seems a possible explanation for the perception of ancient authors about swift and agile Illyrian ships.

Figure 30. Coins of the Daorsi with images of ships (from Kozličić 1993).

'maritime cultural landscape'. This feature in design would certainly make the perception of ancient authors about swift and agile Illyrian vessels more understandable.

One potential ethnographic parallel comes from the other side of the world. A ship named the *kabang* is used by the semi-nomadic Austronesian Moken population who inhabit the Mergui Archipelago, claimed by both Thailand and Burma. The *kabang* consists of an extended dugout bottom and a superstructure which constitutes the boat's side shell plating.[165] The dugout

---

[165] Ivanoff 1999; Hinshiranan 2001.

4. Archaeological and iconographic evidence in protohistoric eastern Adriatic

Figure 31a-b. Coin of king Gentius, with a representation of a ship
(photo: Z. A. Alajbeg, courtesy of the Archaeological Museum of Split).

bottom is characterised by indentations at the front and back which strongly resemble the form of ships from south Adriatic coinage (Figures 33, 34). The shape of the bow and stern is suitable for climbing the boat and beaching it. At the same time, it has a symbolic meaning for the Moken people, representing the mouth that eats the sea and the anus that ejects it, that is, the ingesting and excreting parts of the human body.[166] Although the *kabang* are distant in space and time, it is possible to hypothesise that these ships provided a similar technological solution in regards to beaching it.

4.3. Protohistoric archaeological and iconographical sources for eastern Adriatic ships

The evidence for Adriatic shipping in the Iron Ages is therefore very modest, and does not help much in looking for the ancestors of the later south Adriatic lemb and the liburnian.[167] The only iconographic evidence which might actually help in the search for the liburnian and Illyrian lemb are the images of ships on local coinage from south Adriatic mints. The visible projections on the sterns of these ships are similar to the sterns of the ships represented on 2nd-century BC coins from the Macedonian region of Bottiaea (MAKEDONΩN; BOTTIATON), which Basch, at the time of his publication, considered unique in the shipbuilding practices of the Graeco-Roman world.[168] Taking into account the time when these Bottiaean coins were minted, it is possible to cautiously hypothesise that they depict vessels which were the outcome of experimentations with south Adriatic lembs conducted by the Macedonian king Philip V in the late 3rd century. These representations of ships correspond in space and time to the writings of ancient authors, such as Polybius, Livy and Appian, who refer to the ships used by these

---

[166] Ivanoff 1999: 118.
[167] Cf. Medas 2004: 137-38.
[168] Basch 1983: 407-10; 1987: 129-30. Basch uses the mythological origin of the Bottians to suggest their possible connection with the Minoans, and their stern appendage tradition, which is very problematic.

Figure 32. Coins from south-Illyrian mints (from Kozličić 1981).

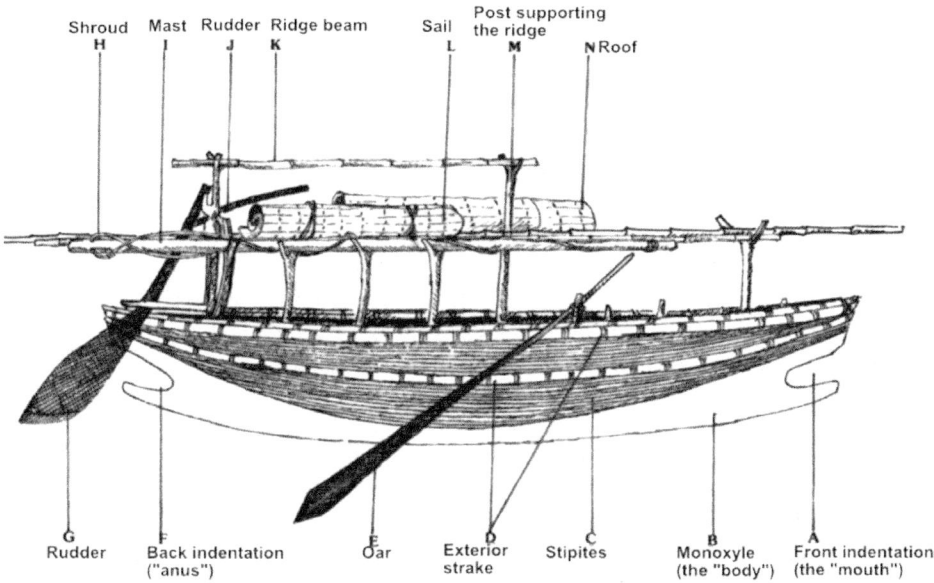

Figure 33. The Moken *kabang* (after J. Ivanoff, M. Bountry, http://www.lampipark.org/wp-content/uploads/2017/01/Moken-Sea-gypsies.pdf last accessed 9/7/2020).

Figure 34. A Moken man builds a miniature *kabang* (from Hinshiranan 2001).

communities in the 3rd and 2nd centuries BC as 'lembs' (see p. 175). Thus, it is reasonable to assume that the images of ships from these coins and the Illyrian lembs from the written sources are the same ships. These were certainly an important and powerful symbol for south Adriatic communities, and it is no accident that they are displayed on their coins. The ships from the Prozor belt-buckle belong to the same chronological framework and could also be hypothetically seen as a stylised representation of these same Illyrian lembs.

The other visual representations of the ships from the Iron Age Adriatic are problematic to interpret. Some scholars think that elements of distinct indigenous Adriatic nautical traditions can be recognised in the images of the boats from Novilara, Glasinac and Nesactium (a recent find from Cattolica should also be added to this list).[169] While

---

[169] Medas 1997: 104-05; 2004: 137-38; 2016; 2019: 59-66

there are indeed some similarities between the Glasinac, Novilara, and perhaps Cattolica graffiti images, which are chronologically the earliest, it is difficult to connect them with the ships shown on the Nesactium and Varvaria images. It is not clear which ships are represented on those images, as there are no indications to help precisely establish their origins. The only exception is the ship from Varvaria, which indeed could have represented the local Liburnian ship, but which is not possible to date with certainty at this moment. It has been suggested that the images of ships from Nesactium and Glasinac show the Etruscan boats, because of similarities with a bulky ship from the krater of Aristonothos from Caere, and other representations of Etruscan boats.[170] The Novilara ships have also been seen as being made in the Greek (or Mediterranean) shipbuilding traditions, in the same way as the Nesactium ship was at one stage compared with the Greek merchant ships, from second half of the 6th century, depicted in the Tomba della nave from Tarquinia.[171] So, as we can see, there are many different opinions in interpreting these images, which cannot be reconciled easily, and it is even more difficult to match them with the images of ships from 3rd century BC onwards, which are discussed below.

The context and meaning of the images also vary. The Glasinac image was a re-purposed foreign artefact used as prestige goods in the elite burial. The Novilara stele was also used in a local funerary and commemorative context, but was most certainly a local product of unknown age.[172] The Nesactium *situla* might commemorate a real event in which its owner or his ancestor participated, but it could also have been an image transmitted from mythology. Its visual aesthetics suited the local community, regardless of whether the artefacts was locally made, or produced somewhere in the neighbouring north Adriatic/south Alpine area. Finally, the relief from Bribirska glavica looks like a locally produced image, initially intended for public display, as it was found in the centre of the settlement rather than in a more private funerary context. The motif of a phallus-shaped boat symbolically entering a port shaped as a woman displays visual symbols which cannot be compared with other similar examples in Liburnia, so it is also difficult to discuss it beyond pure speculation. The craftsmen who made these images in all certainty were not experienced sailors, and it is highly questionable if they had any intention to represent the boats realistically. The images could also have been copied from other works of art, rather than being a direct model for the craftsmen who made these images.

Saying that, it is also important not to fully reject the idea that at least some distinct north Adriatic naval traditions persisted throughout the Iron Ages. The social networks linking both sides of the north Adriatic must have affected the shipbuilders on both coasts. Good evidence for this is the continued use of sewn plank boats in the Roman era, something attested on both sides of north Adriatic and in the written sources. In the 2nd century AD, Aulus Gellius, quoting M. Terentius Varro from two centuries earlier, wrote:

> *I believe that σπάρτα in Homer does not mean spartum, or 'Spanish broom,' but rather σπάρτοι, a kind of broom which is said to grow in the Theban territory. In Greece there has only recently been a*

– recognising these traditions only in the early Iron Ages. Jurišić (1983: 7-10) and Kozličić (1993: 20-22) connect these traditions with Illyrian lemb and liburnian.
[170] Nesactium: Mihovilić 1992: 74; 1996: 46; 2004: 101; Glasinac: De Boer 1992/93: 15.
[171] Novilara: Tiboni 2009; Nesactium: Mihovilić 1992: 74.
[172] Baldelli 1982: 27-28; Tiboni 2009: 405.

supply of spartum, *imported from Spain. The Liburnians did not make use of that material either, but as a rule fastened their ships together with thongs, while the Greeks made more use of hemp, tow, and other cultivated plants* (sativis), *from which ropes got their name of* sparta.[173]

Gellius' contemporary, Sextus Pompeius Festus, wrote a 20-volume epitome of the (at the time) 150-year-old encyclopaedic treatise, *De verborum significatione*, written by M. Verrius Flaccus. The text gives etymologies and meanings for the most common Latin words, omitting exotic or obsolete terms. Describing the term *serilia*, Festus cites Verrius saying:

> Serilia. Verrius thinks that this is a name given to Istrian and Liburnian ships whose grain is thickened with flax and broom; and that the name is derived from 'consero' and 'contexo'. His reason for believing this is that Pacuvius says in The Washing: 'And no tenon held fast the framework of the hull, but it was sewn with flax and plaitage of broom'; whereas it was a coined word which Pacuvius used, as a periphrastic turn, for ropes which are entwined, 'conseruntur', out of broom.[174]

It is clear that the term *serilia* for Verrius marked the rush-rope boats of the Histri and Liburni, with a literal meaning of 'cordage or ropes'. In support of this etymological explanation, Verrius seemingly quoted an early Roman tragic poet, Pacuvius (*c.* 220 - *c.* 130 BC). In his tragedy, Odysseus also built a boat to escape from the island of Ogigia. As the tragedy is lost, we cannot know whether Pacuvius referred to the Liburni in his text. The remarks of Varro cited by Gellius, and those of Verrius cited by Festus are convincing pieces of evidence that the inhabitants of Istria and Liburnia used sewn boats in the period preceding the 1st century BC, and continued to do so until at least the 2nd century AD. Verrius also informs us that the Latin term for these boats was *serilia*. This means that those ships were not called lembs or liburnians but had their own proper Latin name that distinguished them from others.

There are no sewn plank boat shipwrecks discovered in the southern Adriatic, as has been confirmed by recent extensive surveys of the Montenegrin and Albanian coastal waters as part of the Illyrian Coastal Exploration Program. This survey identified 27 ancient shipwrecks preliminarily dating from the 6th century BC up until the Late Roman period.[175] So, it seems very certain that, with the present state of the evidence, north Adriatic traditions of sewn ships had no parallels in the southern Adriatic. North Adriatic sewn ships belonged to several different regional traditions, which were recently defined by Pomey and Boetto as the northeast Adriatic (Istro-Liburnian), north/northwest Adriatic (Ravenna, Po delta, Venetian area and Ravenna), and continental traditions of the Danube basin. These authors convincingly connect the persisting tradition of using sewn-plank boats in the northern Adriatic with notions of a shared 'maritime cultural landscape', closed confined spaces, and the adaptation of shipbuilding methods to a particular geo-ecological zone. In the case of the Istro-Liburnian tradition, the persistence of using sewn plank boats is explained through a shared local indigenous tradition connected to strong cultural identity and the archipelagic characteristic of the eastern Adriatic coast, which: 'has specific sailing conditions, which are quite different to those of open-sea navigation, creating a defined sea space'.[176] This certainly

---
[173] Aul. Gell. *Noct. Att.*, 17.3.
[174] Festus, *Epitoma*, s.v. serilia, 508.33.
[175] Not all of the ships are published – these numbers were gathered from https://rpmnautical.org/projects/montenegro/ and https://rpmnautical.org/projects/albanian-coastal-survey/ (last accessed 6/2/2020). For the publications see: Royal 2012: 411-31; 2013; 2015.
[176] Pomey, Boetto 2019: 18-19, quote from 19. Boetto, Rousse (2012) in an earlier publication recognised only two

could be related to the unique 'maritime cultural landscape' of the area, with the ship-design its visible product. However, the naval traditions overall must have been affected in some degree by Mediterranean shipbuilding traditions which were relayed through complex processes of technological transfer, so the evidence for shipping from the 3rd century BC onwards should be seen in that context.[177]

Fig. 35. A representation of a *liburna* from the 16th-century manuscript of *De rebus bellicis*.

---

distinct Adriatic traditions of making stitched-boats, they called 'Roman-Padane' and 'Roman-Illyrian'.
[177] Dzino, Boršić 2020.

5. Written sources on lembs and Liburnians from the 4th c. BC to Late Antiquity

This section will present a collection of texts in which the terms *lembos* and *liburnica*, with their grammatical and etymological derivatives, appear in ancient Greek and Roman written sources. As we are going to demonstrate, the term *lembos* appears only in the 4th century BC, while the functional explanation remains uncertain. The term *liburnica*, on the other hand, appears in the written sources two centuries later. The Greek and the Latin sources are kept separate, i.e. first the Greek sources are listed, then the Latin. The Greek language authors are numbered with Roman numerals, and the Latin ones with Arabic numerals. The prefix A denotes mentions of lemb, and the prefix B the mention of liburnian. The examples are organised chronologically according to the author's time of birth – known or approximated. If the terms appear in more than one work in one author, they are again put in a chronological order according to the time of the composition of the work, if it can be established. Within the same work, all the instances in which the terms appear are put according to the position in the text in modern editions, and, finally, the texts are presented with their literary and historical contexts. In the tables reproduced in section 6.4 we combined the Greek and the Latin sources so that the reader is provided with a clearer overview of the historical development and usage of both terms.

5.1. Introduction

The terms searched for in this section are:

Greek 'lemb':
    λέμβος, noun
    λεμβῶδες, adjective, 'lemb-like'
Latin 'lemb':
    lembus, noun
    lembulus, noun, grammatical diminutive of lemb
    lembarius, noun, 'boatman on a lemb'
    lenunculus, noun, grammatical diminutive of lemb
Greek 'liburnian':
    λιβύρνα
    λίβυρνος
    λιβυρνίς
    λύβερνα
Latin 'liburnian':
    liburna
    liburnica

Some ethnic terms metonymically used for ships were taken into consideration as well.

The period considered in the selection of our texts was the period from the oldest known textual materials mentioning lembs and liburnians, up through to approximately the 6th century AD. Special attention was placed on early sources, since one of primary goals of this study is to establish the original connection between the two types of ships, and their relation to the eastern Adriatic population. Later sources were not always taken into account because in Late Antiquity and the Middle Ages both words became removed from their original meaning so that the analysis of their usage would not significantly contribute to a better understanding of their mutual relationships, and relation to the eastern Adriatic population. This is also the reason why some Greek and Latin words for lemb were omitted: they appear in later sources and are not relevant for our present research.

For the primary texts, the following online databases were the main sources:

- Thesaurus Linguae Graecae: http://www.tlg.uci.edu
- Centre Traditio Litterarum Occidentalium and Brepols. Library of Latin texts. Series A. Series A: http://clt.brepolis.net/LLTA/
- Centre Traditio Litterarum Occidentalium and Brepols. Library of Latin texts. Series A. Series B: http://clt.brepolis.net/LLTB/
- Bibliotheca Teubneriana Latina Online: https://www.degruyter.com/view/db/btl
- Thesaurus Linguae Latinae (TLL) Online: https://www.degruyter.com/view/db/tll
- Papyri.info, ed. by R. Bagnall, J. Sosin, et al.: http://papyri.info
- Loeb Classical Library (LCL): https://www.loebclassics.com/
- Digital Fragmenta Historicorum Graecorum: http://www.dfhg-project.org/
- The Latin Library: https://www.thelatinlibrary.com/
- Corpus corporum: http://mlat.uzh.ch/

For translation, the following dictionaries were used:

- H. G. Liddell, R. Scott, H. S. Jones, Greek-English Lexicon (LSJ): http://stephanus.tlg.uci.edu/lsj/#eid=1
- P. G. W. Glare, Oxford Latin Dictionary (OLD), https://www.oxfordscholarlyeditions.com/page/the-oxford-latin-dictionary.

These databases, if not otherwise indicated, were used as search machines: the relevant passages were found and reproduced here with the reference to the source from where the original text was reproduced as used in the databases. Any modifications of English translations are indicated in square brackets or in footnotes. If there is no reference to English translation, it means the translation was done originally by Luka Boršić for this book. The main source of information about the authors and some general information on the texts is the online edition of *Brill's Neue Pauly*. All other primary sources are indicated in their complete bibliographical description.

Finally, a few words on the completeness of the sources should be mentioned. We would like to be able to say that *all* places in which there is a mention of either lembs or liburnians are listed here; however, it would almost certainly not be true – not only for the obvious reason that browsing through ancient texts we might have missed some of the occurrences of these terms. This is purely our error and oversight. There are also other factors that render this whole project open to further *addenda*: the limitations of available literature, the limitations of otherwise excellent internet searching engines, but, above all, the fact that some sources were used with some limitations, especially *epigraphica* and *papyri*, which are still somehow

5. Written sources on lembs and Liburnians from the 4th c. BC to Late Antiquity

unsystematic and not easily available sources. However, this collection of texts presents a solid basis which should be amplified in some future research.

5.2. Lemb

5.2.1. Ancient Greek sources

A-I Demosthenes (384/383-322 BC)

A-I.1 *Plea of Demo against Zenothemis*

(6) ὡς δ' ἡλίσκεθ' ὁ Ἡγέστρατος καὶ δίκην δώσειν ὑπέλαβεν, φεύγει καὶ διωκόμενος ῥίπτει αὑτὸν εἰς τὴν θάλατταν, διαμαρτὼν δὲ τοῦ **λέμβου** διὰ τὸ νύκτ' εἶναι, ἀπεπνίγη. ἐκεῖνος μὲν οὕτως, ὥσπερ ἄξιος ἦν, κακὸς κακῶς ἀπώλετο, ἃ τοὺς ἄλλους ἐπεβούλευσε ποιῆσαι, ταῦτα παθὼν αὐτός. (7) οὑτοσὶ δ' ὁ κοινωνὸς αὐτοῦ καὶ συνεργὸς τὸ μὲν πρῶτον εὐθὺς ἐν τῷ πλοίῳ παρά τ' ἀδικήματα, ὡς οὐδὲν εἰδώς, ἀλλ' ἐκπεπληγμένος καὶ αὐτός, ἔπειθεν τὸν πρῳρέα καὶ τοὺς ναύτας εἰς τὸν **λέμβον** ἐκβαίνειν καὶ ἐκλιπεῖν τὴν ναῦν τὴν ταχίστην, ὡς ἀνελπίστου τῆς σωτηρίας οὔσης καὶ καταδυσομένης τῆς νεὼς αὐτίκα μάλα, ἵν', ὅπερ διενοήθησαν, τοῦτ' ἐπιτελεσθείη καὶ ἡ ναῦς ἀπόλοιτο καὶ τὰ συμβόλαι' ἀποστερήσαιεν.
6.4-7.5 (=882-891)

(6) Hegestratus, being caught in the act, and expecting to pay the penalty, took to flight, and, hotly pursued by the others, flung himself into the sea. It was dark, and he missed **the ship's boat**, and so was drowned. Thus, miserable as he was, he met a miserable end as he deserved, suffering the fate which he purposed to bring about for others. (7) As for this fellow, his associate and accomplice, at the first on board the ship immediately after the attempted crime, just as though he knew nothing of it but was himself in utter consternation, he sought to induce the sailing-master and the seamen to embark in **the boat** and abandon the vessel with all speed, declaring that there was no hope of safety and that the ship would presently sink; thinking that thus their design might be accomplished, the ship be lost, and the creditors thus be robbed of their money.
pp. 181-183

A-I.2 *Plea of Chrysippus and Partner against Phormio*

(10) Μετὰ ταῦτα τοίνυν, ὦ ἄνδρες Ἀθηναῖοι, οὗτος μὲν ἐν τῷ Βοσπόρῳ κατελέλειπτο, ὁ δὲ Λάμπις ἀναχθεὶς ἐναυάγησεν οὐ μακρὰν ἀπὸ τοῦ ἐμπορίου· γεγεμισμένης γὰρ ἤδη τῆς νεώς, ὡς ἀκούομεν, μᾶλλον τοῦ δέοντος, προσανέλαβεν ἐπὶ τὸ κατάστρωμα χιλίας βύρσας, ὅθεν καὶ ἡ διαφθορὰ τῇ νηὶ συνέβη. καὶ αὐτὸς μὲν ἀπεσώθη ἐν τῷ **λέμβῳ** μετὰ τῶν ἄλλων παίδων τῶν Δίωνος, ἀπώλεσεν δὲ πλέον ἢ τριάκοντα σώματα ἐλεύθερα χωρὶς τῶν ἄλλων. πολλοῦ δὲ πένθους ἐν τῷ Βοσπόρῳ ὄντος ὡς ἐπύθοντο τὴν διαφθορὰν τῆς νεώς, ηὐδαιμόνιζον Φορμίωνα πάντες τουτονί, ὅτι οὔτε συνανήχθη οὔτε ἐνέθετο εἰς τὴν ναῦν οὐδέν. συνέβαινεν δὲ παρά τε τῶν ἄλλων καὶ παρὰ τούτου ὁ αὐτὸς λόγος. καί μοι ἀνάγνωθι ταύτας τὰς μαρτυρίας.
10.7 (=907-921)

After this, men of Athens, the defendant was left in Bosporus, while Lampis put to sea, and was shipwrecked not far from the port; for although his ship was already overloaded, as we learn, he took on an additional deck-load of one thousand hides, which proved the cause of the loss of the vessel. He himself made his escape **in the boat** with the rest of Dio's servants, but he lost more than thirty lives besides the cargo. There was much mourning in Bosporus when they learned of the loss of the ship, and everybody deemed this Phormio lucky in that he had not sailed with the others, nor put any goods on board the ship. The same story was told by the others and by Phormio himself.
pp. 243-245

Greek: *Demosthenis orationes 2.2*, ed. W. Rennie. Clarendon Press, Oxford 1921.

English: *Demosthenes. Orations, Volume IV: Orations 27-40: Private Cases*, transl. A. T. Murray. LCL 318 (1936).

Besides his excellent education and rhetorical influence, Demosthenes dedicated several of his sixty preserved speeches to maritime problems. He was evidently quite familiar with the intricacies of maritime law system: his speech *Against Lacritus* contains the only surviving maritime loan contract from the period.[178] Thus, there no reason to doubt that his naval terminology was contemporary, quite carefully chosen and justified. So, besides being the earliest source of the name of the ship lemb, both citations suggest that lemb was a sort of a smaller vessel that accompanies a larger ship, that could serve also as a salvage vessel. However, it might have been bigger than a dinghy since it could carry more than thirty people, according to the passage in *Against Phormio*.

A-II Lycurgus (before 383-324 BC)

Against Leocrates:

17. Λεωκράτης δὲ τούτων οὐδενὸς φροντίσας, συσκευασάμενος ἃ εἶχε χρήματα μετὰ τῶν οἰκετῶν **ἐπὶ τὸν λέμβον** κατεκόμισε, τῆς νεὼς ἤδη περὶ τὴν Ἀκτὴν ἐξορμούσης, καὶ περὶ δείλην ὀψίαν αὐτὸς μετὰ τῆς ἑταίρας Εἰρηνίδος κατὰ μέσην τὴν Ἀκτὴν διὰ τῆς πυλίδος ἐξελθὼν πρὸς τὴν ναῦν προσέπλευσε καὶ ᾤχετο φεύγων, οὔτε τοὺς λιμένας τῆς πόλεως ἐλεῶν, ἐξ ὧν ἀνήγετο, οὔτε τὰ τείχη τῆς πατρίδος αἰσχυνόμενος, ὧν τὴν φυλακὴν ἔρημον τὸ καθ' αὑτὸν μέρος κατέλιπεν, οὔτε τὴν ἀκρόπολιν καὶ τὸ ἱερὸν τοῦ Διὸς τοῦ Σωτῆρος καὶ τῆς Ἀθηνᾶς τῆς Σωτείρας ἀφορῶν καὶ προδιδοὺς ἐφοβήθη, οὓς αὐτίκα σώσοντας ἑαυτὸν ἐκ τῶν κινδύνων ἐπικαλέσεται.
17 (=33-90)

Leocrates ignored all these provisions. He collected what belongings he had and with his slaves' assistance placed them **in the ship's boat**, the ship itself being already anchored off the shore. Late in the evening he went out himself with his mistress Irenis through the postern gate on to the open beach and sailed out to the ship. And so he disappeared, a deserter, untouched by pity for the city's harbours from which he was putting out to sea, and unashamed in face of the walls which, for his own part, he left undefended. Looking back at the Acropolis and the temple of Zeus the Saviour and Athena the Protectress, which he had betrayed, he had no fear, though he will presently call upon these gods to save him from danger.
p. 27

Greek: *Lycurgi oratio in Leocratem*, eds C. Scheibe, F. Blass, N. C. Conomis. Teubner, Leipzig 1970.

English: *Minor Attic Orators, Volume II: Lycurgus. Dinarchus. Demades. Hyperides*, transl. J. O. Burtt. LCL 395 (1954).

Lycurgus' elite education, as well as his political interests, especially linked to the Athenian fleet, could serve as an indication that his use of naval terminology is carefully chosen. The

---

[178] McCabe 2012.

speech *Against Leocrates* is the only preserved speech we have today. Within it, Lycurgus confronts the Greek terms λέμβος and ναῦς – for which reason we have the translation of *lembos* as 'the ship's boat', whereas *naus* presents a larger vessel. Contextually the relationship between the two is clear: one uses a *lembos* to reach a *naus*, thus it is safe to assume that lemb, in an educated Athenian context, was understood to be a smaller boat.

A-III Aristotle (384-322 BC)

De incessu animalium (*Progression of Animals*)

ὑπεναντίως δ' ἔχουσιν οἱ ὄρνιθες τοῖς ὁλοπτέροις τὴν τῶν πτερῶν φύσιν, μάλιστα δ' οἱ τάχιστα αὐτῶν πετόμενοι. τοιοῦτοι δ' οἱ γαμψώνυχες· τούτοις γὰρ ἡ ταχυτὴς τῆς πτήσεως χρήσιμος πρὸς τὸν βίον. ἀκόλουθα δ' αὐτῶν ἔοικεν εἶναι καὶ τὰ λοιπὰ μόρια τοῦ σώματος πρὸς τὴν ὠκεῖαν κίνησιν, κεφαλὴ μὲν ἁπάντων μικρὰ καὶ αὐχὴν οὐ παχύς, στῆθος δ' ἰσχυρὸν καὶ ὀξύ, ὀξὺ μὲν πρὸς τὸ εὔτονον εἶναι, καθάπερ ἂν εἰ πλοίου **πρῶρα λεμβώδους**, ἰσχυρὸν δὲ τῇ φύσει τῆς σαρκός, ἵν' ἀπωθεῖν τε [710b] δύνηται τὸν προσπίπτοντα ἀέρα· καὶ τοῦτο δρᾷ ῥᾳδίως καὶ μὴ μετὰ πόνου· τὰ δ' ὄπισθεν κοῦφα καὶ συνήκοντα πάλιν εἰς στενόν, ἵν' ἐπακολουθῇ τοῖς ἔμπροσθεν, μὴ σύροντα τὸν ἀέρα διὰ τὸ πλάτος.
710a32–710b4

But birds are in general at the opposite pole to flying insects as regards their feathers, but especially the swiftest flyers among them. (These are the birds with curved talons, for swiftness of wing is useful to their mode of life.) The rest of their bodily structure is in harmony with their swift movement, the small head, the slight neck, the strong and acute breastbone (acute like **the prow of a clipper-built vessel**, so as to be compact, and strong by dint of its mass of flesh), in order to be able to push away the air that beats against it, and that easily and without exhaustion. The hind-quarters, too, are light and taper again, in order to conform to the movement of the front and not by their breadth to sweep the air.
pp. 1104-1105

Greek: *Aristotelis de animalium motione et de animalium incessu. Ps.-Aristotelis de spiritu libellus*, ed. W. Jaeger. Teubner, Leipzig 1913.

English: *Complete Works of Aristotle*, ed. J. Barnes. Princeton University Press, Princeton 1984.

In this text Aristotle suggest that birds' breastbones have the form of a lemb so that they could cut the air which resists them while flying. The presupposition of this comparison to elucidate something less known by comparing it to something more known. The less known in this case must be the birds' breastbones, which implies that something being λεμβώδης 'lemb-like' was presumably known to the audience. Since the word itself – λεμβώδης – is a *hapax legomenon*, i.e. it appears only once in all known Greek texts, we cannot suppose that Aristotle expected his readers to know this term. However, we can conclude that they would recognise the etymological connection with a lemb, meaning that lemb must have been a well-known boat with a standardised shape in his time.

A-IV Anaxandrides of Rhodes (works between 376 and 349 BC)

A-IV.1 Kassel-Austin, Fragment 12 (from *Helen*)

Α. ἄγκυρα, **λέμβος**, σκεῦος ὅ τι βούλει λέγε.
Β. ὦ Ἡράκλεις, ἀβελτερείου τεμενικοῦ.
Α. ἀλλ' οὐδ' ἂν εἰπεῖν τὸ μέγεθος δύναιτό τις.
=Suidas, s.v. ἀβέλτερος

A. An anchor, **a boat**, a vessel, name anything you want.
B. O Heracles, what a silly priest!
A. However, nobody would be able to say the size.
p. 105

A-IV.2 Kassel-Austin, Fragment 35 (from *Odysseus*)

ὑμεῖς γὰρ ἀλλήλους ἀεὶ χλευάζετ', οἶδ' ἀκριβῶς·
ἂν μὲν γὰρ ᾖ τις εὐπρεπής, ἱερὸν γάμον
καλεῖτε· ἐὰν δὲ μικρὸν παντελῶς ἀνθρώπιον,
σταλαγμόν. λαμπρός τις ἐξελήλυθ', . . ὅλους
οὗτός ἐστι· λιπαρὸς περιπατεῖ Δημοκλῆς, ζωμὸς
κατωνόμασται· χαίρει τις αὐχμῶν ἢ ῥυπῶν,
κονιορτὸς ἀναπέφηνεν· ὄπισθεν ἀκολουθεῖ
κόλαξ τῳ, **λέμβος** ἐπικέκληται.
=Athen. *Deinosoph.* 6.242d-f

Because I'm well aware you always kid each other. If a guy's handsome, you call him Sacred Marriage, whereas if he's a midget, you call him Drop. Someone emerges from his house glistening; he's Pussy. Democles walks around covered with oil; he's dubbed Meat-broth. Somebody likes being dry or dirty; he's proclaimed Dust-cloud. A flatterer follows behind another person; he's nicknamed **Dinghy**.
p. 105

Greek: *Poetae Comici Graecae*: Vol. 2, eds R. Kassel, C. Austin. De Gruyter, Berlin–New York 1991.

English: *Athenaeus. The Learned Banqueters, Volume III: Books 6-7*, ed. and transl. S. Douglas Olson. LCL 224 (2008).

A-IV.2 is the only place where the term λέμβος is used in a derogative meaning, based on the obvious comparison between the lemb, a small boat following a bigger one, and the flatterer following another person. According to Harpocration (**A-XV**), this must have been a common practice in Greek comedies to describe flatterers as lembs. This implies that the reader was expected to be familiar with this specific function of lembs as smaller boats following bigger ones, which is in vein with Alexandrides being a comic writer. However, it is also interesting to note that in **A-IV.1**, the term *lembos* denotes a vessel of undeterminable size, more of a generic name for the boat.

5. Written sources on lembs and Liburnians from the 4th c. BC to Late Antiquity

A-V Cairo Zenon Papyrus 59015 (259/58 BC)

1
[. .]
ευτ[-ca.?-]ας
μετρ[ητὰς -ca.?-] . Α
τὸ δ[ὲ κερ(άμιον) ὡς] ἑκκαι-
δεκά[χουν ἐστίν, ὥσ]τε γίνεσθαι
τομ . [-ca.?-] τῶι δωδεκα-
κοτύλ[ωι -ca.?- τ]οὺς δὲ
πάντ[ας μετρητὰς πη]η χ(οίνικας) ι κο(τύλας) η
——
[-ca.?-]γγραφει ο[-ca.?-](*)
[κε]ρά[μια Μιλήσια] υνθ (*)
[Σ]άμια ξζ
ἡμικάδια Μιλήσια ριε
Σάμια ρξγ
——
τούτω[ν -ca.?- με]θα(*)
ἐκ τοῦ Θέωνος **λέμβο[υ]**

2
[κεράμια Μιλήσια .]
[Σάμια .]
[ἡ]κμικάδια Μ[ιλήσια . .]
[Σ]άμια ἡμικά[δια . .]
ἐκ τοῦ Ἀερόπου **λέ[μβου]**
κεράμια Σάμια [ξβ]
Μιλήσια [ξ]
ἡμικάδια Μιλήσι[α . .]
ἡμικάδια Σάμια [. .]
ἐκ τοῦ η[-ca.?- κέλητος]
κερ[ά]μ[ια] Μιλήσια ιδ[γ]
ἡ[μικ]άδια λδ
καὶ ἀγὴν φέρουσιν ἐν τῶι
Θέωνος **λέμβωι** κερ(άμια) [Μιλήσ(ια) γ]
ἡμικάδιον Σάμιον [α]
ἐν **τῶι** Ἀερόπου κερ(άμιον) Σάμιον α
ἡμικάδιον Σάμιον α
ἐν **τῶι** κέλ(ητι) κεράμια Μιλήσ(ια) β
ἡμικάδιον Μιλήσιον α

3,ctr
ἐν οἷς ὑποτιθέμεθα μετρ(ητὰς(?)) ͵α {χ} (τούτου)
σύνηκται μετρ(ηταὶ) β χ(όες) ε
λοιποὶ μετρηταὶ ωπα χ(όες) δ
ὧν τίμησις ἐγ(*) (δραχμῶν) νβ (τάλαντα) ζ
(δραχμαὶ) Γωιβ

'We have received a consignment of oil, amounting nominally to 1000 *metretae*. But as we assume the average contents of a *keramion* to be 16 *choes*, and not 18, the 1000 *metretae* are reduced to 888 8/9. Of the total quantity shipped we have received so many jars from **the vessel** of Theon, so many from that of Aeropos, and so many from the *keles*. Breakages are reported (ἀγὴν φέρουσιν) to the amount of so many jars, making a total of 18 *metretae*, of which 10 *metr.*, 5 *choes* have been recovered. This leaves 881 *metr.*, 4 *ch.*, which at the valuation of 52 *drachmae* the *metretae* are worth 7 talents, 3,812 *drachmae*. Deduct from this the import duty of 50 per cent, the minor taxes, and the charges for freight, amounting altogether to 4 talents, 637 *drachmae*. The balance is 3 talents, 3,175 *drachmae*.'

3,md
ὧν τιμὴ \παραμετρουμένων εἰς [τ]ὸ βα[σι]λικὸν/
τὸμ(*) με(τρητὴν) ἀν(ὰ) (δραχμὰς) μς (γίνονται) (τάλαντα)
λοι(ποὶ) με(τρηταὶ) Ϡϙζ χ(οὲς) ζ ὧν <τιμὴ> (τάλαντα) η Γωνζ \Αωοθ/
3, ctr
τούτου τέλος τὸ ἥμυ[σ]υ(*) (τάλαντα) γ (δραχμαὶ) ΔϞς
διακοσιαστὴ [(δραχμαὶ)] σκθ
εὐπλοίας τοῦ μετρ(ητοῦ) (ἡμιωβέλιον) (γίνονται) (δραχμαὶ) ογ (ὀβολοὶ) β
τριηράρχημα
κεραμίων φκ⟦δ⟧(*) ἀν(ὰ) (ὀβολὸν) (γίνονται) (δραχμαὶ) πς (τετρώβολον)
ἡμικαδίων σοε ἀν(ὰ) (ἡμιωβέλιον) (γίνονται) (δραχμαὶ) κβ (ὀβολοὶ) ε

3,ms
φκα (ὀβολοὶ) γ
ἐπωβελία ὡσαύτως (δραχμαὶ) ρθ (ὀβολοὶ) γ

——

ναῦλον τῶι Θέωνος κερ(αμίων) σνε ἀν(ὰ) (δραχμὴν) α (τετρώβολον) (γίνονται) (δραχμαὶ) υκε
ἡμικαδίων ρα ἀν(ὰ) (δραχμὴν) α (ὀβολὸν) α (γίνονται) (δραχμαὶ) ρκς (ὀβολὸς) α τῶι Ἀερόπου κερ(αμίων) ρκβ (δραχμαὶ) σγ (διώβολον)
ἡμικαδίων ρμ (δραχμαὶ) ροε

4,ctr
τ[ῶι -ca.?-] κεˊλ(ητι) κερ(αμίων) ρμγ (δραχμαὶ) σλζ (διώβολον)

4,ms
Ασθ (ὀβολοὶ) β 4,ctr
ἡμικαδίων λδ (δραχμαὶ) μβ (τριώβολον)
τὸ πᾶν ἀνήλωμα (τάλαντα) δ (δραχμαὶ) χλζ
λοιπὰ (τάλαντα) γ (δραχμαὶ) Γροε

4,minf
ρις χ(όες) γ
(τάλαντον) α (δραχμαὶ) με
4,md
οὗ τιμὴ παραμετρουμένου εἰς τὸ βασιλικὸν
τὸμ(*) με(τρητὴν) (δραχμῶν) μς (γίνονται) (τάλαντα) ζ Γχνα (ὀβολὸς)

5. Written sources on lembs and Liburnians from the 4th c. BC to Late Antiquity

((parens-deletion-opening)) τῶν Α με(τρητῶν)
τι(μὴ) (τάλαντα) ζ Δ ἀφαιρουμένης δὲ ((parens-deletion-closing))

((parens-deletion-opening)) τῆς ἀγῆς με(τρητῶν)
ζ χ(οῶν) ζ οὗ τιμὴ (δραχμαὶ) τμη (πεντώβολον)
((parens-deletion-closing))
((parens-deletion-opening)) λοι(πὰ) ἀρ(γυρίου)
(τάλαντα) ζ Βχοα (ὀβολὸς) ((parens-deletion-closing))
ἀπὸ τούτου τέλος ἐκ τιμήσεως
με(τρητῶν) ⋌ρβ χ(οῶν) ε τοῦ μετρητοῦ (δραχμαὶ)
νβ (γίνονται) (τάλαντα) η (δραχμαὶ) Γφοε
(διώβολον)
τούτου τὸ ἥμυσυ(*) (τάλαντα) δ Αψπζ
[(τετρώβολον)]
διακοσιαστὴ σνζ (ὀβολοὶ) ε
εὐπλοίας τοῦ με(τρητοῦ) (ἡμιωβέλιον) πβ
(ὀβολοὶ) δ χ(αλκοῦς) β

Greek: P. Cair. Zen. 1.59015r V (=*Sammelbuch griech. Urkunden aus Ägypten* 3.6781). http://papyri.info/ddbdp/p.cair.zen;1;59015

English: Edgar 1923: 87.

The papyri associated with the name of Zenon, were found on the eastern edge of the Fayoum in Egypt. Zenon, who wrote most of these documents and collected them, was the secretary of a certain Apollonios who held the post of *dioiketes*, an administrator of finance, during the last fifteen years of the reign of Ptolemy II. The text reproduces a modern edition of the four columns of this badly damaged papyrus. The translation on the right is Edgar's attempt at reconstructing the text. The papyrus contains the list of the value of the shipment after payment of custom duty and customs, for the importer, Apollonios. Most likely the merchandise was oil imported from the Aegean to Egypt. For us, the importance of this papyrus lies in two things. First, there is the question of the size of lembs: there was one lemb belonging to Theon and another belonging to Aeropos, with an addition of one *keles* that carried all that load. The second question is the question of the origins of the lembs.

As for the first question, the question of size, Casson brings this papyrus as an evidence for the lemb being a ship of a considerable size.[179] To support this, Casson cites relatively large amounts of merchandise carried on a lemb:

> *P. Cairo Zen. 59015 (259/8 B.C.): one lembos carried 258 18-chous jars (= ca. 12,900 liters) and 102 half-jars (= ca. 2,550 liters) of oil from Samos and Miletus to Alexandria, while another carried 122 18-chous jars and 140 half-jars. The cargo of the bigger, in the neighborhood of 25 tons [...] is almost twice that carried by the* keles *which appears in the same document ...*[180]

---
[179] Casson 1971: 162-63 particularly, n.36.
[180] Casson 1971: 160, n.36.

However, as we see from the text above, the papyrus is very badly damaged, and the reading is but conjectural. This reading of F. Bilabel (reproduced above), already differs on many points from Edgar's edition, which was the basis for Casson's text. Thus, Casson's judgment on the size of lembs, which has received wide reception, should be taken *cum grano salis*. Moreover, there are some discrepancies between the numbers Edgar supposed, and Casson accepted on the one hand, and the modern reading of the text on the other, provided here in the Greek column.

Secondly, in relation to the origin of the lembs, it is impossible to say anything with certainty. Since the merchandised oil was imported in Milesian and Samian jars, it is reasonable to suppose that the oil, as well as the lembs carrying them, were also from the Aegean. However, this cannot be established with any certainty. Edgar comments this portion of the papyrus in the following way: 'The fact that the oil was carried in Milesian and Samian jars indicates its origin, and a comparison of the account with no. 59012, leads us to believe that it came by sea to Alexandria and was unloaded at that sea-port.'[181]

A-VI The Petrie Papyrus 2.20 iv (around 218 BC)

(ἔτους) ε Φαῶφι ιζ.
Ἡρακλείδηι οἰκονόμωι παρὰ Θεοφίλου τοῦ παρ' Ἀντικλέους πρὸς τῆι ἐξαγωγῆι τοῦ βα(σιλικοῦ) σίτου τῶν δι' αὐτοῦ πλ[οί]ων τοῦ ὑπάρχοντος **λέμβου** Ἀντικλεῖ ἐν τῶι βα(σιλικῶι) ὑποδοχίωι ἀγ(ωγῆς) Ϡ ἐφ' οὗ κυ(βερνήτης) Πόρτις ἀγγαρευθέντος ὑπὸ σοῦ ἐν Πτολεμαίδι \τῆι ι τοῦ αὐτοῦ μηνός/ συνέμειξά σοι ἐπὶ τοῦ Λαβυρίνθου \τῆι ια/ ἀξιῶν ἀφεῖναι, \σοῦ/ 〚. . .〛 δὲ φήσαντος χρείαν αὐτοῦ εἶναι πρὸς τῆι καταγωγῆι \ὥστε ἀπάγειν . . . τον τοῖς ἐν Μέμφει ἐλέφασιν/ ((parens-deletion-opening)) καὶ ὅτι λυσιτελέστερον ἀπαλλά ((ξει πλεῖον γὰρ λήμψε\ται/ τοῦ σιτικῶν πλοίων παρὰ τῶν ἠργολαβηκότων)) ἐμοῦ δὲ προενεγκαμένου σοι ἐπειδὴ ἐργολαβίας γεγενημένης περὶ τοῦ . . . του καταγωγῆς τοῦ ἀποστελλομένου εἰς Μέμφιν ὥσ\τε/ τοῖς ἐλέ[φ]ασιν σὺ ἀγγαρεύσας τὸν Ἀντικλέους **λέμβον** ἀγ(ωγῆς) Ϡ ἐφ' οὗ [κυ(βερνήτης)] Πό]ρτις παρεστηκας αὐτοῖς. ἀξιῶ σε ἐάν σοι φαίνηται ἐπαναπ[έμπειν -ca.?-] [-ca.?-] σιτικὰ τὰ συνχρησθ[έντα -ca.?-]

Year 5, Phaophi 17 (= 2 December 218 B.C.) To Herakleides, oikonomos, from Theophilos, of the office of Antikles who is in charge of the shipping out of the royal grain in the boats under his supervision. Because **the lembos** of Antikles in the royal dockyard, 900 (sc. artabs) burden and commanded by Portis, was requisitioned by you in Ptolemais on the 10th of this month, I contacted you in the Labyrinth on the 11th with the request that you release it, but you said that you had need of it for transportation downstream to deliver hay to the elephants in Memphis [*in erasure*: and (pointed out) that it will come out more profitably, for it will obtain more than the freightage of grain cargoes from the contractors], and I stated the case to you: since you, because there is a contract for the transportation downstream of the hay that is being sent to Memphis for the elephants, have requisitioned Antikles' lembos, 900 (sc. artabs) burden and commanded by Portis, and turned it over to them (sc. the contractors), I (now) ask of you, if you please, to send back Antikles' **lembos** [...].

Greek and English: Casson 1993.

---

[181] Edgar 1923: 88.

5. Written sources on lembs and Liburnians from the 4th c. BC to Late Antiquity

Casson describes the circumstances of this document in the following way:

> *There was a cargocarrying version of the* lembos *as well: an account in the Zenon papyri lists two* lemboi *that transported oil from Asia Minor to Pelusium for delivery to Apollonios, Zenon's employer (P. Cair. Zen. I 59015.19 [lembos commanded by Aeropos], 28 [lembos commanded by Theon]). Both were small: one had aboard about 25 tons and the other about 15.*
>
> *In the papyri the term* lembos *occurs in two documents only, the account just mentioned and P.Petr. II 20 iv. In the former the ships sailed the Mediterranean waters; in the latter the ship was on the Nile - the sole instance of a* lembos *there. In other words, Antikles' boat in Ptolemais Hormos must have been as startling a sight, say, as a Chinese junk in New York harbor. And clearly this stranger to the Nile offered something for which shippers who had contracted to deliver hay downriver to Memphis were willing to pay a premium; that is the clear implication of the remark of Herakleides that Theophilos recorded and then deleted. It was not carrying capacity; Antikles' boat was 900 artabs burden, ca. 22.5 tons; like the two* lemboi *of the Zenon papyrus it was small, ten times smaller, e.g., than the smallest in a fleet of* kerkouroi *that in 171 BC was mobilized to transport the grain revenues downriver from Ptolemais Hormos. The only thing a* lembos *could have offered over any ordinary Nile cargo galley was what induced navies to use it-speed.*[182]

Casson here, contrary what we have seen above in the commentary for **A-V**, claims that the lembs were *not* big ships.

A-VII Theocritus (3rd century BC)

Idyll 21

Ἰχθύος ἀγρευτῆρες ὁμῶς δύο κεῖντο γέροντες / στρωσάμενοι βρύον αὖον ὑπὸ πλεκταῖς καλύβαισι, / κεκλιμένοι τοίχῳ τῷ φυλλίνῳ· ἐγγύθι δ' αὐτοῖν / κεῖτο τὰ ταῖν χειροῖν ἀθλήματα, τοὶ καλαθίσκοι, / τοὶ κάλαμοι, τἄγκιστρα, τὰ φυκιόεντα δέλητα, / ὁρμιαὶ κύρτοι τε καὶ ἐκ σχοίνων λαβύρινθοι, / μήρινθοι κῶπαί τε **γέρων τ' ἐπ' ἐρείσμασι λέμβος**· / νέρθεν τᾶς κεφαλᾶς φορμὸς βραχύς, εἵματα, πῖλοι. / οὗτος τοῖς ἁλιεῦσιν ὁ πᾶς πόρος, οὗτος ὁ πλοῦτος· / οὐ κλεῖδ', οὐχὶ θύραν ἔχων, οὐ κύνα· πάντα περισσά / ταῦτ' ἐδόκει τήνοις· ἁ γὰρ πενία σφας ἐτήρει.
6-15

Two old fishermen were lying down together on a bed of dried seaweed which they had strewn in their plaited hut, and they were reclining against the leafy wall. Near them lay the tools of their trade—baskets, rods, hooks, seaweed-covered bait, lines, weels, traps made from rushes, cords, oars, **an old boat on props**, a little mat for a pillow, their clothes, their caps. This was the fishermen's only resource, their only wealth; they had no key, no door, no guard dog: all these things seemed unnecessary to them, because poverty was their safeguard.
p. 282

Greek: *Theocritus 1*, ed. S. F. Gow, 2nd ed. CUP, Cambridge 1952.

English: *Theocritus. Moschus. Bion*, ed. and transl. N. Hopkinson. LCL 28 (2015).

---
[182] Casson 1993: 90-91.

The context of the verses is the following: Two old fishermen are resting in a lonely hut. Asphalion recounts his dream of catching a golden fish and hopes that he will become rich, but his companion responds that only hard work can keep him from starving. The moral of the story is that poverty is (or should be) the stimulus to hard work. These verses are usually taken to show that lemb was also a type of fishing boat, with the second meaning of the term λέμβος in *LSJ* rendered as '*fishing-boat*', with reference to the above verse. However, this would be the only instance in the corpus of early Greek texts in which the term *lembos* would have such a specific meaning. Due to *licentia poetica* and the constringency to fit words into metric scheme, it is possible that Theocritus used the term *lembos* as a generic name for a small boat that would fit the purpose of this poem.

A-VIII Philo of Byzantium (*c.* 280 - *c.* 220 BC)

Parasceuastica et poliorcetica

A-VIII.1
(C54) ἢ πλοῖα ἐναντία <ὁρμίζεται> πολεμιστήρια ὅπλα ἔχοντα, εἰ δὲ μή, **λέμβοι** καὶ ὧν ἂν ἔχῃς τὰ πλεῖστα προσορμισθέντα (95) πρὸς ἄλληλα συναναρτᾶται καὶ συμβολαὶ κατασκευάζονται αὐτοῖς δοκῶν παχεῶν τετραγώνων πρὸ τῆς πρώρας τεθεισῶν, καὶ τούτων συγγομφωθεισῶν καὶ συνδεθεισῶν εἰς τὸ αὐτό, καὶ ἐπ' ἄκρῳ ἐμβόλου περὶ αὐτὰ καθαρμοσθέντος.

(94) Or the war ships are anchored on the opposite side, and if not, then **the lembs** or any other kinds of ships you can have are anchored and (95) connected to one another. The connections between them are prepared by thick, four-sided beams put in front of prows which are tied and fastened together with nails and on top they are fastened with a peg.

A-VIII.2
(D21) ὡσαύτως δὲ καὶ ἐκ θαλάσσης ἐὰν <ποιῇ τὴν> προσαγωγήν, ἐπί τε τῶν ὁλκάδων καὶ **τῶν λέμβων** στήσας μηχανήματα πρόσαγε.

(21) Similarly if you attack from the sea, proceed by putting war machines on ships of burthen or on **lembs**.

A-VIII.3
(D38) αἱ δ' ἐπὶ **τῶν λέμβων** χελῶναι κατασκευάζονται περιφερεῖς ἄνωθεν ἐκ σανίδων ἰσχυρῶν συμπηγνύμεναι, ὑπόφαυσιν κάτωθεν ἔχουσαι, ὅθεν οἱ λιθοβόλοι ἀφίενται.

(38) The protection sheds ('turtles') on **the lembs** are built from strong timbers curved from above and put together, with an opening from below, from which stone-throwers discharge.

Greek: *Exzerpte aus Philons Mechanik b. VII und VIII (vulgo fünftes Buch) Griechisch und Deutsch*, eds H. Diels, E. Schramm [Abhandlungen der preussischen Akademie der Wissenschaften, Philosoph.-hist. Kl. 12]. Reimer, Berlin 1920.

Philo wrote a work on mechanics (Μηχανικὴ σύνταξις) in nine volumes. The present text comes from volumes seven and eight (Παρασκευαστικά, 'Military Weapons' and Πολιορκητικά, 'Art of Siege warfare'), which deal with tactics and weapons for defending and besieging towns. These passages suggests that lembs could be of a considerable size if they were able to be equipped with catapults and other war machines.

5. Written Sources on Lembs and Liburnians from the 4th c. BC to Late Antiquity

A-IX Polybius (before 199 - *c.* 120 BC)

Historiae

A-IX.1

[...] τότε δὴ πρῶτον ἐν νῷ λαμβάνοντες οὕτως τολμηρῶς ἐνεχείρησαν ὥστε πρὶν ἢ πειραθῆναι τοῦ πράγματος, εὐθὺς ἐπιβαλέσθαι Καρχηδονίοις ναυμαχεῖν τοῖς ἐκ προγόνων ἔχουσι τὴν κατὰ θάλατταν ἡγεμονίαν ἀδήριτον. μαρτυρίῳ δ᾽ ἄν τις χρήσαιτο πρὸς τὴν ἀλήθειαν τῶν νῦν ὑπ᾽ ἐμοῦ λεγομένων καὶ πρὸς τὸ παράδοξον αὐτῶν τῆς τόλμης· ὅτε γὰρ τὸ πρῶτον ἐπεχείρησαν διαβιβάζειν εἰς τὴν Μεσσήνην τὰς δυνάμεις, οὐχ οἷον κατάφρακτος αὐτοῖς ὑπῆρχε ναῦς, ἀλλ᾽ οὐδὲ καθόλου μακρὸν πλοῖον **οὐδὲ λέμβος οὐδ᾽ εἷς**, ἀλλὰ παρὰ Ταραντίνων καὶ Λοκρῶν ἔτι δ᾽ Ἐλεατῶν καὶ Νεαπολιτῶν συγχρησάμενοι πεντηκοντόρους καὶ τριήρεις ἐπὶ τούτων παραβόλως διεκόμισαν τοὺς ἄνδρας.
1.20.13-14

[...] yet when they [the Romans] once had conceived the project, they took it in hand so boldly, that before gaining any experience in the matter they at once engaged the Carthaginians who had held for generations undisputed command of the sea. Evidence of the truth of what I am saying and of their incredible luck is this. When they first undertook to send their forces across to Messene not only had they not any decked ships, but no long warships at all, **not even a single boat**, and borrowing fifty-oared boats and triremes from the Tarentines and Locrians, and also from the people of Elea and Naples they took their troops across in these at great hazard.
p. 61

A-IX.2

ὁμοίως δὲ καὶ τοῖς ἐκ τῶν Συρακουσῶν προαπεσταλμένοις ταμίαις ἀνήγγειλαν **οἱ προπλεῖν εἰθισμένοι λέμβοι** τὸν ἐπίπλουν τῶν ὑπεναντίων.
1.53.9–10

The approach of the enemy was also announced by **the light boats that usually sail in front of a fleet** to the Quaestors who had been sent on in advance from Syracuse.
p. 163

A-IX.3

δεδογμένων δὲ τούτων, καὶ δέον τῇ κατὰ πόδας ἡμέρᾳ γενέσθαι τὴν αἵρεσιν καὶ τὴν παράληψιν τῆς ἀρχῆς, καθάπερ ἔθος ἐστὶν Αἰτωλοῖς, προσπλέουσιν τῆς νυκτὸς **ἑκατὸν λέμβοι** πρὸς τὴν Μεδιωνίαν κατὰ τοὺς ἔγγιστα τόπους τῆς πόλεως, **ἐφ᾽ ὧν ἦσαν Ἰλλυριοὶ πεντακισχίλιοι**. καθορμισθέντες δὲ καὶ τῆς ἡμέρας ἐπιγενομένης ἐνεργὸν καὶ λαθραίαν ποιησάμενοι τὴν ἀπόβασιν καὶ χρησάμενοι τῇ παρ᾽ αὑτοῖς εἰθισμένῃ τάξει προῆγον κατὰ σπείρας ἐπὶ τὴν τῶν Αἰτωλῶν στρατοπεδείαν. [...] καὶ πολλοὺς μὲν αὐτῶν ἀπέκτειναν, ἔτι δὲ πλείους αἰχμαλώτους ἔλαβον· τῶν δ᾽ ὅπλων καὶ τῆς ἀποσκευῆς ἐγένοντο πάσης ἐγκρατεῖς. οἱ μὲν οὖν Ἰλλυριοὶ πράξαντες τὸ συνταχθὲν ὑπὸ τοῦ βασιλέως καὶ διακομίσαντες τὴν ἀποσκευὴν καὶ τὴν ἄλλην ὠφέλειαν **ἐπὶ τοὺς λέμβους** εὐθέως ἀνήγοντο, ποιούμενοι τὸν πλοῦν εἰς τὴν οἰκείαν.
2.3

This decree had been passed, and next day the election was to be held, and the new Strategus was to enter at once into office, as is the practice of the Aetolians, when that night **a hundred boats containing a force of five thousand Illyrians** arrived at the nearest point on the coast to Medion. Anchoring there they landed, as soon as it was daylight, with promptitude and secrecy, and forming in the order customary in Illyria, advanced by companies on the Aetolian camp. [...] They killed many Aetolians and took a still larger number of prisoners, capturing all their arms and baggage. The Illyrians, having thus executed the orders of their king, carried off **to their boats** the baggage and other booty and at once set sail for home.
pp. 267-269

A-IX.4

Ὁ δὲ βασιλεὺς Ἄγρων, ἐπεὶ κατέπλευσαν **οἱ λέμβοι**, διακούσας τῶν ἡγεμόνων τὰ κατὰ τὸν κίνδυνον καὶ περιχαρὴς γενόμενος ἐπὶ τῷ δοκεῖν Αἰτωλοὺς τοὺς μέγιστον ἔχοντας τὸ φρόνημα νενικηκέναι, πρὸς μέθας καί τινας τοιαύτας ἄλλας εὐωχίας τραπεὶς ἐνέπεσεν εἰς πλευρῖτιν· ἐκ δὲ ταύτης ἐν ὀλίγαις ἡμέραις μετήλλαξε τὸν βίον.
2.4.6

King Agron, when **the flotilla** returned and his officers gave him an account of the battle, was so overjoyed at the thought of having beaten the Aetolians, then the proudest of peoples, that he took to carousals and other convivial excesses, from which he fell into a pleurisy that ended fatally in a few days.
p. 271

A-IX.5

ἡ δὲ Τεύτα, καταπλευσάντων πρὸς αὐτὴν τῶν ἐκ τῆς Ἠπείρου **λέμβων**, καταπλαγεῖσα τὸ πλῆθος καὶ τὸ κάλλος τῆς ἀγομένης κατασκευῆς [...].
2.8.4

Teuta, on the return of **the flotilla** from Epirus, was so struck with admiration by the quantity and beauty of the spoils they brought back [...].
p. 281

A-IX.6

Ἡ δὲ Τεύτα, τῆς ὥρας ἐπιγενομένης, ἐπισκευάσασα **λέμβους** πλείους τῶν πρότερον ἐξαπέστειλε πάλιν εἰς τοὺς κατὰ τὴν Ἑλλάδα τόπους. ὧν οἱ μὲν διὰ πόρου τὸν πλοῦν ἐπὶ τὴν Κέρκυραν ἐποιοῦντο, μέρος δέ τι προσέσχε τὸν τῶν Ἐπιδαμνίων λιμένα, λόγῳ μὲν ὑδρείας καὶ ἐπισιτισμοῦ χάριν, ἔργῳ δ' ἐπιβουλῆς καὶ πράξεως ἐπὶ τὴν πόλιν.
2.9.1-2

Teuta, when the season came, fitted out a larger number of **boats** than before and dispatched them to the Greek coasts. Some of them sailed straight across the high sea to Corcyra, while a part put in to the harbor of Epidamnus, professedly to water and provision, but really with the design of surprising and seizing the town.
p. 285

A-IX.7

οἱ δ' Ἰλλυριοὶ συμπαραλαβόντες Ἀκαρνάνων ναῦς κατὰ τὴν συμμαχίαν, οὔσας ἑπτὰ καταφράκτους, ἀνταναχθέντες συνέβαλλον τοῖς τῶν Ἀχαιῶν σκάφεσιν περὶ τοὺς καλουμένους Παξούς. Οἱ μὲν οὖν Ἀκαρνᾶνες καὶ τῶν Ἀχαϊκῶν νεῶν αἱ κατὰ τούτους ταχθεῖσαι πάρισον ἐποίουν τὸν ἀγῶνα καὶ διέμενον ἀκέραιοι πάρισον ἐποίουν τὸν ἀγῶνα καὶ διέμενον ἀκέραιοι κατὰ τὰς συμπλοκὰς πλὴν τῶν εἰς αὐτοὺς τοὺς ἄνδρας γινομένων τραυμάτων. Οἱ δ' Ἰλλυριοὶ ζεύξαντες **τοὺς παρ' αὐτῶν λέμβους** ἀνὰ τέτταρας συνεπλέκοντο τοῖς πολεμίοις Καὶ τῶν μὲν ἰδίων ὠλιγώρουν καὶ παραβάλλοντες πλαγίους συνήργουν ταῖς ἐμβολαῖς τῶν ὑπεναντίων. ὅτε δὲ τρώσαντα καὶ δεθέντα κατὰ τὰς ἐμβολὰς δυσχρήστως (διέκειτο πρὸς τὸ παρόν) τὰ τῶν ἀντιπάλων σκάφη, προσκρεμαμένων αὐτοῖς περὶ τοὺς ἐμβόλους **τῶν ἐζευγμένων λέμβων**, τότ' ἐπιπηδῶντες ἐπὶ τὰ καταστρώματα τῶν Ἀχαϊκῶν νεῶν κατεκράτουν διὰ τὸ πλῆθος τῶν ἐπιβατῶν. Καὶ τούτῳ τῷ τρόπῳ τεττάρων μὲν πλοίων ἐκυρίευσαν τετρηρικῶν, μίαν δὲ πεντήρη σὺν αὐτοῖς τοῖς ἀνδράσιν

The Illyrians, now reinforced by seven decked ships sent by the Acarnanians in compliance with the terms of their treaty, put to sea and encountered the Achaean ships off the islands called Paxi. The Acarnanians and those Achaean ships which were told off to engage them fought with no advantage on either side, remaining undamaged in their encounter except for the wounds inflicted on some of the crew. The Illyrians lashed **their boats** together in batches of four and thus engaged the enemy. They sacrificed their own boats, presenting them broadside to their adversaries in a position favoring their charge, but when the enemy's ships had charged and struck them and getting fixed in them, found themselves in difficulties, as in each case **the four boats lashed together** were hanging on to their beaks, the marines leapt on to the decks of the Achaean ships and overmastered them by their numbers. In this way they captured four quadriremes and sunk with all hands a quinquereme, on board of which was Margus

5. Written sources on lembs and Liburnians from the 4th c. BC to Late Antiquity

ἐβύθισαν, ἐφ' ἧς ἔπλει Μάργος ὁ Καρυνεύς, ἀνὴρ πάντα τὰ δίκαια τῷ κοινῷ τῶν Ἀχαιῶν πολιτεύματι πεποιηκὼς μέχρι τῆς καταστροφῆς. Οἱ δὲ πρὸς τοὺς Ἀκαρνᾶνας διαγωνιζόμενοι, συνιδόντες τὸ κατὰ τοὺς Ἰλλυριοὺς προτέρημα καὶ πιστεύοντες τῷ ταχυναυτεῖν, ἐπουρώσαντες. ἀσφαλῶς τὴν ἀποχώρησιν εἰς τὴν οἰκείαν ῥώσαντες ἀσφαλῶς τὴν ἀποχώρησιν εἰς τὴν οἰκείαν ποιήσαντο.
2.10

of Caryneia, a man who up to the end served the Achaeans most loyally. The ships that were engaged with the Acarnanians, seeing the success of the Illyrians, and trusting to their speed, made sail with a fair wind and escaped home in safety.
pp. 287-289

A-IX.8
ἐκυρίευσαν δὲ καὶ **λέμβων** εἴκοσι τῶν ἀποκομιζόντων τὴν ἐκ τῆς χώρας ὠφέλειαν.
2.11.14

They [the Romans] also captured twenty [Illyrian] **boats** which were conveying the plunder from the country.
p. 293

A-IX.9
ὑπὸ δὲ τὴν ἐαρινὴν ὥραν ἡ Τεύτα διαπρεσβευσαμένη πρὸς τοὺς Ῥωμαίους ποιεῖται συνθήκας, ἐν αἷς εὐδόκησε φόρους τε τοὺς διαταχθέντας οἴσειν πάσης τ' ἀναχωρήσειν τῆς Ἰλλυρίδος πλὴν ὀλίγων τόπων, καὶ τὸ συνέχον, ὃ μάλιστα πρὸς τοὺς Ἕλληνας διέτεινε, μὴ πλεύσειν πλέον ἢ **δυσὶ λέμβοις** ἔξω τοῦ Λίσσου, καὶ τούτοις ἀνόπλοις.
2.12.3-4

In the early spring Teuta sent an embassy to the Romans and made a treaty, by which she consented to pay any tribute they imposed, to relinquish all Illyria except a few places, and, what mostly concerned the Greeks, undertook not to sail beyond Lissus with more than **two unarmed vessels**.
pp. 293-295

A-IX.10
συνέβαινε γὰρ κατ' ἐκείνους τοὺς καιροὺς Δημήτριον τὸν Φάριον, ἐπιλελησμένον μὲν τῶν Δημήτριον τὸν Φάριον, ἐπιλελησμένον μὲν τῶν προγεγονότων εἰς αὐτὸν εὐεργετημάτων ὑπὸ Ῥωμαίων, καταπεφρονηκότα δὲ πρότερον μὲν διὰ τὸν ἀπὸ Γαλατῶν τότε δὲ διὰ τὸν ἀπὸ Καρχηδονίων φόβον περιεστῶτα Ῥωμαίους, πάσας δ' ἔχοντα τὰς ἐλπίδας ἐν τῇ Μακεδόνων οἰκίᾳ διὰ τὸ συμπεπολεμηκέναι καὶ μετεσχηκέναι τῶν πρὸς Κλεομένη κινδύνων Ἀντιγόνῳ, πορθεῖν μὲν καὶ καταστρέφεσθαι τὰς κατὰ τὴν Ἰλλυρίδα πόλεις τὰς ὑπὸ Ῥωμαίους ταττομένας, πεπλευκέναι δ' ἔξω τοῦ Λίσσου παρὰ τὰς συνθήκας **πεντήκοντα λέμβοις** καὶ πεπορθηκέναι πολλὰς τῶν Κυκλάδων νήσων.
3.16.2-3

It so happened that at that time in Illyria Demetrius of Pharos, oblivious of the benefits that the Romans had conferred on him, contemptuous of Rome because of the peril to which she was exposed first from the Gauls and now from Carthage, and placing all his hopes in the Royal House of Macedon owing to his having fought by the side of Antigonus in the battles against Cleomenes, was sacking and destroying the Illyrian cities subject to Rome, and, sailing beyond Lissus, contrary to the terms of the treaty, **with fifty boats**, had pillaged many of the Cyclades.
p. 43

A-IX.11

ἐξ οὗ τῶν μὲν κατὰ πρόσωπον τῶν δὲ κατὰ νώτου πονούντων, τέλος οἱ περὶ τὸν Δημήτριον ἐτράπησαν· καί τινες μὲν αὐτῶν ἔφυγον ὡς πρὸς τὴν πόλιν, οἱ δὲ πλείους ἀνοδίᾳ κατὰ τῆς νήσου διεσπάρησαν. ὁ δὲ Δημήτριος ἔχων ἑτοίμους **λέμβους** πρὸς τὸ συμβαῖνον ἔν τισι τόποις ἐρήμοις ὑφορμοῦντας ἐπὶ τούτους ἐποιήσατο τὴν ἀποχώρησιν. εἰς οὓς ἐμβὰς ἐπιγενομένης τῆς νυκτὸς ἀπέπλευσε καὶ διεκομίσθη παραδόξως πρὸς τὸν βασιλέα Φίλιππον, παρ' ᾧ τὸ λοιπὸν διέτριβε τοῦ βίου μέρος.
3.19.7-9

At the end, being hard pressed both in front and in the rear, Demetrius' troops turned and fled, some escaping to the city, but the greater number dispersing themselves over the island across country. Demetrius had some **boats** lying ready for such a contingency at a lonely spot, and retreating there and embarking sailed away at nightfall and managed to cross and reach King Philip [Philip V], at whose court he spent the rest of his life.
p. 53

A-IX.12

Ἀννίβας δὲ προσμίξας τοῖς περὶ τὸν ποταμὸν τόποις εὐθέως ἐνεχείρει ποιεῖσθαι τὴν διάβασιν κατὰ τὴν ἁπλῆν ῥύσιν, σχεδὸν ἡμερῶν τεττάρων ὁδὸν ἀπέχων στρατοπέδῳ τῆς θαλάττης. καὶ φιλοποιησάμενος παντὶ τρόπῳ τοὺς παροικοῦντας τὸν ποταμὸν ἐξηγόρασε παρ' αὐτῶν τά τε μονόξυλα πλοῖα πάντα καὶ τοὺς **λέμβους**, ὄντας ἱκανοὺς τῷ πλήθει διὰ τὸ ταῖς ἐκ τῆς θαλάττης ἐμπορίαις πολλοὺς χρῆσθαι τῶν παροικούντων τὸν Ῥοδανόν. ἔτι δὲ τὴν ἁρμόζουσαν ξυλείαν ἐξέλαβε πρὸς τὴν κατασκευὴν τῶν μονοξύλων· ἐξ ὧν ἐν δυσὶν ἡμέραις πλῆθος ἀναρίθμητον ἐγένετο πορθμείων, ἑκάστου σπεύδοντος μὴ προσδεῖσθαι τοῦ πέλας, ἐν αὑτῷ δ' ἔχειν τὰς τῆς διαβάσεως ἐλπίδας.
3.42.1-3

Hannibal, on reaching the neighborhood of the river, at once set about attempting to cross it where the stream is single at a distance of about four days' march from the sea. Doing his best to make friends with the inhabitants of the bank, he bought up all their canoes and **boats**, amounting to a considerable number, since many of the people on the banks of the Rhone engage in maritime traffic. He also got from them the logs suitable for making the canoes, so that in two days he had an innumerable quantity of ferryboats, every one doing his best to dispense with any assistance and relying on himself for his chance of getting across.
p. 109

A-IX.13

(1) Οὐ μὴν ἀλλ' ἐπιγενομένης τῆς πέμπτης νυκτὸς οἱ μὲν προδιαβάντες ἐκ τοῦ πέραν ὑπὸ τὴν ἑωθινὴν προῆγον παρ' αὐτὸν τὸν ποταμὸν ἐπὶ τοὺς (2) ἀντίπερα βαρβάρους, ὁ δ' Ἀννίβας ἑτοίμους ἔχων τοὺς στρατιώτας ἐπεῖχε τῇ διαβάσει, **τοὺς μὲν λέμβους** πεπληρωκὼς τῶν πελτοφόρων ἱππέων, τὰ δὲ (3) μονόξυλα τῶν εὐκινητοτάτων πεζῶν. εἶχον δὲ τὴν μὲν ἐξ ὑπερδεξίου καὶ παρὰ τὸ ῥεῦμα τάξιν **οἱ λέμβοι**, τὴν δ' ὑπὸ τούτους τὰ λεπτὰ τῶν πορθμείων, **ἵνα τὸ πολὺ τῆς τοῦ ῥεύματος βίας ἀποδεχομένων τῶν λέμβων** ἀσφαλεστέρα γίνοιτο τοῖς μονοξύλοις (4) ἡ παρακομιδὴ διὰ τοῦ πόρου. κατὰ δὲ τὰς πρύμνας **τῶν λέμβων** ἐφέλκειν διενοοῦντο τοὺς ἵππους νέοντας, τρεῖς ἅμα καὶ τέτταρας τοῖς ἀγωγεῦσιν ἑνὸς ἀνδρὸς ἐξ

On the fifth night, however, the force which had already crossed began a little before dawn to advance along the opposite bank against the barbarians there, while Hannibal had got his soldiers ready and was waiting till the time for crossing came. He had filled **the boats** with his light horse and the canoes with his lightest infantry. **The large boats** were placed highest up stream and the lighter ferryboats farther down, **so that the heavier vessels receiving the chief force of the current** the canoes should be less exposed to risk in crossing. They" hit on the plan of towing the horses astern of **the boats** swimming, one man at each side of the stern guiding three or four horses by their

ἑκατέρου τοῦ μέρους τῆς πρύμνης οἰακί- ἀνδρὸς ἐξ ἑκατέρου τοῦ μέρους τῆς πρύμνης οἰακίζοντος, ὥστε πλῆθος ἱκανὸν ἵππων συνδιακομίζεσθαι (5) κατὰ τὴν πρώτην εὐθέως διάβασιν. οἱ δὲ βάρβαροι θεωροῦντες τὴν ἐπιβολὴν τῶν ὑπεναντίων ἀτάκτως ἐκ τοῦ χάρακος ἐξεχέοντο καὶ σποράδην, πεπεισμένοι κωλύειν εὐχερῶς τὴν ἀπόβασιν (6) τῶν Καρχηδονίων. Ἀννίβας δ' ἅμα τῷ συνιδεῖν ἐν τῷ πέραν ἐγγίζοντας ἤδη τοὺς παρ' αὐτοῦ στρατιώτας, σημηνάντων ἐκείνων τὴν παρουσίαν τῷ καπνῷ κατὰ τὸ συντεταγμένον, ἐμβαίνειν ἅπασιν ἅμα παρήγγελλε καὶ βιάζεσθαι πρὸς τὸ ῥεῦμα τοῖς (7) ἐπὶ τῶν πορθμείων τεταγμένοις. ταχὺ δὲ τούτου γενομένου, καὶ τῶν ἐν τοῖς πλοίοις ἁμιλλωμένων μὲν πρὸς ἀλλήλους μετὰ κραυγῆς, διαγωνιζομένων (8) δὲ πρὸς τὴν τοῦ ποταμοῦ βίαν, τῶν δὲ στρατοπέδων ἀμφοτέρων ἐξ ἑκατέρου τοῦ μέρους παρὰ τὰ χείλη τοῦ ποταμοῦ παρεστώτων, καὶ τῶν μὲν ἰδίων συναγωνιώντων καὶ παρακολουθούντων μετὰ κραυγῆς, τῶν δὲ κατὰ πρόσωπον βαρβάρων παιανιζόν των καὶ προκαλουμένων τὸν κίνδυνον, ἦν τὸ γινόμενον (9) ἐκπληκτικὸν καὶ παραστατικὸν ἀγωνίας.
3.43.1-9

leading reins, so that a considerable number were got across at once in the first batch. The barbarians seeing the enemy's project poured out of their camp, scattered and in no order, feeling sure that they would easily prevent the Carthaginians from landing. Hannibal, as soon as he saw that the force he had previously sent across was near at hand on the opposite bank, they having announced their approach by a smoke signal as arranged, ordered all in charge of the ferryboats to embark and push up against the current. He was at once obeyed, and now with the men in the boats shouting as they vied with one another in their efforts and struggled to stem the current, with the two armies standing on either bank at the very brink of the river, the Carthaginians following the progress of the boats with loud cheers and sharing in the fearful suspense, and the barbarians yelling their war cry and challenging to combat, the scene was in the highest degree striking and thrilling.
pp. 111-113

A-IX.14

(5) ῥύματα δὲ καὶ πλείω ταύταις ἐνῆψαν, οἷς ἔμελλον **οἱ λέμβοι** ῥυμουλκοῦντες οὐκ ἐάσειν φέρεσθαι κατὰ ποταμοῦ, βίᾳ δὲ πρὸς τὸν ῥοῦν κατέχοντες παρα (6) κομιεῖν καὶ περαιώσειν ἐπὶ τούτων τὰ θηρία. μετὰ δὲ ταῦτα χοῦν ἔφερον ἐπὶ πάσας πολύν, ἕως ἐπιβάλλοντες ἐξωμοίωσαν, ὁμαλὴν καὶ σύγχρουν ποιοῦντες τῇ διὰ τῆς χέρσου φερούσῃ πρὸς τὴν διάβασιν (7) ὁδῷ. τῶν δὲ θηρίων εἰθισμένων τοῖς Ἰνδοῖς μέχρι μὲν πρὸς τὸ ὑγρὸν ἀεὶ πειθαρχεῖν, εἰς δὲ τὸ ὕδωρ ἐμβαίνειν οὐδαμῶς ἔτι τολμώντων, ἦγον διὰ τοῦ χώματος δύο προθέμενοι θηλείας, πειθαρχούντων (8) αὐταῖς τῶν θηρίων. ἐπεὶ δ' ἐπὶ τὰς τελευτων αὐταῖς τῶν θηρίων. ἐπεὶ δ' ἐπὶ τὰς τελευταίας ἐπέστησαν σχεδίας, διακόψαντες τοὺς δεσμούς, οἷς προσήρτηντο πρὸς τὰς ἄλλας, καὶ **τοῖς λέμβοις** ἐπισπασάμενοι τὰ ῥύματα ταχέως ἀπέσπασαν ἀπὸ τοῦ χώματος τά τε θηρία καὶ τὰς ὑπ' αὐτοῖς σχεδίας.
3.46.5-8

They attached to these several towing lines by which **boats** were to tow them, not allowing them to be carried down stream, but holding them up against the current, and thus were to convey the elephants which would be in them across. After this they piled up a quantity of earth on all the line of rafts, until the whole was on the same level and of the same appearance as the path on shore leading to the crossing. The animals were always accustomed to obey their mahouts up to the water, but would never enter it on any account, and they now drove them along over the earth with two females in front, whom they obediently followed. As soon as they set foot on the last rafts the ropes which held these fast to the others were cut, and **the boats** pulling taut, the towing lines rapidly tugged away from the pile of earth the elephants and the rafts on which they stood.
p. 121

A-IX.15
(6) Ἤδη δ' ἐπιλελεγμένων τῶν Ἀχαϊκῶν νεανίσκων καὶ συντεταγμένων ὑπὲρ τῆς βοηθείας τῶν Λακεδαιμονίων καὶ Μεσσηνίων, Σκερδιλαΐδας ὁμοῦ καὶ Δημήτριος ὁ Φάριος ἔπλευσαν ἐκ τῆς Ἰλλυρίδος ἐν ἐνενήκοντα **λέμβοις** ἔξω τοῦ Λίσσου παρὰ τὰς πρὸς (7) Ῥωμαίους συνθήκας. οἳ τὸ μὲν πρῶτον τῇ Πύλῳ προσμίξαντες καὶ ποιησάμενοι προσβολὰς ἀπέπεσον·(8) μετὰ δὲ ταῦτα Δημήτριος μὲν ἔχων τοὺς πεντήκοντα **τῶν λέμβων** ὥρμησεν ἐπὶ νήσων, καὶ περιπλέων τινὰς μὲν ἠργυρολόγει, τινὰς δ' ἐπόρθει τῶν (9) Κυκλάδων· Σκερδιλαΐδας δὲ ποιούμενος τὸν πλοῦν ὡς ἐπ' οἴκου προσεῖχε πρὸς Ναύπακτον μετὰ τετταράκοντα **λέμβων**, πεισθεὶς Ἀμυνᾷ τῷ βασιλεῖ τῶν Ἀθαμάνων, ὃς ἐτύγχανε κηδεστὴς ὑπάρχων αὐτοῦ. (10) ποιησάμενος δὲ συνθήκας πρὸς Αἰτωλοὺς δι' Ἀγε(10) ποιησάμενος δὲ συνθήκας πρὸς Αἰτωλοὺς δι' Ἀγελάου περὶ τοῦ μερισμοῦ τῶν λαφύρων, ὑπέσχετο συνεμβαλεῖν ὁμόσε τοῖς Αἰτωλοῖς εἰς τὴν Ἀχαΐαν.
4.16.6-10

The Achaean levy of young men had been enrolled, and the Lacedaemonians and Messenians had contracted to send their contingents, when Scerdilaïdas, together with Demetrius of Pharos, sailed from Illyria with a fleet of ninety **boats** and passed Lissus, thus breaking the treaty with Rome. They touched first at Pylos and made some attacks on it which failed. Demetrius now with fifty of **the boats** started for the islands, and sailing through the Cyclades pillaged or levied blackmail on some of them. Scerdilaïdas on his voyage home touched at Naupactus with his forty **boats** at the request of Amynas, the king of Athamania, who was his connexion by marriage. Here, having come to terms with the Aetolians through Agelaus about the division of the spoil, he promised to join them in invading Achaea.
pp. 369-371

A-IX.16
(7) ὁ δὲ Ταυρίων, πυνθανόμενος τὴν τῶν Αἰτωλῶν εἰσβολὴν καὶ τὰ περὶ τὴν Κύναιθαν πεπραγμένα, θεωρῶν δὲ τὸν Δημήτριον τὸν Φάριον ἀπὸ τῶν νήσων εἰς τὰς Κεγχρεὰς καταπεπλευκότα, παρεκάλει τοῦτον βοηθῆσαι τοῖς Ἀχαιοῖς καὶ διισθμίσαντα **τοὺς λέμβους** ἐπιτίθεσθαι τῇ τῶν Αἰτωλῶν (8) διαβάσει. ὁ δὲ Δημήτριος λυσιτελῆ μέν, οὐκ εὐσχήμονα δὲ πεποιημένος τὴν ἀπὸ τῶν νήσων ἐπάνοδον, διὰ τὸν τῶν Ῥοδίων ἐπ' αὐτὸν ἀνάπλουν, ἄσμενος ὑπήκουσε τῷ Ταυρίωνι, προσδεξαμένου 'κείνου τὴν εἰς τὴν ὑπέρβασιν **τῶν λέμβων** δαπάνην.
4.19.7-8

Taurion had learnt of the Aetolian invasion and the fate of Cynaetha; and seeing that Demetrius of Pharos had sailed back from the islands to Cenchreae, begged him to assist the Achaeans, and after conveying his **boats** across the Isthmus, to fall upon the Aetolians during their crossing. Demetrius, whose return from his expedition to the islands had been much to his advantage indeed, but somewhat ignominious, as the Rhodians were sailing to attack him, lent a ready ear to Taurion, who had engaged to meet the expense of transporting **the boats**.
p. 379

A-IX.17
(7) διόπερ ὑποκαθημένης ἐκ τούτων αὐτῷ τῆς ὀργῆς, βραχέα προσαναμνήσαντος τοῦ Φιλίππου, ταχέως ὑπήκουσε καὶ συνέθετο μεθέξειν τῆς κοινῆς συμμαχίας, ἐφ' ᾧ λαμβάνειν μὲν εἴκοσι τάλαντα κατ' ἐνιαυτόν, πλεῖν δὲ **λέμβοις τριάκοντα** καὶ πολεμεῖν τοῖς Αἰτωλοῖς κατὰ θάλατταν.
4.29.7

As he [Scerdilaïdas] had been nursing anger against them [the Aetolians] for this ever since, it only required a brief mention by Philip of this grievance to make him at once consent and agree to become a member of the general alliance on condition of receiving an annual sum of twenty talents, in consideration of which he was to attack the Aetolians by sea **with thirty boats**.
p. 407

5. Written sources on lembs and Liburnians from the 4th c. BC to Late Antiquity

A-IX.18
μέλλοντος δ' αὐτοῦ διαβαίνειν τὸν Ἀμβρακικὸν κόλπον ἐξ Ἀκαρνανίας εἰς Ἤπειρον, παρῆν **ἐφ' ἑνὸς λέμβου** Δημήτριος ὁ Φάριος, ἐκπεπτωκὼς ὑπὸ Ῥωμαίων ἐκ τῆς Ἰλλυρίδος·[...].
4.66.4

As he [king Philip] was about to cross the Gulf of Ambracia from Acarnania to Epirus, Demetrius of Pharus appeared **in a single frigate**, having been driven by the Romans from Illyria [...].
p. 501

A-IX.19
(3) ὁ μὲν οὖν βασιλεὺς περὶ ταῦτα καὶ πρὸς τούτοις ἦν. κατὰ δὲ τὸν καιρὸν τοῦτον πεντεκαίδεκα μὲν ἦκον **λέμβοι** παρὰ Σκερδιλαΐδου—τοὺς γὰρ πλείστους ἐκωλύθη πέμψαι διὰ τὰς γενομένας ἐπιβουλὰς καὶ ταραχὰς περὶ τοὺς κατὰ τὴν Ἰλλυρίδα (4) πολιδυνάστας—ἧκον δὲ καὶ παρ' Ἠπειρωτῶν καὶ παρ' Ἀκαρνάνων ἔτι δὲ Μεσσηνίων οἱ διαταχθέν (5) τες σύμμαχοι· τῆς γὰρ τῶν Φιαλέων πόλεως ἐξαιρεθείσης ἀπροφασίστως τὸ λοιπὸν ἤδη μετεῖχον Μεσσήνιοι τοῦ πολέμου.
5.4.3-5

While the king was thus occupied, fifteen **boats** arrived from Scerdilaïdas, who had been prevented from sending the major part of his fleet owing to plots and disturbances among the city despots throughout Illyria, and there came also the contingents ordered from Epirus, Acarnania, and Messene; for now that Phigaleia had been taken, the Messenians had no longer any hesitation in taking part in the war.
p. 13

A-IX.20
(1) Ἅμα δὲ τοῖς προειρημένοις Σκερδιλαΐδας, νομίζων ὑπὸ τοῦ βασιλέως ἀδικεῖσθαι διὰ τό τινα τῶν χρημάτων ἐλλείπειν αὐτῷ τῶν κατὰ τὰς συντάξεις ὁμολογηθέντων, ἃς ἐποιήσατο πρὸς Φίλιππον, ἐξαπέστειλε **λέμβους** πεντεκαίδεκα, μετὰ δόλου ποιούμενος τὴν ἐπιβολὴν τῆς κομιδῆς τῶν χρημάτων.
5.95.1

Simultaneously with these events Scerdilaïdas, considering himself wronged by the king, as the sum due to him by the terms of their agreement had not been paid in full, sent out fifteen **galleys** with the design of securing payment by trickery.
p. 249

A-IX.21
(1) Διόπερ ἀκούων τοὺς Σκερδιλαΐδου **λέμβους** περὶ Μαλέαν λῄζεσθαι καὶ πᾶσι τοῖς ἐμπόροις ὡς πολεμίοις χρῆσθαι, παρεσπονδηκέναι δὲ καὶ τῶν ἰδίων (2) τινὰ πλοίων ἐν Λευκάδι συνορμήσαντα, καταρτίσας δώδεκα μὲν καταφράκτους ναῦς, ὀκτὼ δ' ἀφράκτους, τριάκοντα δ' ἡμιολίους, ἔπλει δι' Εὐρίπου, σπεύδων μὲν καταλαβεῖν καὶ τοὺς Ἰλλυριούς, καθόλου δὲ μετέωρος ὢν ταῖς ἐπιβολαῖς ἐπὶ τὸν κατὰ τῶν Αἰτωλῶν πόλεμον διὰ τὸ μηδέν πω συνεικέναι τῶν (3) ἐν Ἰταλίᾳ γεγονότων. συνέβαινε δέ, καθ' οὓς καιροὺς ἐπολιόρκει τὰς Θήβας Φίλιππος, ἡττῆσθαι Ῥωμαίους ὑπ' Ἀννίβου τῇ περὶ Τυρρηνίαν μάχῃ, τὴν δὲ φήμην ὑπὲρ τῶν γεγονότων μηδέπω προσπεπτωκέναι (4) τοῖς Ἕλλησιν. ὁ δὲ Φίλιππος, **τῶν λέμβων**

Hearing, therefore, that the **galleys** of Scerdilaïdas were committing acts of piracy off Cape Malea and treating all merchants as enemies, and that he had treacherously seized some Macedonian ships which were anchored near him at Leucas, he manned twelve decked ships, eight undecked ones, and thirty *hemiolii*, and sailed through the Euripus, being anxious to capture the Illyrians also, and altogether in high hopes of success in the war with the Aetolians, as he had hitherto had no news of what was going on in Italy. It was while Philip was besieging Thebes that the Romans were defeated by Hannibal in Etruria, but the report of this event had not yet reached Greece. Philip missed the Illyrian **galleys**, and, anchoring off Cenchreae, sent

ὑστερήσας καὶ καθορμισθεὶς πρὸς Κεγχρεαῖς τὰς μὲν καταφράκτους ναῦς ἐξαπέστειλε, συντάξας περὶ Μαλέαν ποιεῖσθαι τὸν πλοῦν ὡς ἐπ' Αἰγίου καὶ Πατρῶν, τὰ δὲ λοιπὰ τῶν πλοίων ὑπερισθμίσας ἐν (5) Λεχαίῳ παρήγγελλε πᾶσιν ὁρμεῖν. αὐτὸς δὲ κατὰ σπουδὴν ἧκε μετὰ φίλων ἐπὶ τὴν τῶν Νεμέων πανήγυριν εἰς Ἄργος. 5.101.1-5.

off his decked ships with orders to sail round Cape Malea toward Aegium and Patrae: the rest of his vessels he dragged over the Isthmus, ordering them all to anchor at Lechaeum; and himself with his friends hastened to Argos to be present at the celebration of the Nemean festival.
pp. 263-265

A-IX.22
109. Φίλιππος δὲ κατὰ τὴν παραχειμασίαν ἀναλογιζόμενος ὅτι πρὸς τὰς ἐπιβολὰς αὐτοῦ χρεία πλοίων ἐστὶ καὶ τῆς κατὰ θάλατταν ὑπηρεσίας, καὶ ταύτης (2) οὐχ ὡς πρὸς ναυμαχίαν—τοῦτο μὲν γὰρ οὐδ' ἂν ἤλπισε δυνατὸς εἶναι, Ῥωμαίοις διαναυμαχεῖν—ἀλλὰ μᾶλλον ἕως τοῦ παρακομίζειν στρατιώτας καὶ θᾶττον διαίρειν οὗ πρόθοιτο καὶ παραδόξως ἐπὶ (3) φαίνεσθαι τοῖς πολεμίοις διόπερ ὑπολαβὼν ἀρίστην εἶναι πρὸς ταῦτα τὴν τῶν Ἰλλυριῶν ναυπηγίαν ἑκατὸν ἐπεβάλετο **λέμβους** κατασκευάζειν, σχεδὸν πρῶτος (4) τῶν ἐν Μακεδονίᾳ βασιλέων. καταρτίσας δὲ τούτους συνῆγε τὰς δυνάμεις ἀρχομένης θερείας, καὶ βραχέα προσασκήσας τοὺς Μακεδόνας ἐν ταῖς (5) εἰρεσίαις ἀνήχθη. κατὰ δὲ τὸν αὐτὸν καιρὸν Ἀντίοχος μὲν ὑπερέβαλε τὸν Ταῦρον, Φίλιππος δὲ ποιησάμενος τὸν πλοῦν δι' Εὐρίπου καὶ [τοῦ] περὶ Μαλέαν ἧκε πρὸς τοὺς περὶ Κεφαλληνίαν καὶ Λευκάδα τόπους, ἐν οἷς καθορμισθεὶς ἐκαραδόκει πολυπραγμονῶν (6) τὸν τῶν Ῥωμαίων στόλον. πυνθανόμενος δὲ περὶ τὸ Λιλύβαιον αὐτοὺς ὁρμεῖν, θαρρήσας νος δὲ περὶ τὸ Λιλύβαιον αὐτοὺς ὁρμεῖν, θαρρήσας ἀνήχθη, καὶ προῆγε ποιούμενος τὸν πλοῦν ὡς ἐπ' Ἀπολλωνίας.
110. ἤδη δὲ συνεγγίζοντος αὐτοῦ τοῖς περὶ τὸν Ἄων ποταμὸν τόποις, ὅς ῥεῖ παρὰ τὴν τῶν Ἀπολλωνιατῶν πόλιν, ἐμπίπτει πανικὸν παραπλήσιον (2) τοῖς γινομένοις ἐπὶ τῶν πεζικῶν στρατοπέδων. τῶν γὰρ ἐπὶ τῆς οὐραγίας πλεόντων τινὲς **λέμβοι**, καθορμισθέντες εἰς τὴν νῆσον, ἣ καλεῖται μὲν Σάσων, κεῖται δὲ κατὰ τὴν εἰσβολὴν τὴν εἰς τὸν Ἰόνιον πόρον, ἧκον ὑπὸ νύκτα πρὸς τὸν Φίλιππον, φάσκοντες συνωρμηκέναι τινὰς αὐτοῖς πλέοντας ἀπὸ πορθμοῦ, (3) τούτους δ' ἀπαγγέλλειν, ὅτι καταλείποιεν ἐν Ῥηγίῳ πεντήρεις Ῥωμαϊκὰς πλεούσας ἐπ' Ἀπολλωνίας (4) καὶ πρὸς Σκερδιλαΐδαν. ὁ δὲ Φίλιππος, ὑπολαβὼν ὅσον οὔπω τὸν στόλον ἐπ' αὐτὸν παρεῖναι, περίφοβος γενόμενος

109. During the winter Philip took into consideration that for his enterprise he would require ships and crews to man them, not it is true with the idea of fighting at sea—for he never thought he would be capable of offering battle to the Roman fleet—but to transport his troops, land where he wished, and take the enemy by surprise. Therefore, as he thought the Illyrian shipwrights were the best, he decided to build a hundred **galleys**, being almost the first king of Macedonia who had taken such a step. Having equipped these fleets he collected his forces at the beginning of summer and, after training the Macedonians a little in rowing, set sail. It was just at the time that Antiochus crossed the Taurus, when Philip sailing through the Euripus and round Cape Malea reached the neighborhood of Cephallenia and Leucas, where he moored and awaited anxiously news of the Roman fleet. Hearing that they were lying off Lilybaeum, he was encouraged to put to sea again and advanced sailing toward Apollonia.
110. Just as he was approaching the mouth of the river Aoüs, which runs past Apollonia, his fleet was seized by a panic such as sometimes overtakes land forces. For some of the **galleys** in the rear, which had anchored off an island called Sason lying at the entrance to the Ionian Sea, came in the night and informed Philip that some vessels which had crossed from the Sicilian Strait had anchored in the same roadstead and announced to them that they had left at Rhegium some Roman quinqueremes which were on their voyage to Apollonia to join Scerdilaïdas. Philip, in the belief that the Roman fleet would be upon him in less than no time, was seized by fear, and at once weighed anchor and gave orders to sail back. Quitting his anchorage and making the return voyage in thorough disorder he reached Cephallenia on the second day, travelling

5. Written sources on lembs and Liburnians from the 4th c. BC to Late Antiquity

καὶ ταχέως ἀνασπάσας τὰς ἀγκύρας (5) αὖτις εἰς τοὐπίσω παρήγγειλε πλεῖν. οὐδενὶ δὲ κόσμῳ ποιησάμενος τὴν ἀναζυγὴν καὶ τὸν ἀνάπλουν δευτεραῖος εἰς Κεφαλληνίαν κατῆρε, συνεχῶς (6) ἡμέραν καὶ νύκτα τὸν πλοῦν ποιούμενος. βραχὺ δέ τι θαρρήσας ἐνταῦθα κατέμεινε, ποιῶν ἔμφασιν ὡς τι θαρρήσας ἐνταῦθα κατέμεινε, ποιῶν ἔμφασιν ὡς ἐπί τινας τῶν ἐν Πελοποννήσῳ πράξεων ἐπεστροφώς. (7) συνέβη δὲ ψευδῶς γενέσθαι τὸν ὅλον φόβον (8) περὶ αὐτόν. ὁ γὰρ Σκερδιλαΐδας, ἀκούων κατὰ χειμῶνα **λέμβους** ναυπηγεῖσθαι τὸν Φίλιππον πλείους, καὶ προσδοκῶν αὐτοῦ τὴν κατὰ θάλατταν παρουσίαν, διεπέμπετο πρὸς τοὺς Ῥωμαίους διασαφῶν (9) ταῦτα καὶ παρακαλῶν βοηθεῖν, οἱ δὲ Ῥωμαῖοι δεκαναΐαν ἀπὸ τοῦ περὶ τὸ Λιλύβαιον ἐξαπέστειλαν στόλου, (10) ταύτην τὴν περὶ τὸ Ῥήγιον ὀφθεῖσαν· ἣν Φίλιππος εἰ μὴ πτοηθεὶς ἀλόγως ἔφυγε, τῶν περὶ τὴν Ἰλλυρίδα πράξεων μάλιστ' ἂν τότε καθίκετο διὰ τὸ τοὺς Ῥωμαίους πάσαις ταῖς ἐπινοίαις καὶ παρασκευαῖς περὶ τὸν Ἀννίβαν καὶ τὴν περὶ Κάνναν μάχην γίνεσθαι, τῶν τε πλοίων ἐκ τοῦ κατὰ λόγον ἐγκρατὴς (11) ἂν ἐγεγόνει. νῦν δὲ διαταραχθεὶς ὑπὸ τῆς προσαγγελίας ἀβλαβῆ μέν, οὐκ εὐσχήμονα δ' ἐποιήσατο τὴν ἀναχώρησιν εἰς Μακεδονίαν.
5.109–110

continuously by day and night. Plucking up a little courage he remained there pretending that he had returned to undertake some operations in the Peloponnese. As it turned out, the whole had been a false alarm. For Scerdilaïdas, hearing that Philip had been building a considerable number of **galleys** in the winter and expecting him to arrive by sea, sent to inform the Romans and beg for help, upon which the Romans sent a squadron of ten ships from their fleet at Lilybaeum, these being the ships that had been sighted off Rhegium. Had Philip not taken alarm so absurdly and fled before this squadron, now was the opportunity for him to make himself master of Illyria, the whole attention and all the resources of the Romans being concentrated on Hannibal and the situation connected with the battle of Cannae; and most probably the ships would have fallen into his hands also. But as it was the news upset him so much, that he made his way back to Macedonia without suffering any loss indeed but that of prestige.
pp. 283-287

A-IX.23

Φίλιππος δὲ περικαταλαμβανόμενος τοῖς καιροῖς, δοὺς τὸ σύνθημα τοῖς ἐπὶ τοῦ δεξιοῦ καὶ παραγγείλας ἀντιπρώρρους ποιεῖν τὰς ναῦς καὶ συμπλέκεσθαι τοῖς πολεμίοις ἐρρωμένως, αὐτὸς ὑπὸ τὰς νησίδας ἀναχωρήσας μετά τινων **λέμβων**, τὰς μεταξὺ τοῦ πόρου κειμένας, ἀπεκαραδόκει τὸν (9) κίνδυνον. ἦν δὲ τῶν μὲν τοῦ Φιλίππου νεῶν τὸ πλῆθος τὸ συγκαταστὰν εἰς τὸν ἀγῶνα κατάφρακτοι τρεῖς καὶ πεντήκοντα, σὺν δὲ τούτοις ἄφρακτα *** **λέμβοι** δὲ σὺν ταῖς πρίστεσιν ἑκατὸν καὶ πεντήκοντα· τὰς γὰρ ἐν τῇ Σάμῳ ναῦς οὐκ ἠδυνήθη (10) καταρτίσαι πάσας. τὰ δὲ τῶν πολεμίων σκάφη κατάφρακτα μὲν ἦν ἑξήκοντα καὶ πέντε σὺν τοῖς τῶν Βυζαντίων, μετὰ δὲ τούτων ἐννέα τριημιολίαι καὶ τριήρεις τρεῖς ὑπῆρχον.
16.2.8–10.

Philip, thus anticipated, after signaling to those on his right orders to turn their ships directly toward the enemy and engage him vigorously, retired himself with a few **light vessels** to the islands in the middle of the strait and awaited the result of the battle. Philip's fleet which took part in the battle consisted of fifty-three decked warships, ... undecked ones, and a hundred and fifty **galleys** and beaked ships, for he had not been able to fit out all the ships which were at Samos. The enemy had sixty-five decked warships, including those of the Byzantines, nine trihemioliae, and three triremes.
pp. 7-9

A-IX.24

(1) Τῶν δὲ λοιπῶν νεῶν τοῦ πλήθους ὁ κίνδυνος (2) ἐφάμιλλος ἦν· καθ' ὅσον γὰρ ἐπλεόναζον οἱ παρὰ τοῦ Φιλίππου **λέμβοι**, κατὰ τοσοῦτον διέφερον οἱ περὶ τὸν Ἄτταλον τῷ τῶν καταφράκτων νεῶν πλήθει. (3) καὶ τὰ μὲν περὶ τὸ δεξιὸν κέρας τοῦ Φιλίππου τοιαύτην εἶχε τὴν διάθεσιν ὥστ' ἀκμὴν ἄκριτα μένειν τὰ ὅλα, πολὺ δὲ τοὺς περὶ τὸν Ἄτταλον ἐπικυδεστέρας ἔχειν τὰς ἐλπίδας. (4) οἱ δὲ Ῥόδιοι κατὰ μὲν τὰς ἀρχὰς εὐθέως ἐκ τῆς ἀναγωγῆς ἀπεσπάσθησαν τῶν πολεμίων, καθάπερ ἀρτίως εἶπα, τῷ δὲ ταχυναυτεῖν παρὰ πολὺ διαφέροντες τῶν ἐναντίων συνῆψαν τοῖς ἐπὶ τῆς οὐραγίας Μακεδόσι. (5) καὶ τὸ μὲν πρῶτον ὑποχωροῦσι τοῖς σκάφεσι κατὰ πρύμναν ἐπι-πρῶτον ὑποχωροῦσι τοῖς σκάφεσι κατὰ πρύμναν ἐπι-φερόμενοι τοὺς ταρσοὺς παρέλυον· (6) ὡς δ' οἱ μὲν παρὰ τοῦ Φιλίππου συνεπιστρέφειν ἤρξαντο παραβοηθοῦντες τοῖς κινδυνεύουσι, τῶν δὲ Ῥοδίων οἱ καθυστεροῦντες ἐκ τῆς ἀναγωγῆς συνῆψαν τοῖς περὶ (7) τὸν Θεοφιλίσκον, τότε κατὰ πρόσωπον ἀντιπρώρρους τάξαντες τὰς ναῦς ἀμφότεροι συνέβαλον εὐψύχως, ὁμοῦ ταῖς σάλπιγξι καὶ τῇ κραυγῇ παρακαλοῦντες (8) ἀλλήλους. εἰ μὲν οὖν μὴ μεταξὺ τῶν καταφράκτων νεῶν ἔταξαν οἱ Μακεδόνες τοὺς **λέμβους**, ῥᾳδίαν ἂν καὶ σύντομον ἔλαβε κρίσιν ἡ ναυμαχία· νῦν δὲ ταῦτ' ἐμπόδια πρὸς τὴν χρείαν τοῖς Ῥοδίοις ἐγίνετο (9) κατὰ πολλοὺς τρόπους. μετὰ γὰρ τὸ κινηθῆναι τὴν ἐξ ἀρχῆς τάξιν ἐκ τῆς πρώτης συμβολῆς πάντες (10) ἦσαν ἀναμὶξ ἀλλήλοις, ὅθεν οὔτε διεκπλεῖν εὐχερῶς οὔτε στρέφειν ἐδύναντο τὰς ναῦς οὔτε καθόλου χρῆσθαι τοῖς ἰδίοις προτερήμασιν, ἐμπιπτόντων αὐτοῖς τῶν **λέμβων** ποτὲ μὲν εἰς τοὺς ταρσούς, ὥστε δυσχρηστεῖν ταῖς εἰρεσίαις, ποτὲ δὲ πάλιν εἰς τὰς πρώρρας, ἔστι δ' ὅτε κατὰ πρύμναν, ὥστε παραποδίζεσθαι καὶ τὴν τῶν κυβερνητῶν καὶ τὴν παραποδίζεσθαι καὶ τὴν τῶν κυβερνητῶν καὶ τὴν (11) τῶν ἐρετῶν χρείαν. κατὰ δὲ τὰς ἀντιπρώρρους (12) συμπτώσεις ἐποίουν τι τεχνικόν· αὐτοὶ μὲν γὰρ ἔμπρωρρα τὰ σκάφη ποιοῦντες ἐξάλους ἐλάμβανον τὰς πληγάς, τοῖς δὲ πολεμίοις ὕφαλα τὰ τραύματα (13) διδόντες ἀβοηθήτους ἐσκεύαζον τὰς πληγάς. σπανίως δ' εἰς τοῦτο συγκατέβαινον· καθόλου γὰρ ἐξέκλινον τὰς συμπλοκὰς διὰ τὸ γενναίως ἀμύνεσθαι τοὺς Μακεδόνας ἀπὸ τῶν καταστρωμάτων ἐν (14) ταῖς

Among the other ordinary ships of the fleet the contest was equal; for the advantage that Philip had in the number of his **galleys** was balanced by Attalus' superiority in decked ships. The position of affairs on Philip's right wing was such that the result was still doubtful; but Attalus was by far the most sanguine of success. The Rhodians, as I just said, were at first from the moment that they put out to sea very widely separated from the enemy, but as they sailed a great deal faster they caught up the rear of the Macedonian fleet. At first they attacked the ships which were retreating before them from the stern, breaking their banks of oars. But as soon as the rest of Philip's fleet began to put about and come to the assistance of their comrades in peril and those of the Rhodians who had been the last to put to sea joined Theophiliscus, then both fleets directing their prows against each other engaged gallantly, cheering each other on with loud cries and the peal of trumpets. Now had not the Macedonians interspersed their **galleys** among their decked ships the battle would have been quickly and easily decided, but as it was these **galleys** impeded the action of the Rhodian ships in many ways. For, once the original order of battle had been disturbed in their first charge, they were utterly mixed up, so that they could not readily sail through the enemy's line nor turn their ships round, in fact could not employ at all the tactics in which they excelled, as the **galleys** were either falling foul of their oars and making it difficult for them to row, or else attacking them in the prow and sometimes in the stern, so that neither the could serve efficiently. But in the direct charges prow to prow they employed a certain artifice. For depressing the ships toward the prow themselves they received the enemy's blow above water, but piercing him below water produced breaches which could not be repaired. It was but seldom, however, that they resorted to this mode of attack; for as a rule they avoided closing with the enemy, as the Macedonian soldiers offered a valiant resistance from the deck in such close combats. For the most part they cut the enemy's line and put his banks of oars out of action, afterward turning and sailing round again

5. Written sources on lembs and Liburnians from the 4th c. BC to Late Antiquity

συστάδην γινομέναις μάχαις. τὸ δὲ πολὺ κατὰ μὲν τοὺς διέκπλους παρασύροντες τῶν πολεμίων νεῶν τοὺς ταρσοὺς ἠχρείουν· μετὰ δὲ ταῦτα πάλιν ἐκπεριπλέοντες, καὶ τοῖς μὲν κατὰ πρύμναν ἐμβάλλοντες, τοῖς δὲ πλαγίοις καὶ στρεφομένοις ἀκμὴν προσπίπτοντες οὓς μὲν ἐτίτρωσκον, οἷς δὲ παρέλυον (15) ἀεί τι τῶν πρὸς τὴν χρείαν ἀναγκαίων. καὶ δὴ τῷ τοιούτῳ τρόπῳ μαχόμενοι παμπληθεῖς τῶν πολεμίων ναῦς διέφθειραν.

and charging him sometimes in the stern and sometimes in flank while he was still turning; thus they made breaches in some of the ships and in others damaged some part of the necessary gear. Indeed by this mode of fighting they destroyed quite a number of the enemy's ships.
pp. 11-15

16.4

A-IX.25
(4) ἐν ᾧ καιρῷ Θεοφιλίσκος, βοηθήσας μετὰ τριῶν πεντήρων, τὴν μὲν ναῦν οὐκ ἠδυνήθη σῶσαι διὰ τὸ πλήρη θαλάττης εἶναι, δύο δὲ ναῦς πολεμίας τρώ (5) σας τοὺς ἐπιβάτας ἐξέβαλε. ταχὺ δὲ περιχυθέντων αὐτῷ **λέμβων** πλειόνων καὶ καταφράκτων νεῶν, τοὺς μὲν πλείστους ἀπέβαλε τῶν ἐπιβατῶν ἐπιφανῶς (6) ἀγωνισαμένους, αὐτὸς δὲ τρία τραύματα λαβὼν καὶ παραβόλως τῇ τόλμῃ κινδυνεύσας μόλις ἐξέσωσε τὴν ἰδίαν ναῦν ἐπιβοηθήσαντος αὐτῷ Φιλοστράτου καὶ συναναδεξαμένου τὸν ἐνεστῶτα κίνδυνον εὐψύχως.
16.5.4–6

At this moment Theophiliscus came up to help with three quinqueremes, and though he could not save the ship as she was full of water, rammed two of the enemy's ships and forced the troops on board to take to the water. He was rapidly surrounded by a number of **galleys** and decked ships, and after losing most of his soldiers, who fought splendidly, and receiving himself three wounds and displaying extraordinary courage, just managed to save his own ship, Philostratus coming up to his succor and taking a gallant part in the struggle.
pp. 15-17

A-IX.26
(4) ὁ δὲ Φίλιππος, συνθεασάμενος ἀπεσπασμένον πολὺ τὸν Ἄτταλον ἀπὸ τῶν ἰδίων, παραλαβὼν τέτταρας πεντήρεις καὶ τρεῖς ἡμιολίας, ἔτι δὲ τῶν **λέμβων** τοὺς ἐγγὺς ὄντας, ὥρμησε, καὶ διακλείσας τὸν Ἄτταλον ἀπὸ τῶν οἰκείων νεῶν ἠνάγκασε μετὰ μεγάλης ἀγωνίας εἰς τὴν γῆν ἐκβαλεῖν τὰ σκάφη. (5) τούτου δὲ συμβάντος αὐτὸς μὲν ὁ βασιλεὺς μετὰ τῶν πληρωμάτων εἰς τὰς Ἐρυθρὰς ἀπεχώρησε, τῶν δὲ πλοίων καὶ τῆς βασιλικῆς κατασκευῆς ἐγκρατὴς ὁ (6) Φίλιππος ἐγένετο. καὶ γὰρ ἐποίησάν τι τεχνικὸν ἐν τούτοις τοῖς καιροῖς οἱ περὶ τὸν Ἄτταλον· τὰ γὰρ ἐπιφανέστατα τῆς βασιλικῆς κατασκευῆς ἐπὶ τὸ (7) κατάστρωμα τῆς νεὼς ἐξέβαλον. ὅθεν οἱ πρῶτοι τῶν (7) κατάστρωμα τῆς νεὼς ἐξέβαλον. ὅθεν οἱ πρῶτοι τῶν Μακεδόνων, συνάψαντες ἐν τοῖς **λέμβοις**, συνθεασάμενοι ποτηρίων πλῆθος καὶ πορφυρῶν ἱματίων καὶ τῶν τούτοις παρεπομένων σκευῶν, ἀφέμενοι τοῦ (8) διώκειν ἀπένευσαν ἐπὶ τὴν τούτων ἁρπαγήν.
16.6.4–8

Philip now, seeing that Attalus was widely separated from his own fleet, took four quinqueremes and three hemioliae and such **galleys** as were near him and, intercepting the return of Attalus to his own fleet, compelled him in great disquietude to run his ships ashore. After this the king and the crews escaped to Erythrae, but Philip gained possession of the ships and the royal furniture. Attalus indeed resorted to an artifice on this occasion by causing the most splendid articles of his royal furniture to be exposed on the deck of his ship, so that the Macedonians who were the first to reach it in their **galleys**, when they saw such a quantity of cups, purple cloaks, and other objects to match, instead of continuing the pursuit turned aside to secure this booty, so that Attalus made good his retreat to Erythrae.
p. 19

A-IX.27
(1) Ἐφθάρησαν δὲ τοῦ μὲν Φιλίππου ναῦς ἐν μὲν τῇ πρὸς Ἄτταλον ναυμαχίᾳ δεκήρης, ἐννήρης, ἑπτήρης, ἑξήρης, τῶν δὲ λοιπῶν κατάφρακτοι μὲν δέκα καὶ τριημιολίαι τρεῖς, **λέμβοι** δὲ πέντε καὶ εἴκοσι καὶ (2) τὰ τούτων πληρώματα· ἐν δὲ τῇ πρὸς Ῥοδίους διεφθάρησαν κατάφρακτοι μὲν δέκα, **λέμβοι** δὲ περὶ τετταράκοντα τὸν ἀριθμόν· ἥλωσαν δὲ δύο τετρήρεις (3) καὶ **λέμβοι** σὺν τοῖς πληρώμασιν ἑπτά. τῶν δὲ παρ' Ἀττάλου κατέδυσαν μὲν τριημιολία μία καὶ δύο πεντήρεις, (ἥλωσαν δὲ δύο τετρήρεις) καὶ τὸ (4) τοῦ βασιλέως σκάφος.
16.7.1-3

Of Philip's ships there were sunk in the battle with Attalus one 'ten,' one 'nine,' one 'seven,' and one 'six,' and of the rest of his fleet ten decked ships, three trihemioliae, and twenty-five **galleys** with their crews. In his battle with the Rhodians he lost ten decked ships and about forty **galleys** sunk and two quadriremes and seven **galleys** with their crews captured. Out of Attalus' fleet one trihemiolia and two quinqueremes were sunk, two quadriremes and the royal ship were taken.
p. 21

A-IX.28
(1) Ἐπελθόντος δὲ τοῦ τεταγμένου καιροῦ παρῆν ὁ μὲν Φίλιππος ἐκ Δημητριάδος ἀναχθεὶς εἰς τὸν Μηλιέα κόλπον, πέντε **λέμβους** ἔχων καὶ μίαν πρίστιν, ἐφ' ἧς αὐτὸς ἐπέπλει.
18.1.1

When the time fixed for the conference came, Philip arrived, having sailed from Demetrias to the Malian Gulf with five **galleys** and a beaked ship in which he traveled himself.
p. 99

A-IX.29
καὶ γὰρ ἡ τῶν **λέμβων** παρουσία καὶ τὸ πλῆθος τῶν ἀπὸ λωλότων ἱππέων καὶ *** ἡ τοῦ Γενθίου μετάθεσις συνέτριβεν αὐτούς.
29.11.3

For the presence of the **galleys**, the large losses of the Roman cavalry, and Genthius' change of attitude weighed on their spirits.
p. 73

A-IX.30
(3) παραστήσας οὖν **λέμβον** καὶ παραδοὺς αὐτὸν Δημητρίῳ τινὶ τῶν φίλων ἐξαπέστειλεν. [...] (9) καταπλαγεὶς τὴν περίστασιν ἐνέβη πάλιν τὸν **λέμβον** πρὸς τὸν Δημήτριον.
30.9.3 and 9

(3) He [the king Ptolemy] therefore procured a **galley**, and putting him in charge of Demetrius, one of the royal friends, sent him off [...]. (9) Polyaratus, alarmed at his dangerous situation, went on board the **galley** again to Demetrius [...].
pp. 127-129

Greek: *Polybii historiae* 1-4, ed. T. Büttner-Wobst. Teubner, Leipzig 1889-1904.

English: *Polybius. The Histories, Volume I: Books 1-2; Volume II: Books 3-4; Volume III: Books 5-8; Volume IV: Books 9-15; Volume V: Books 16-27*, transl. W.R. Paton, rev. F.W. Walbank, Ch. Habicht. LCL 128, 137-38, 159-60 (2010-12).

Polybius. The Histories, Volume VI: Books 28-39. Fragments, ed. and transl. S. Douglas Olson, transl. W.R.Paton, Rev. F.W. Walbank, Ch. Habicht. LCL 161 (2012).

With 51 mentions of lembs, Polybius is by far the most extensive source we have about this ship from the ancient Greek world. If we summarise the passages above, several things come out as immediately evident. Paton/Walbank/Halbicht's translations vary significantly, for the Greek term λέμβος has been translated variously as:

5. Written sources on lembs and Liburnians from the 4th c. BC to Late Antiquity

- boat: **A-IX.1-3, 6-8, 10-17, 19**
- flotilla (λέμβοι): **A-IX.4-5**
- vessel: **A-IX.9**
- light vessel: **A-IX.23**
- frigate: **A-IX.18**
- galley: **A-IX.20-30** (in connection with the Macedonians and Rhodians)

Different builders/users/owners of the lembs were mentioned:

- the Romans: **A-IX.1-2**
- the Carthaginians (built by local inhabitants on banks of Rhone, used by Hannibal): **A-IX.12-14**
- the 'Illyrians': **A-IX.3-4, 7** (with a strong self-reference used in the phrase τοὺς παρ' αὐτῶν λέμβους, 'properly their own lembs'), **A-IX.8**
- of Demetrius of Pharos: **A-IX.10-11, 15-16, 18**
- of Scerdilaïdas: **A-IX.17, 19-21**
- the Macedonians: **A-IX.22-30**

From Polybius we can also learn the following details about the lembs.

- Lembs are used to sail in front [of the fleet – translator's addition] (**A-IX.2**: οἱ προπλεῖν εἰθισμένοι λέμβοι). The word used here – εἰθισμένοι – is a form of ἐθίζω, stemming from ἔθος, 'to accustom', a relatively rare word that denotes someone's or something's natural property. If Polybius was careful in using his vocabulary here, this would imply that the original function of the lembs was to sail in front of other ships.
- 100 lembs can carry 5000 people, **A-IX.4**
- Lembs are bigger than 'canoes' (μονόξυλα πλοῖα – vessels made of a single wood), **A-IX.12-13**
- Although it is originally used at sea, it can be used also on rivers (e.g. Rhone, **A-IX.12**)
- There are usually many lembs in an expedition (**A-IX.13, 27**)
- The Macedonian king Philip ordered 100 lembs to be built by Illyrian shipwrights of Illyrian fame for having good shipbuilding techniques. Polybius comments that this was the first time that a Macedonian king could do such a thing (**A-IX.22**). Polybius however does not claim that the lemb was originally an Illyrian ship.
- It is used in association with the word πρίστις, probably denoting a saw-fish (*LSJ*), and in relation to the naval idiom, 'ship of war, prob. from its shape'. However, the place in the text is corrupt and it is hard to say whether Polybius had another sort of ship in mind, or he described lembs as having a specific form (**A-IX.23**).
- It is a very fast boat: it can make a trip from Sason to Cephallenia, 160 nautical miles, in less than two days, which, according to Casson, makes the lemb travel at an average speed of four knots (**A-IX.22**).[183]

---

[183] Casson 1951: 147.

A-X Posidonius (135 - c. 51 BC)

Fragmenta

(§12) Ἐκ δὲ τούτου συμβαλόντα τὸν Εὔδοξον ὡς δυνατὸς εἴη ὁ περίπλους ὁ Λιβυκός, πορευθέντα οἴκαδε τὴν οὐσίαν ἐνθέμενον πᾶσαν ἐξορμῆσαι. καὶ πρῶτον μὲν εἰς Δικαιαρχίαν, εἶτ' εἰς Μασσαλίαν ἐλθεῖν, καὶ τὴν ἑξῆς παραλίαν μέχρι Γαδείρων, πανταχοῦ δὲ διακωδωνίζοντα ταῦτα καὶ χρηματιζόμενον κατασκευάσασθαι πλοῖον μέγα καὶ **ἐφόλκια δύο λέμβοις λῃστρικοῖς ὅμοια**, ἐμβιβάσαι τε μουσικὰ παιδισκάρια καὶ ἰατροὺς καὶ ἄλλους τεχνίτας, ἔπειτα πλεῖν ἐπὶ τὴν Ἰνδικὴν μετέωρον ζεφύροις συνεχέσι. καμνόντων δὲ τῷ πλῷ τῶν συνόντων, ἄκοντα ἐπουρίσαι πρὸς γῆν, δεδοικότα τὰς πλημμυρίδας καὶ τὰς ἀμπώτεις. καὶ δὴ καὶ συμβῆναι ὅπερ ἐδεδίει·καθίσαι γὰρ τὸ πλοῖον, ἡσυχῇ δέ, ὥστε μηδ' ἀθροῦν διαλυθῆναι, ἀλλὰ φθῆναι τὰ φορτία σωθέντα εἰς γῆν καὶ τῶν ξύλων τὰ πλεῖστα·ἐξ ὧν **τρίτον λέμβον** συμπηξάμενον πεντηκοντόρῳ πάρισον πλεῖν, ἕως ἀνθρώποις συνέμιξε τὰ. αὐτὰ ῥήματα φθεγγομένοις, ἅπερ πρότερον ἀπεγέγραπτο. ἅμα δὲ τοῦτό τε γνῶναι, ὅτι τε οἱ ἐνταῦθα ἄνθρωποι ὁμοεθνεῖς εἶεν τοῖς Αἰθίοψιν ἐκείνοις, καὶ ὅτι ὁμοροῖεν τῇ Βόγου βασιλείᾳ. Ἀφέντα δὴ τὸν ἐπὶ Ἰνδοὺς πλοῦν ἀναστρέφειν· ἐν δὲ τῷ παράπλῳ νῆσον εὔυδρον καὶ εὔδενδρον ἐρήμην ἰδόντα σημειώσασθαι. (§13) σωθέντα δὲ εἰς τὴν Μαυρουσίαν, διαθέμενον **τοὺς λέμβους** πεζῇ κομισθῆναι πρὸς τὸν Βόγον καὶ συμβουλεύειν αὐτῷ τὴν ναυστολίαν ἐπανελέσθαι ταύτην. ἰσχῦσαι δ' εἰς τἀναντία τοὺς φίλους ὑποτείνοντας φόβον, μὴ συμβῇ τὴν χώραν εὐεπιβούλευτον γενέσθαι, δειχθείσης παρόδου τοῖς ἔξωθεν ἐπιστρατεύειν ἐθέλουσιν.
BNJ, 87 F28 §12-13 (=Strabo, 2.3.4)

(§12) From this Eudoxos deduced that it was possible to circumnavigate Africa; so he went home, put everything he had on board, and sailed out. First he came to Dikaiarchia (Puteoli), then to Marseilles and the successive coastlines as far as Cádiz. Advertising this everywhere he went and making money, he had a large ship built **and two tow-boats similar to pirate lemboi**, on which he put girl musicians, doctors, and other professionals, and then sailed on the high sea towards India with a steady west wind. But as those who were with him on the ship grew ill, he reluctantly tacked towards land though he feared the ebb and flow. And indeed what he feared actually happened: the ship ran aground, if gently and without it breaking it up all at once. He was able to get the cargo safely to land first and most of the timbers, from which he built a **third lembos** much like a pentekonter and sailed. until he was with men who spoke the same words that he had previously written down. Whereupon he realised that the men here were of the same people as those Aithiopians and that they bordered on the kingdom of Bogos (Bocchus) So, he gave up the idea of the voyage to the Indians and returned. As he sailed along, he saw, and made a note of, an uninhabited island with a good water supply and plenty of trees. (§13) He got back safely to Maurousia, disposed **of the lemboi**, travelled on foot to Bogos and recommended him to take on this expedition. However, the king's friends prevailed against him, making out that there would be a worry that the place would attract hostile attention once the route there had been revealed to foreigners wishing to invade [i.e., Maurousia]

Greek and English: K. Dowden, 'Poseidonios (87)', *BNJ*.

In these fragments, taken from Strabo 2.3.4, where he reports on Posidonius' account of Eudoxus of Cyzikus attempting to circumnavigate Africa, we see that the term λέμβος is being comparted to two different vessels. First, an ἐφόλκιον, a boat that is towed or a '*small boat towed after a ship*' (LSJ), is claimed to be similar to a pirate lemb. Posidonius' need to add '*pirate*' to the term *lembos* (λέμβοις λῃστρικοῖς) suggest that this was not an obvious allusion, and it may refer to a certain shape of lembs that pirates used to employ. In the second comparison the size of the third lemb is compared to a penteconter, a considerably larger boat with fifty oars. However, Strabo, in his comment on Posidonius' account casts doubt on the possibility of building such a large ship in the desert.[184] Both instances of lembs in this passage suggest that this word was used in a more generic sense, like a certain shape or function rather than a specific type of ship.

A-XI Diodorus Siculus (before 60 - after 36 BC)

Library of History

ἐν ὅσῳ δὲ ταῦτα τὴν συντέλειαν ἐλάμβανεν, ἀθροίσας τοὺς ἁδροτάτους **τῶν λέμβων** καὶ τούτους καταφράξας σανίσι καὶ θυρίδας κλειστὰς κατασκευάσας ἐνέθετο μὲν τῶν τρισπιθάμων ὀξυβελῶν τοὺς πορρωτάτω βάλλοντας καὶ τοὺς τούτοις κατὰ τρόπον χρησομένους, ἔτι δὲ τοξότας Κρῆτας, τὰς δὲ ναῦς προσαγαγὼν ἐντὸς βέλους κατετίτρωσκε τοὺς κατὰ τὴν πόλιν ὑψηλότερα τὰ παρὰ τὸν λιμένα τείχη κατασκευάζοντας. 20.85.3	In the interval while these were receiving their finishing touches, he [Demetrios] collected the strongest of **the light craft**, fortified them with planks, provided them with ports that could be closed, and placed upon them those of the catapults for bolts three palms long which had the longest range and the men to work them properly, and also Cretan archers; then, sending the boats within range, he shot down the men of the city who were building higher the walls along the harbour. p. 367

Greek: *Diodori bibliotheca historica*, 5 vols., eds I. Bekker, L. Dindorf, K. T. Fischer and F. Vogel 3rd ed. Teubner, Leipzig 1888-96.

English: *Diodorus Siculus. Library of History, Volume X: Books 19.66-20*, transl. R. M. Geer. LCL 390 (1954).

In this passage, the Macedonian king Demetrios Poliorcretes' siege of Rhodes is described. He uses lembs as war ships, adding fortifications with planks, and putting *'windows/doors that can be closed'* as a sort of closed cabin or a structure on the vessel (in the text above translated as *'ports that could be closed'*). He then added catapults, men who would operate them, and archers, creating rather potent war ships. It is worth noting that it is the Macedonians who took advantage of lembs here. In an earlier book (15.14.2), Diodorus mentions *'the light craft of the Illyrians'* (see p. 20 above), but in this context he does not use the expected lembs, but rather a more generic term πλοιάριον, a deminutive of πλοῖον, 'boat' (ἐπέπλευσε τοῖς τῶν Ἰλλυριῶν πλοιαρίοις – he sailed against Illyrian light vessels).

---

[184] Strabo, 2.3.5: 'And when he had gone off to Gadeira and built himself ships and sailed away in regal fashion, how did he manage, when his boat was wrecked, to build himself a third cutter in the desert?'

A.XII Plutarch of Chaeronea (*c.* 45 - before 125)

Life of Aemilius Paulus

(1) Γναῖος δ' Ὀκτάβιος ὁ ναυαρχῶν Αἰμιλίῳ προσορμισάμενος τῇ Σαμοθρᾴκῃ τὴν μὲν ἀσυλίαν παρεῖχε τῷ Περσεῖ διὰ τοὺς θεούς, ἔκπλου (2) δὲ καὶ φυγῆς εἶργεν. οὐ μὴν ἀλλὰ λανθάνει πως ὁ Περσεὺς Ὀροάνδην τινὰ (3) Κρῆτα **λέμβον** ἔχοντα συμπείσας μετὰ χρημάτων ἀναλαβεῖν αὐτόν. ὁ δὲ κρητισμῷ χρησάμενος, τὰ μὲν χρήματα νύκτωρ ἀνέλαβεν, ἐκεῖνον δὲ τῆς ἑτέρας νυκτὸς ἥκειν κελεύσας ἐπὶ τὸν πρὸς τῷ Δημητρείῳ λιμένα μετὰ τῶν τέκνων καὶ θεραπείας ἀναγκαίας, εὐθὺς ἀφ' ἑσπέρας ἀπέπλευσεν.
23.1-3

But to resume, Gnaeus Octavius, the admiral of Aemilius, came to anchor off Samothrace, and while he allowed Perseus to enjoy asylum, out of respect to the gods, he took means to prevent him from escaping by sea. However, Perseus somehow succeeded in persuading a certain Cretan named Oroandes, the owner of **a small skiff**, to take him on board with his treasures. So Oroandes, true Cretan that he was, took the treasures aboard by night, and after bidding Perseus to come during the following night to the harbour adjoining the Demetrium, with his children and necessary attendants, as soon as evening fell sailed off.
p. 423

Greek: *Plutarchi vitae parallelae* 2.1, ed. K. Ziegler, 2nd ed. Teubner, Leipzig 1964.

English: *Plutarch. Lives, Volume VI: Dion and Brutus. Timoleon and Aemilius Paulus*, transl. B. Perrin. LCL 98 (1918).

Plutarch attributes the ownership of a lemb to a Cretan living in Samothrace. The context suggests that it was not a big ship, and that it was a sort of a vessel used for various purposes.

A.XIII Appian (*c.* 90-160)

A-XIII.1 *The Punic Wars* (*Libyca*)

(50) καὶ Μασσανάσσης μὲν Καρχηδονίοις οὕτως ἐπολέμει, ὁ δὲ Σκιπίων, ἐπεί οἱ πάντα εὐτρεπῆ γεγένητο ἐν Σικελίᾳ, ἔθυε Διὶ καὶ Ποσειδῶνι καὶ ἐς Λιβύην ἀνήγετο ἐπὶ νεῶν μακρῶν μὲν δύο καὶ πεντήκοντα, φορτίδων δὲ τετρακοσίων· κέλητές τε καὶ **λέμβοι** πολλοὶ (51) συνείποντο αὐτῷ. καὶ στρατιὰν ἦγε πεζοὺς μὲν ἑξακισχιλίους ἐπὶ μυρίοις, ἱππέας δὲ χιλίους καὶ ἑξακοσίους. ἐπήγετο δὲ καὶ βέλη καὶ ὅπλα καὶ μηχανήματα ποικίλα (52) καὶ ἀγορὰν πολλήν.
50–51

III.13 In this way Masinissa was making war on the Carthaginians. In the meantime Scipio, having completed his preparations in Sicily, and sacrificed to Jupiter and Neptune, set sail for Africa with fifty-two warships and 400 transports, with a great number of **smaller craft** following behind. His army consisted of 16,000 foot and 1600 horse. He carried also projectiles, arms, and engines of various kinds, and a plentiful supply of provisions.
p. 421

A-XIII.2 *The Illyrian wars*

(17) Ἄγρων ἦν βασιλεὺς Ἰλλυριῶν μέρους ἀμφὶ τὸν κόλπον τῆς θαλάσσης τὸν Ἰόνιον, ὃν δὴ καὶ Πύρρος, ὁ τῆς Ἠπείρου βασιλεύς, κατεῖχε καὶ οἱ τὰ Πύρρου διαδεξάμενοι. Ἄγρων δ' ἔμπαλιν τῆς τε Ἠπείρου τινὰ καὶ Κέρκυραν ἐπ' αὐτοῖς καὶ Ἐπίδαμνον καὶ Φάρον καταλαβὼν ἔμφρουρα εἶχεν. ἐπιπλέοντος δ' αὐτοῦ καὶ τὸν ἄλλον Ἰόνιον, (18) νῆσος, ᾗ ὄνομα Ἴσσα, ἐπὶ Ῥωμαίους κατέφυγεν. οἳ δὲ πρέσβεις τοῖς Ἰσσίοις συνέπεμψαν, εἰσομένους τὰ Ἄγρωνος ἐς αὐτοὺς ἐγκλήματα. τοῖς δὲ πρέσβεσιν ἔτι προσπλέουσιν ἐπαναχθέντες **Ἰλλυρικοὶ λέμβοι** τῶν μὲν Ἰσσίων πρεσβευτὴν Κλεέμπορον, τῶν δὲ Ῥωμαίων Κορογκάνιον (19) ἀναιροῦσιν· οἱ δὲ λοιποὶ διέδρασαν αὐτούς. καὶ ἐπὶ τῷδε Ῥωμαίων ἐπ' Ἰλλυριοὺς ναυσὶν ὁμοῦ καὶ πεζῷ στρατευόντων [...]. (20) μετὰ ταῦτα δὲ ἡ Ἄγρωνος γυνὴ πρέσβεις ἐς Ῥώμην ἔπεμψε τά τε αἰχμάλωτα ἀποδιδόντας αὐτοῖς καὶ τοὺς αὐτομόλους ἄγοντας, καὶ ἐδεῖτο συγγνώμης τυχεῖν τῶν οὐκ ἐφ' ἑαυτῆς, (21) ἀλλ' ἐπὶ Ἄγρωνος γενομένων. οἳ δὲ ἀπεκρίναντο Κέρκυραν μὲν καὶ Φάρον καὶ Ἴσσαν καὶ Ἐπίδαμνον καὶ Ἰλλυριῶν τοὺς Ἀτιντανοὺς ἤδη Ῥωμαίων ὑπηκόους εἶναι, Πίννην δὲ τὴν ἄλλην Ἄγρωνος ἀρχὴν ἔχειν καὶ φίλον εἶναι Ῥωμαίοις, ἢν ἀπέχηται τῶν προλελεγμένων, καὶ τὴν Λίσσον μὴ παραπλέωσιν Ἰλλυρικοὶ λέμβοι δυοῖν (22) πλείονες, καὶ τούτοιν δὲ ἀνόπλοιν. ἡ μὲν δὴ ταῦτα πάντα ἐδέχετο [...]. [...]
(25) Ῥωμαῖοι Μακεδόσιν ἐπολέμουν, καὶ Περσεὺς ἦν ἤδη Μακεδόνων βασιλεὺς μετὰ Φίλιππον· Περσεῖ δὲ Γένθιος, Ἰλλυριῶν ἑτέρων βασιλεύς, ἐπὶ χρήμασι συνεμάχει καὶ ἐς τοὺς Ῥωμαίων Ἰλλυριοὺς ἐνέβαλε καὶ πρέσβεις Ῥωμαίων πρὸς αὐτὸν ἐλθόντας ἔδησεν, αἰτιώμενος οὐ (26) πρέσβεις ἀλλὰ κατασκόπους ἐλθεῖν. Ἀνίκιος δέ, Ῥωμαίων στρατηγός, **λέμβους** τε τοῦ Γενθίου τινὰς εἷλεν ἐπιπλεύσας καὶ κατὰ γῆν αὐτῷ συνενεχθεὶς ἐκράτει τὴν μάχην καὶ συνέκλεισεν ἔς τι χωρίον, [...].
17–26

Chap. II.7 Agron was king of that part of Illyria which borders the Adriatic Sea, over which sea Pyrrhus, king of Epirus, and his successors held sway. Agron in turn captured a part of Epirus and also Corcyra, Epidamnus, and Pharus in succession, and established garrisons in them. When he threatened the rest of the Adriatic with his fleet, the isle of Issa implored the aid of the Romans. The latter sent ambassadors to accompany the Issii and to ascertain what offences Agron imputed to them. **The Illyrian light vessels** attacked the ambassadors as they sailed up, and slew Cleemporus, the envoy of Issa, and the Roman Coruncanius; the remainder escaped by flight. Thereupon the Romans invaded Illyria by land and sea. [...] After these events the widow of Agron sent ambassadors to Rome to surrender the prisoners and deserters into their hands. She begged pardon also for what had been done, not by herself, but by Agron. They received for answer that Corcyra, Pharus, Issa, Epidamnus, and the Illyrian Atintani were already Roman subjects, that Pinnes might have the remainder of Agron's kingdom and be a friend of the Roman people if he would keep hands off the aforesaid territory, and agree not to sail beyond Lissus with more than two Illyrian **pinnaces**, both unarmed. She accepted all these conditions. [...]
When the Romans were at war with the Macedonians during the reign of Perseus, the successor of Philip, Genthius, the king of another Illyrian tribe, made an alliance with Perseus for money and attacked Roman Illyria, and put the ambassadors sent by the Romans in chains, charging them with coming not as ambassadors, but as spies. The Roman general, Anicius, in a naval expedition, captured some of Genthius' **pinnaces** and then engaged him in battle on land, defeated him, and shut him up in a fortress.
pp. 63-69

Greek: *Appiani Historia Romana 1*, eds E. Gabba, A. G. Roos, P. Viereck. Teubner, Leipzig 1939.

English: *Appian. Roman History, Volumes I-II*, ed. and transl. B. McGing. LCL 2-3 (1912).

A-XIII.3 Civil Wars

ὁ δὲ πολὺς ἐς Σικελίαν ᾔει, γειτονεύουσαν τῆς Ἰταλίας, καὶ Πομπηίου σφᾶς προθύμως ὑποδεχομένου. λαμπροτάτην γὰρ δὴ σπουδὴν ἐς τοὺς ἀτυχοῦντας ὁ Πομπήιος ἐν καιρῷ τότε ἔδειξε, κήρυκάς τε περιπέμπων, οἳ πάντας ἐς αὑτὸν ἐκάλουν, καὶ τοῖς περισῴζουσιν αὐτοὺς ἐλευθέροις τε καὶ θεράπουσι προλέγων διπλάσια τῶν διδομένων τοῖς αἱροῦσι· **λέμβοι** τε αὐτοῦ καὶ στρογγύλα ὑπήντα τοῖς πλέουσι, καὶ τριήρεις τοὺς αἰγιαλοὺς ἐπέπλεον, σημεῖά τε ἀνίσχουσαι τοῖς ἀλωμένοις, καὶ τὸν ἐντυγχάνοντα περισῴζουσαι.
4.6.36

The greater number, however, went to Sicily because of its nearness to Italy, where Sextus Pompeius received them gladly. The latter showed the most admirable zeal in behalf of the unfortunate at this crisis, sending heralds who invited all to come to him, and offered to those who should save the proscribed, both slaves and free persons, double the rewards that had been offered for killing them. His **small boats** and merchant ships met those who were escaping by sea, and his war-ships sailed along the shore and made signals to those wandering there and saved such as they found.
p. 201

Greek: *Appiani historia Romana 2*, eds L. Mendelssohn, P. Viereck. Teubner, Leipzig 1905.

English: *Appian, Roman History, Volume IV: The Civil Wars,* transl. H. White. LCL 5 (1913).

There are three main points to make here. First (**A-XIII.1**), the Romans used lembs during the Punic wars, confirming the evidence given by Polybius (1.53.9). Second, a lemb was a type of ship that was typically used by the Illyrians (**A-XIII.2**). That explains the terms of the treaty between Agron's widow, Teuta, and the Romans, in which only two unarmed lembs were allowed to sail beyond Lissus (Lezhë), which would make it reasonable to suppose that lembs were typical and not dangerous vessels. Finally, Sextus Pompeius had lembs in his fleet in 36 BC, but these clearly were not warships (**A-XIII.3**).

A-XIV Hamburg Papyrus *P. Hamb.* 4 248 (24 March, 145)

(ἔτους ὀγδόου Αὐτοκράτορος Καίσαρος Τίτου Αἰλίου Ἀδριανοῦ Ἀντωνείνου Σεβαστοῦ Εὐσεβοῦς Φαμενὼθ κη. διέγρ(αψαν) Θέωνι ἐγλήμ(πτορι) ἀγορανομίας νομοῦ καὶ ἄλλω(ν) ὠνῶν δι(ὰ) Ἡρακλ( ) Ὥρος Σαταβοῦτος καὶ οἱ λοιπ(οὶ) ἱερεῖς Σοκνοπ(αίου) Νήσου ὑπὲρ ἀποτάκτου λέμβων ἠθμῶ[ν] τοῦ διε[λ]ηλυθότος ἑβδόμου ἔτους ἀργ(υρίου) (δραχμὰς) ἑκατόν, (γίνονται) (δραχμαὶ) ρ, ὁμοίως οἱ αὐτοὶ ὑπὲρ διπλώματ(ος) ὧν ἔχουσι ὄνων τοῦ διεληλυθ(ότος) ἑβδόμου ἔτους ἀργ(υρίου) (δραχμὰς) ἑκατόν, (γίνονται) (δραχμαὶ) ρ.

In the eighth year of the Emperor Caesar Titus Aelius Hadrianus Antoninus Augustus Pius, on the 28th of Phamenoth (24th March AD 145?). [The following sum] should be transferred to Theon, the tax leaseholder [or manager] of the province's tax and other taxes, [represented] by Herakl[...], Horos, the son of Satabus and the other priests of Soknopaiu Nesos, for the rent of lembs in the previous, seventh year [of the government]: hundred silver drachmae which makes 100 [Dr]. The same [should be given to] the aforementioned priests for the permits for the donkeys they possess, for the previous seventh year [of the government]: 100 silver drachmae, which makes 100 [Dr].

5. Written sources on lembs and Liburnians from the 4th c. BC to Late Antiquity

Greek: *Griechische papyri der Staats- und Universitätsbibliothek Hamburg (P. Hamb. IV)*, eds B. von Kramer *et al.* [Archiv für Papyrusforschung und verwandte Gebiete – Beihefte 4]. Teubner, Stuttgart–Leipzig 1998. http://papyri.info/ddbdp/p.hamb;4;248

Besides the δίπλωμα ὄνων, a permit for the use of donkeys, which is not relevant for our topic, the subject of this papyrus is the ἀπότακτος λέμβων. The phrase ἀπότακτος λέμβων is somehow ambiguous: '*... it is uncertain whether this [ἀπότακτος] represents rent upon boats leased from the state or a payment for an exclusive concession to use boats (owned by the temple) for fishing on the waterways of the state, in this case Lake Moeris.*'[185] In any case it is about a certain sum of money that has to be paid for the rent of the lembs. The whole transaction happens in the village of Soknopaiou Nesus, modern Dīmā. This village is situated on the Lake Moeris, modern Lake Qārūn in Egypt, so that the usage of boats on this fairly large lake is understandable.

A-XV (Valerius) Harpocration (2nd century)

Lexicon in decem oratores

(8) **Λέμβος**· Δημοσθένης ἐν τῷ Πρὸς Ζηνόθεμιν. εἶδος νεὼς ὁ **λέμβος**· πολλάκις δὲ εἴρηται ἐν τῇ κωμῳδίᾳ.

Lemb: Demosthenes mentions it in *Against Zenothemis*[186] **A lemb** is a kind/form/shape of a ship: it is often mentioned in comedy.

Greek: *Harpocration: Lexeis of the Ten Orators*, ed. J. J. Keaney. Hakkert, Amsterdam 1991.

This brief lexicon entry, taken from the first lexicographic work in strict alphabetical order (R. Tosi, *NP*), that contains glosses of the ten orators of the Hellenistic Canon, is important for our purposes for it emphatically states that the lemb is an εἶδος of ship. The word εἶδος employed here by Harpocration is typically associated with Aristotle's philosophy of which Harpocration must have been quite conscious since he otherwise quotes abundantly from Aristotle. However, the Aristotelian εἶδος has a double meaning: it can mean both 'kind' and 'species' as well as 'form' or 'shape'. From this brief entry it is impossible to establish exactly which meaning Harpocration had in mind. One meaning would imply that lembs were, so to say, a species of the genus 'ship', a separate group of vessels that differ from other species of ship by some specific differences. The other meaning would be more 'philosophical' in the following way: lemb would be a term of a specific form – or a function – and any kind of vessel that has this specific form or serves that specific purpose would be called a lemb.

---

[185] Wallace 1969: 219.
[186] See above **A-I.1**.

A-XVI Dionysius of Byzantium (2nd century)

Per Bosporum navigatio

(102) Πλησίον δ' ἄκρα **Λέμβος** ὄνομα· κέκληται δ' ἀπὸ τοῦ σχήματος· καὶ συνεχὴς αὐτῷ αἰγιαλός· κατὰ στόμα δ' αὐτοῦ νῆσος πάνυ βραχεῖα, καθ' ἣν λευκαινόμενος ὁ βυθὸς ὑφάλοις ῥαχίαις ἐπὶ τὴν Εὐρώπην ἀποτρέπει τῶν ἰχθύων τὸν δρόμον· πτοούμενοι γὰρ δὴ τὴν ὄψιν ἐπιφόρῳ τῷ ῥεύματι τέμνουσι τὸν πόρον· Βλάβην αὐτὸ Χαλκηδόνιοι καλοῦσιν, ἑτοῖμον ὄνομα θέμενοι καὶ τῇ πείρᾳ τοῦ συμβαίνοντος οἰκεῖον.
§102

102. Nearby is a high point with a name **Lemb**, thusly called from the shape. Adjoining it is a shore, at its mouth there is a very small island, where the bottom is whitened by submerged reefs that divert the path of fish toward Europe. Frightened by the sight they direct their course with a favoring current. The Chalcedonians call it *Blabe* (*Harm*), giving it an appropriate name suitable to what happens in their experience.

Greek: *Dionysii Byzantii anaplus Bospori una cum scholiis x saeculi*, ed. R. Güngerich, 2nd ed. Weidmann, Berlin 1958.

Dionysius of Byzantium is the 2nd-century geographer known for his Ἀνάπλους Βοσπόρου (*Voyage through the Bosporus*), which describes the coastline of the Bosporus. In this passage he describes high point named Λέμβος, named after its shape. This coincides with other sources (e.g. **A-III**, **A-XV**) which have an understanding that their audience is familiar with distinctive shape of lemb.

A-XVII Aelius Aristides (2nd century AD)

A-XVII.1 *Sacred Tales*

καὶ φράζει δὴ τὸ πᾶν, ὡς εἱμαρμένον τε εἴη ναυαγῆσαί μοι καὶ τούτου ἄρα ἕνεκα καὶ ταῦτα συμβαίη· καὶ νῦν ἔτι δέοι ὑπὲρ ἀσφαλείας καὶ τοῦ παντάπασιν ἐκπλῆσαι τὸ χρεών, ἐμβάντα εἰς **λέμβον** ἐν τῷ λιμένι οὕτω ποιῆσαι, ὡς τὸν μὲν **λέμβον** ἀνατραπῆναι καὶ καταδῦναι, αὐτὸν δὲ ἐξάραντός τινος ἐξενεχθῆναι πρὸς τὴν γῆν· ἐν γὰρ τούτῳ τελεῖσθαι τὰ ἀναγκαῖα.
2.13 (=Jebb, p. 294.2-11)

He [god Asclepius] told me everything: that it was my destiny to be shipwrecked, that this occurred for a reason, and that now, for my safety and the complete fulfilment of what was necessary, I had to climb a **lemb** in the harbor and arrange it in such way that the **lemb** overturns and goes down, but I am brought ashore by someone who pulls me out. For then, in that case, what happens must necessarily take place.

Greek: *Aristides*, Vol. 1, ed. W. Dindorf. Reimer, Leipzig 1829. The pagination in Greek text also follows *Aelius Aristides. Opera Omnia Graece & Latine*, ed. S. Jebb, Oxford 1722.

5. Written sources on lembs and Liburnians from the 4th c. BC to Late Antiquity

A-XVII.2 *Oratio 36: The Egyptian Discourse*

καὶ προσέτι αὐτὸς ἐπεθύμησα εἰς τὸν **λέμβον** ἐμβὰς πειραθῆναι τοῦ πλοῦ, οὐ μόνον διὰ τῶν αὐτῶν δι' ὧνπερ ἐκείνους εἶδον κατενηνεγμένους, ταῦτα δ' ἦν τὰ πρὸς ἕω τῆς νήσου, ἀλλ' ἀρξάμενος αὐτόθεν περιπλεῦσαι κύκλῳ πᾶν τὸ ὁρώμενον, καὶ κατὰ τὴν ἑτέραν τῆς νήσου πλευρὰν ἀφεῖναι κατὰ ῥοῦν ἐπὶ τὰς πόλεις.
(=Jebb, p. 344.15-23)

And then when I entered the **lemb** I wished to try to sail not only over those [cataracts] that I saw flowing down, as they were on the East side of the island [Elephantine], but also starting from here I tried to sail around everything that can be seen and from the other sind of the island, sailing down the stream, I reached the cities.

Greek: *Aristides 2*, ed. W. Dindorf. Reimer, Leipzig 1829. The pagination in Greek text follows *Aelius Aristides. Opera Omnia Graece & Latine*, ed. S. Jebb, Oxford 1722.

The lemb to which Aristides refers in **A-XVII.1** must be a small boat, so that it could be overturned by one person.

A-XVIII Pausanias of Halicarnassus (2nd century)

Collection of Attic Words

(7) λέμβος καὶ ἐφόλκιον· σκάφος ἢ πλοῖον.
s.v. λέμβος

(7) Lemb and "epholkion" [small boat towed after a ship]: a boat or a vessel.

Greek: *Untersuchungen zu den attizistischen Lexika*, ed. H. Erbse [Abhandlungen der deutschen Akademie der Wissenschaften zu Berlin, Philosoph.-hist. Kl. 2]. Akademie Verlag, Berlin 1950.

In this lexicographic entry λέμβος is put in the same category as ἐφόλκιον (cf. **A-X** and **A-XXIX**). This former name still lives today in the Italian name of the smaller ship *felucca* and Croatian *filjuga*.[187] The lexicon *Collection of Attic Words* 'not only provides instructions for good Attic expression, but also was intended to promote the appreciation of the Attic writers through the inclusion of cultural-historical information.' (S. Matthaios, *NP*). This means that Pausanias saw the terms *lembos* and *epholkion* as Attic expressions which would require some explanation those who were not familiar with elevated Atticised parlance.

---

[187] The etymology goes as ital. *feluca* < ar. pl. *faluk* from sing. *fűik* < gr. ἐφόλκιον, Skok 1971: 517.

A-XIX Achilles Tatius (fl. late 2nd century)

Leucippe and Clitophon

A-XIX.1

(1) Ναῦν δὲ εἶχεν ἰδίαν, τοῦτο προκατασκευάσας οἴκοθεν εἰ τύχοι τῆς ἐπιχειρήσεως. οἱ μὲν δὴ ἄλλοι θεωροὶ ἀπέπλευσαν, αὐτὸς δὲ μικρὸν ἀπεσάλευε τῆς γῆς, ἅμα μὲν ὡς δοκοίη τοῖς πολίταις ἕπεσθαι, ἅμα δὲ ἵνα μή, πλησίον τῆς Τύρου τοῦ σκάφους ὄντος, κατάφωρος γένοιτο μετὰ τὴν ἁρπαγήν. (2) ἐπεὶ δὲ ἐγένετο κατὰ Σάραπτα κώμην Τυρίων ἐπὶ θαλάττῃ κειμένην, ἐνταῦθα προσπορίζεται **λέμβον**, δίδωσι δὲ τῷ Ζήνωνι· τοῦτο γὰρ ἦν ὄνομα τῷ οἰκέτῃ, ὃν ἐπὶ τὴν ἁρπαγὴν παρεσκευάκει. (3) ὁ δὲ (ἦν γὰρ καὶ ἄλλως εὔρωστος τὸ σῶμα καὶ φύσει πειρατικός) ταχὺ μὲν ἐξεῦρε λῃστὰς ἁλιεῖς ἀπὸ τῆς κώμης ἐκείνης καὶ δῆτα ἀπέπλευσεν ἐπὶ τὴν Τύρον. ἔστι δὲ μικρὸν ἐπίνειον Τυρίων, νησίδιον ἀπέχον ὀλίγον τῆς Τύρου (Ῥοδόπης αὐτὸ τάφον οἱ Τύριοι λέγουσιν), ἔνθα **ὁ λέμβος** ἐφήδρευεν. [...] 2.18. (2) ἄρτι δὲ γενομένων ἡμῶν ἐπὶ τῷ χείλει τῆς θαλάσσης, ὁ μὲν τὸ συγκείμενον ἀνέτεινε σημεῖον, **ὁ δὲ λέμβος** ἐξαίφνης προσέπλει, καὶ ἐπεὶ πλησίον ἐγένετο, ἦσαν ἐν αὐτῷ νεανίσκοι δέκα.
2.17.1-2.18.2

17. He had a vessel of his own—he had made all these preparations at home, in case he should succeed in such an attempt: so when the rest of the envoys sailed off, he weighed anchor and rode a little off the land, waiting in order that he might seem to be accompanying his fellow-citizens on their homeward journey, and that after the carrying off of the girl his vessel might not be too close to Tyre and so himself be taken in the act. When he had arrived at Sarepta, a Tyrian village on the sea-board, he acquired **a small boat** and entrusted it to Zeno; that was the name of the servant in whose charge he had placed the abduction—a fellow of a robust body and the nature of a brigand. Zeno picked up with all speed some fishermen from that village who were really pirates as well, and with them sailed away for Tyre: **the boat** came to anchor, waiting in ambush, in a little creek in a small island not far from Tyre, which the Tyrians call Rhodope's Tomb. [...] Hardly had we arrived at the water's edge, when he hoisted the preconcerted signal; **the boat** rapidly sailed toward the shore, and when it had come close, it was apparent that it contained ten youths.
pp. 91-93

A-XIX.2

(1) Λέγει δὴ καὶ ὁ Σώστρατος· "Ἐπεὶ τοίνυν τοὺς ὑμετέρους μύθους, ὦ παιδία, κατελέξατε, φέρε ἀκούσατε," ἔφη, "καὶ παρ' ἐμοῦ τὰ οἴκοι πραχθέντα περὶ Καλλιγόνην τὴν σήν, ὦ Κλειτοφῶν, ἀδελφήν, ἵνα μὴ ἀσύμβολος ὦ μυθολογίας παντάπασι." (2) κἀγὼ ἀκούσας τὸ τῆς ἀδελφῆς ὄνομα πάνυ τὴν γνώμην ἐπεστράφην καί, "Ἄγε, πάτερ," εἶπον, "λέγε· μόνον περὶ ζώσης λέγοις." ἄρχεται δὴ λέγειν ἃ φθάνω προειρηκὼς ἅπαντα, τὸν Καλλισθένην, τὸν χρησμόν, τὴν θεωρίαν, **τὸν λέμβον**, τὴν ἁρπαγήν.
8.17.1–2

Then said Sostratus: 'Now that you, my children, have finished your stories, listen to mine: the story of what happened at home with regard to Calligone—your sister, Clitophon; I shall thus not have contributed absolutely nothing to these excellent recitals.' Hearing the name of my sister, I was all attention: 'Speak on, father,' said I, 'only may your story be of one who is still in the land of the living!' He began by recounting all that I described some time ago—about Callisthenes, and the oracle, and the sacred embassy, and **the boat**, and the abduction.
p. 447

Greek: *Achilles Tatius. Leucippe and Clitophon*, ed. E. Vilborg. Almqvist & Wiksell, Stockholm 1955.

English: *Achilles Tatius. Leucippe and Clitophon*, transl. S. Gaselee. LCL 45 (1969).

5. Written sources on lembs and Liburnians from the 4th c. BC to Late Antiquity

The romantic novel *Leucippe and Cleitophon*, divided into eight books, from which the above fragments are taken, is one of the five surviving Ancient Greek romances. The story itself carries much of an intimate character and the usage of a lemb in it has some private connotations, something that normal, ordinary people could afford.

A-XX Alciphron (probably 2nd or early 3rd century?)

Letters

A-XX.1

(4) [...] καὶ ὁ Ἕρμων ἀφεὶς τὸ φερνίον αὐτοῖς ἰχθύσιν, ἀφεὶς δὲ καὶ ἡμᾶς αὐτῷ σκάφει, ᾤχετο **ἐπὶ λέμβου** κωπήρους Ῥοδίοις τισὶ θαλαττουργοῖς ἀναμιχθείς. καὶ ὁ μὲν δεσπότης οἰκέτην, ἡμεῖς δὲ συνεργὸν ἀγαθὸν ἐπενθήσαμεν.
1.2.4

[...] when Hermon abandoned his creel, fish and all, and deserted us too, with skiff and all, and went off **in a small boat**, lost in a crowd of independent fishermen, Rhodians equipped with oars. So our master had the loss of a servant, and we of a good fellow-worker, to lament.
p. 43

A-XX.2

(2) ἄκουε δὴ ὡς ἔχει καὶ πρὸς ὅ τι σε δεῖ τὴν γνώμην ἐξενεγκεῖν. τὰ ἡμέτερα, ὡς οἶσθα, παντελῶς ἐστιν ἄπορα καὶ βίος κομιδῇ στενός· τρέφει γὰρ οὐδὲν ἡ θάλαττα. **ὁ λέμβος** οὖν οὗτος ὃν ὁρᾷς ὁ κωπήρης, <ὁ> τοῖς πολλοῖς ἐρέταις κατηρτυμένος, Κωρυκιόν ἐστι σκάφος, λῃσταὶ δ' Ἀτταλῆς τὸ ἐν αὐτῷ σύστημα. οὗτοί με κοινωνὸν ἐθέλουσι λαβεῖν τοῦ τολμήματος, πόρους ἐκ πόρων [εὐμεγέθεις] ὑπισχνούμενοι.
1.8.2

Just hear how the matter stands and what the problem is on which you must express your opinion. Our situation, as you know, is desperate, and our way of life is extremely cramped; for the sea is yielding nothing. Now **that boat** you see—the boat with oars, I mean, the one equipped with many rowers—is a Corycian craft and its crew are Attalian pirates. They want to take me as a partner in their wicked venture, and revenues after revenues, huge ones too, they promise from it.
pp. 52-54

Greek: *Alciphronis rhetoris epistularum libri IV*, ed. M. A. Schepers. Teubner, Leipzig 1905.

English: *Alciphron, Aelian, and Philostratus: The Letters*, transl. A. R. Benner, F. H. Fobes. LCL 383 (1949).

In **A-XX.1** we have a syntactically unusual construction which renders the relationship between the σκάφος (translated above as '*skiff*') and λέμβος (translated as '*a small boat*') hard to decipher. As it is suggested, it could be that *skaphos* is *lembos*, but it could also mean that the other fishermen remained in the *skaphos* and Hermon went on a *lembos*. In any case, as the remainder of the passage suggests, taking a boat with him was not considered a major harm, since it is not mentioned as something that was considered a considerable loss. In fact, for the master, losing a servant was a more substantial damage than losing this particular boat. On the other hand, in **A-XX.2** *lembos* must mean a considerably bigger boat, for it is '*equipped with many rowers*' which can hardly be the case in the previous text. The boat is identified as Corycian, from a city on the coast of Lycia, not far south of Attalia in present-day Turkey (modern Kizkalesi) and belonging to the local pirates. Without any further evidence from this author, it would be hard to reconcile those two evidently disparate descriptions. In the former case it must be a smaller boat that can be operated by one person, in the later case it was a much bigger ship that served pirates' purposes. An obvious solution would be to suggest that, at the time this text was written, the term *lembos* signified a boat in general in educated Attic dialect.

A-XXI Michigan Papyrus 1972 (15 November 249)

```
-- -- -- -- -- -- -- --
1 χοντ( ) πρὸς τ[ -ca.?- ] '
ἐμοῦ δὲ παρέχο[ντος -ca.?- ]
πηγου καὶ λεμβ[ -ca.?- ]                        [lemb]
στησῃ α . α .[ -ca.?- ]
5 δώδεκα καὶ δ .[ -ca.?- ἑκά-]
στῳ δὲ τῶν ε[ -ca.?- ]
ὀκτὼ καὶ ὑπὲ[ρ -ca.?- ]
καὶ ἐκτάκτων ναυπ[ηγῶν -ca.?- ]
α καὶ ἑκάστῳ ναυπη[γῷ δώσω κατὰ μῆνα πυ-]
10 ροῦ ἀρτάβην μίαν κ[αὶ - ca.10 - ἑκά-]
στῳ κατὰ μηνα χοίνι[κα(ς) - ca. 10 -]
ὡς ἑκάστ(ῳ) κατὰ μῆν[α δοθήσεται ἐλαίου]
χρηστ(οῦ) κοτύλ(η) μία καὶ ξ[- ca.15 -]
χρόνου κεράμιον ἓν [καὶ ὁμολογῶ ἐσχη-]
15 κέναι εἰς ἀραβῶνος(*) λόγ[ον δραχμὰς -ca.?- ]
αἵπερ κατὰ μέρος ἐ[ξοδιασθήσονται ὑπὲρ ἡ-]
μερησίων μισθ(ῶν). ἀρξόμε[νος δὲ ἀπὸ τοῦ ἐνεστῶ-]
τος μηνὸς Ἀθὺρ τοῦ ἐνε[στῶτος α (ἔτους) σοι ἐργά-]
σομαι μέχρις οὗ συντελίω[σις](*)[ γένηται καὶ παρα-]
20 δώσω τὸ πλοῖον εὐάρεσ[τον καὶ καλὸν ἄνευ]
πάσης μέμψεως τέχνη[ς, οὔσης σοι τῆς πρά-]
ξεως παρά τε ἐμοῦ καὶ τῶν [ὑπαρχόντων μοι πάν-]
των. ἡ ἐπιδοχ(ὴ) κυρία κ[αὶ ἐπερωτηθ(εὶς) ὡμολ(όγησα).]
25 Δεκίου Τραιανοῦ Εὐσεβ[ο]ῦς Ἐ[ὐτυχοῦς Σεβαστοῦ]
Ἀθὺρ ιθ. Αὐρ(ήλιος) Ἥρας Διογᾶτ[ος ἐπεδεξάμην τὴν κα-]
τασκευὴν τοῦ προκειμ[ένου πλοίου καὶ ἔσχηκα]
ἰς(*) λόγον [το]ῦ ἀραβῶνος(*) δραχ[μὰς -ca.?- καὶ μισθὸν δώσω καθ' ἕ-]
καστο[ν μ]ῆνα καὶ παρα[δώσω τὸ πλοῖον καθὼς]
30 πρόκ[ειται] καὶ ἐπερω[τηθ(εὶς) ὡμολόγησα.]
Traces 1 line
31-- -- -- -- -- -- -- -- --
```

and I shall give each shipbuilder monthly one artaba of wheat and - - - each one monthly x choinix (-kes) of - - - each one will be given monthly one kotyle of good olive oil and - - -period one keramion and I acknowledge that I have received as an arrha x drachmas which will be spent in part for daily wages. Starting from the present month Hathyr of the present 1st year I shall start to work for you until a discharge of obligations in full is reached and I shall hand over the boat in acceptable and good condition without any technical blame while you have the right of execution upon me and upon all my possessions. The lease is normative and in answer to the formal question I gave my consent. Year 1 of Imperator Caesar Gaeus Messius Quintus Decius Trajanus Pius Felix Augustus, Hathyr 19. I, Aurelius Heras, son of Diogas, have taken upon me to build the afore mentioned boat and I have received as an arrha x drachmas and I shall provide pay each month and I shall hand the boat over as written above and in answer to the formal questions I gave my consent. - - -

5. Written sources on lembs and Liburnians from the 4th c. BC to Late Antiquity

Greek and English: Sijpesteijn 1996.

This is a badly preserved papyrus. The only time the term *lembos* is mentioned is in line 3 with the initial four letters (λεμβ), whereas the last character, indicated with sublinear dot, is unclear or imperfect in some way. The rest of the text mentions a more generic term πλοῖον (vessel). The text says that a certain Aurelius Heras, son of Diogas, must be an independent contractor who was given the job of building a lemb for a person whose name has been lost. Unfortunately, the amount of *drachmae* spent on this project has not been preserved, so we cannot really say anything about the size of the vessel or the labour required to build it.

A-XXII Diogenes Laertius (3rd century)

Lives of Eminent Philosophers

Γεγόνασι δ' Ἡρακλεῖδαι τεσσαρεσκαίδεκα· [...] πέμπτος Καλλατιανὸς ἢ Ἀλεξανδρεύς, γεγραφὼς τὴν Διαδοχὴν ἐν ἓξ βιβλίοις καὶ Λεμβευτικὸν λόγον, ὅθεν καὶ Λέμβος ἐκαλεῖτο [...]. 5.94	Fourteen persons have borne the name of Heraclides: [...] (5) of Callatis or Alexandria, author of the Succession of Philosophers in six books and a work entitled Lembeuticus, from which he got the surname of Lembus (a fast boat or scout) [...]. pp. 547-549

Greek: *Diogenis Laertii vitae philosophorum*, 2 vols. ed. H.S. Long. Clarendon Press, Oxford 1964.

English: *Diogenes Laertius. Lives of Eminent Philosophers, Volume I: Books 1-5*, transl. R. D. Hicks. LCL 184 (1925).

This fragment is of significance because it mentions the Platonist philosopher, politician and doxographer from the 4th century BC, Heraclides Lembus. *Lembus* was a nickname which Diogenes Laertius attributes to his script called Λεμβευτικός (lemb-like, lembish?). Since we possess no trace of this script, it is impossible to say anything more about the origin of this name. Heraclides was born in Callatis in Pontus (less likely in Oxyrhynchus in Middle Egypt) and later became a citizen of Alexandria (D. T. Runia, *NP*). From this perspective it would be difficult to imagine that his nickname would have anything to do with a boat that was Illyrian in origin.

A-XXIII Heliodorus of Emesa (3rd century)

The Aethiopica

Ἐγὼ μὲν ὡς εἶχον εὐθὺς μετὰ τὴν κρίσιν εἰς τὸν Πειραιᾶ κατέβην καὶ νεὼς ἀναγομένης ἐπιτυχὼν τὸν πλοῦν εἰς Αἴγιναν ἐποιούμην, ἀνεψιοὺς εἶναί μοι τῆς μητρὸς ἐνταῦθα πυνθανόμενος· κατάρας δὲ καὶ τοὺς ἐπιζητουμένους ἀνευρὼν οὐκ ἀηδῶς τὰ πρῶτα διῆγον. Εἰκοστῇ δὲ ὕστερον ἡμέρᾳ συνήθως ἀλύων ἐπὶ λιμένα κατῆλθον· καὶ **λέμβος** ἄρτι κατεφέρετο. Μικρὸν οὖν ἐπιστὰς ὁπόθεν τε εἴη καὶ τίνας ἄγοι περιεσκόπουν. Οὔπω δὲ τῆς ἀποβάθρας ἀκριβῶς κειμένης ἐξήλατό τις καί με προσδραμὼν περιέβαλλεν· [...].
1.14.2–3

I, in such case as I was after the judgment, came to the Piraeus, and finding a ship ready to depart sailed to Aegina, for I knew I had some kinsfolk there, by my mother's side. When I arrived there and had found those I sought for, at first I lived pleasantly enough: about twenty days after, roaming about as I was wont to do, I walked down to the haven, and behold, a **barque** was within kenning. I stayed there a little, and devised with myself whence that barque should come and what manner of people should be in her. The bridge was scarce down when one leaped out and ran and embraced me
pp. 23-24

Greek: *Héliodore. Les Éthiopiques (Théagène et Chariclée)*, 3 Vols., eds T.W. Lumb, J. Maillon, R.M. Rattenbury, 2nd ed. Les Belles Lettres, Paris 1960.

English: *Heliodorus: An Aethiopian Romance (The Aethiopica)*, transl T. Underdowne, rev. F. A. Wright. George Routledge & Sons, London 1923.

A-XXIV Rufinus (*c.* 3rd or 4th century)

Epigrams

εἰς τὰς ἑταίρας **Λέμβιον** καὶ Κερκούριον· εἰσὶ δὲ ταῦτα τὰ ὀνόματα μικρῶν καραβίων, τῶν παρὰ ἡμῖν σανδάλων

Λέμβιον, ἡ δ' ἑτέρα Κερκούριον, αἱ δύ' ἑταῖραι αἰὲν ἐφορμοῦσιν τῷ Σαμίων λιμένι. ἀλλά, νέοι, πανδημὶ τὰ λῃστρικὰ τῆς Ἀφροδίτης φεύγεθ'· ὁ συμμίξας καὶ καταδὺς πίεται.
5.44

On the courtesans **Lembion** and Cercurion; these are names of small vessels (boats to us)

Lembion and Cercurion, two courtesans, always lie moored in the harbor of Samos. Young men, flee en masse from Aphrodite's pirate ships! He who engages them is both sunk and swallowed up.
pp. 230-231

Greek and English: *Greek Anthology, Volume I: Books 1-5.* Transl. W. R. Paton, Rev. M. A. Tueller. LCL 67 (2014).

Rufinus is variously dated between AD 150 and 400, but the most succinct analysis, that of Page, places him in the upper part of this chronological range – most likely the 4th century.

The epigram shows that the term lemb represents a small boat, while the play of words with πίεται meaning 'sea swallowing a ship' sunk by the pirates but also the prostitute Lembion (named after lemb) 'swallowing the money of her customers', would have evidently made sense for Rufinus' audience.[188]

A-XXV Themistius (c. 317-385)

Oration 1: On Love of Mankind (Constantius)

12a [...] Τοῦτ' οὖν ἐστιν ὃ διόλλυσιν ἐκεῖνον, οὐχ ἡ μέση τῶν ποταμῶν, ἀλλ' ἡ βασιλέως ἀρετὴ πλησίον λάμπουσα. [b] καὶ οὐκ ἐπίσταται, ὃ μόνον κέρδος γειτονήσεως, ἐπιτρέψαι τὰ πηδάλια τῆς διανοίας ἐγγὺς ὄντι τῷ κυβερνᾶν ἐπισταμένῳ καὶ ἐξαρτῆσαι τὸ σκάφος τῆς μεγάλης νεώς. τοῦτο γὰρ ἄμεινον, οἶμαι, ἢ **λέμβῳ μικρῷ** ἐπιπλέοντα, δίχα οἰάκων τε καὶ τῆς ἄλλης παρασκευῆς διαναυμαχεῖν πρὸς τριήρη μεγάλην τε καὶ ἰσχυράν, πολλοὺς μὲν ὁπλίτας ἔχουσαν, πολλοὺς δὲ ἐρέτας τε καὶ ἐπιβάτας, κυβερνήτην τε ἐκ σπαργάνων τοῖς πηδαλίοις ἐντεθραμμένον. [c] πονηρὰν γὰρ τὴν ναυμαχίαν ὁ τοιοῦτος ναυτίλλεται, εἰ καὶ βραχύν τινα χρόνον ὑπὸ κουφότητος διαφύγοι τὰς ἐμβολάς.
1.12 a-c

This then is what brings about the ruin of that man: not Mesopotamia, but the virtue of the king shining out next to him. [b] And he [Shapur] does not understand what is the only advantage of proximity, which is to entrust the steering oar of his mind to one who is near by and knows how to steer, and to lash his ship [the Persian state] to the great vessel [the Roman Empire]. For, in my opinion, this is better than to put to sea on a **small pinnace** without steering gear or other equipment, and to fight to the finish with a great and strong trireme carrying many solders, and many oarsmen and marines, and a helmsman reared at the tiller from his infant clothes. Such a man [Shapur] is terribly at sea when it comes to a naval battle, [c] even if for some little time he might escape ramming through his manoeuverability.
p. 90

Greek: *Themistii orationes quae supersunt* 1, eds G. Downey, H. Schenkl. Teubner, Leipzig 1965.

English: *Politics, Philosophy, and Empire in the Fourth Century. Select Orations of Themistius*, transl. P. Heather, D. Moncurt. TTH 36 (2001).

This speech was given during Themistius' initial appearance before the Emperor Constantius II in around AD 350.[189] In the carefully chosen speech on praising the Emperor's philanthropy, Themistius claims that a victory over the Persian king, probably Shapur II, was due to Constantius' character. Themistius used the rhetorical techniques of metaphor and comparison: first he claimed that it would have been much more profitable for the Persian king to join the Roman Emperor if only he knew how to direct his mind ('*to entrust the steering oar of his mind*'). Then he compared Shapur to emperor Constantius: it is like a lemb (Shapur) compared to a trireme (the Emperor). The comparison serves a triple purpose: first, it is about the size: Shapur, the lemb, is much smaller than the Emperor, the fully equipped trireme; second, it is about the function: it would be just natural for the Persian king to follow the lead and be

---
[188] Page 1978: 3-49 (dating), 88-89 (commentary).
[189] Downey 1958: 50.

attached to the Emperor, just like a lemb is naturally attached to and follows a larger boat; and third, lemb is '*without steering gear or other equipment*', which hints at the Shapur's incapacity to control and direct himself properly. This is the only place in which it says that the lemb would not have a steering gear or '*other equipment*' and thus it may be taken to be a mere rhetorical figure of speech, a sort of a hyperbole to accentuate the desired effect.

A-XXVI St Basil the Great (329/330-381/382)

Letters

Ἀλλ' ὅμοιός εἰμι τοῖς ἐν θαλάσσῃ ὑπὸ τῆς κατὰ τὸν πλοῦν ἀπειρίας ἀπολλυμένοις καὶ ναυτιῶσιν· οἳ τῷ μεγέθει τοῦ πλοίου δυσχεραίνουσιν, ὡς πολὺν τὸν σάλον παρεχομένου, κἀκεῖθεν ἐπὶ τὸν **λέμβον** ἢ τὸ ἀκάτιον μεταβαίνοντες πανταχοῦ ναυτιῶσι καὶ ἀποροῦνται, συμμετέρχεται γὰρ αὐτοῖς ἡ ἀηδία καὶ ἡ χολή. Τοιοῦτον οὖν τι καὶ τὸ ἡμέτερον. 2.1.13–19	On the contrary, I am like those who go to sea, and because they have had no experience in sailing are very distressed and sea-sick, and complain of the size of the boat as causing the violent tossing; and then when they leave the ship and take to **the dinghy** or the cock-boat, they continue to be sea-sick and distressed wherever they are; for their nausea and bile go with them when they change. Our experience is something like this. p. 9

Greek: *Saint Basile. Lettres*, 3 vols., ed. Y. Courtonne. Les Belles Lettres, Paris 1957-66.

English: *Basil. Letters, Volume I: Letters 1-58*, transl. R. J. Deferrari. LCL 190 (1926).

In this passage St Basil uses compares a lemb to ἀκάτιον, another type of small boat – here in a diminutive form of ἄκατος. The term *lembos* is obviously a synonym for a small boat.

A-XXVII Libanius (314-393)

A-XXVII.1 *Epistolae*

(3) καὶ σοὶ δὴ συναχθομένων ἡμῶν καὶ πρὸς ἀλλήλους ὀδυρομένων, εἰς οἷον ἄρα χειμῶνα πέπτωκας, τὸν σεισμὸν ἐκεῖνον καὶ τὰ ἀπὸ τοῦ σεισμοῦ, θεῶν τις ἔδωκε καταφυγὴν ἄνδρα φίλον, ἐν ᾧ μέγεθός τε διανοίας καὶ τὸ δύνασθαι πράττειν ἄττα ἂν προθυμηθῇ. (4) παραινῶ δή σοι τῆς παρούσης ἔχεσθαι τύχης καὶ τὸν σαυτοῦ **λέμβον** ἐξαρτῆσαι τῆς Ἀλεξάνδρου νεώς, ὃν ἤδη μοι δοκῶ θεωρεῖν ἐπὶ μείζονος ἀρχῆς καὶ τρίτης πάλιν καὶ τετάρτης. οἶμαι δέ, αὐτὸν οὐδὲ γῆρας ἐξαιρήσεται πρὸς ἡσυχίαν ἀπὸ τῶν ταῖς πόλεσι λυσιτελούντων πόνων. οἶδε γὰρ βασιλεὺς τὸν ἄνδρα καὶ χρώμενος οὐ λήξει. 281.3-4	(3) While we were grieving together and lamenting to each other what kind of tempest has fallen up upon you, and the earthquake and everything after the earthquake, one of the gods gave you as a shelter a friendly man, full of greatness and intelligence, who is able to do everything he desires. (4) I advise you to take advantage of the present luck and to attach your **lemb** to Alexander's ship, who seems to me to be looking at a higher position, again the third or the fourth. However, I think, the old age will not coerce him to retire from working to the advantage of the cities. The king knows this man and he will not stop using him.

5. Written sources on lembs and Liburnians from the 4th c. BC to Late Antiquity

Greek: *Libanii opera*, vols. 10-11, ed. R. Foerster. Teubner, Leipzig 1921-1922.

A-XXVII.2 *Orationes*

(21) "Ὦ λιμένος, ὃν φεύγουσαι νῆες ἀνήγοντο σπουδῇ <τὰ> ἀπόγεια κόπτοντες. ὁ πρὶν ὁλκάδων ἔμπλεως οὐδὲ **λέμβον** εἰσπλέοντα δείκνυσιν, ἀλλ' ἔστι φοβερώτερος ἐμπόροις τοῦ τῆς Σκύλλης οἰκητηρίου.
61.21

O port, from where running ships hastily sailed away cutting the mooring cables! While it was full of trading vessels one could not even see a **lemb** sailing in, but it was too fearful to the merchant of Scilla's dwelling.

Greek: *Libanii opera*, vols. 1-4, ed. R. Foerster. Teubner, Leipzig 1903-08.

A-XXVII.3 *Argumenta orationum Demosthenicarum*

(3) ἡ μὲν οὖν τοῦ Λάμπιδος ναῦς ἀναχθεῖσα διαφθείρεται καὶ μετ' ὀλίγων ὁ Λάμπις ἐν τῷ **λέμβῳ** σῴζεται καὶ ἀφικόμενος Ἀθήναζε μηνύει Χρυσίππῳ τὸ εὐτύχημα τοῦ Φορμίωνος, ὡς ἀπελείφθη τε ἐν τῷ Βοσπόρῳ καὶ εἰς τὴν ναῦν οὐδὲν ἐνέθετο.
43.3

Lampid's ship was destroyed after having sailed out, and Lampid was saved with a few in a **lemb** and, having arrived to Athens, announced the good news to Phormion's Chrysippus that he was left in Bosporos and that he did not take anything aboard.

Greek: *Libanii opera*, vol. 8, ed. R. Foerster. Teubner, Leipzig 1915.

Libanius was an expert in archaic and classical literature who paid special attention to his style and expression for which he gained a reputation of one of the most famous masters of the Greek language. Libanius's linguistic purism suggests to us that the term *lembos* was not perceived as a strange or 'unpure' word that should be avoided. Moreover, his notorious reluctance to accept anything 'new' (i.e. Hellenistic) or coming from Latin sources lets us believe that he perceived the term *lembos* as an integral part of the corpus of the stylistically acceptable Greek words.

A-XXVIII Zosimus (c. 500)

Historia nova

(2) Τοῦ Ῥήνου πρὸς ταῖς ἐσχατιαῖς τῆς Γερμανίας, ὅπερ ἐστὶν ἔθνος Γαλατικόν, εἰς τὸ Ἀτλαντικὸν πέλαγος ἐκδιδόντος, οὗ τῆς ἠόνος ἡ Βρεττανικὴ νῆσος ἐννακοσίοις σταδίοις διέστηκεν, ἐκ τῶν περὶ τὸν ποταμὸν ὑλῶν ξύλα συναγαγὼν ὀκτακόσια κατεσκεύασε πλοῖα μείζονα **λέμβων**, ταῦτά τε εἰς τὴν Βρεττανίαν ἐκπέμψας κομίζεσθαι σῖτον ἐποίει·καὶ τοῦτον τοῖς ποταμίοις πλοίοις ἀνάγεσθαι διὰ τοῦ Ῥήνου παρασκευάζων, τοῦτό τε ποιῶν συνεχέστερον διὰ τὸ βραχὺν εἶναι τὸν πλοῦν, ἤρκεσε τοῖς ἀποδοθεῖσι ταῖς οἰκείαις πόλεσιν εἰς τὸ καὶ τροφῇ χρήσασθαι καὶ σπεῖραι τὴν γῆν καὶ ἄχρις ἀμητοῦ τὰ ἐπιτήδεια ἔχειν.
3.5.2

(2) From where the Rhine flows out into the Atlantic Ocean in the remotest part of Germany, where a Gallic people live, to the island of Britain is a distance of nine hundred stades. In the woods around the river he [Caesar, i.e. Julian (Apostata)] gathered timber and built eight hundred vessels, larger than **fast galleys**, which he sent to Britain to bring back grain. By doing this continuously, owing to the shortness of the voyage, and bringing the grain up the Rhine in river boats, he kept those who had been returned to their own towns well supplied with food, with seed to sow their ground, and with provisions until the harvest.
p. 51

Greek: *Zosime. Histoire nouvelle*, vols. 1-3.2, ed. F. Paschoud. Les Belles Lettres, Paris 1971-1989.

English: *Zosimus. New History*, transl. with comm. R. T. Ridley [Byzantina Australiensia 2]. Australian Association for Byzantine Studies, Canberra 1982.

For our purposes, the curiosity of the above passage is that lembs are mentioned as being used as far north as the Rhine Delta, in the present-day Netherlands. Here the term *lembos* designates a smaller boat.

A-XXIX Hesychius (5th or 6th century AD)

Lexicon

(638) **λέμβος**· τὸ μικρὸν πλοιάριον nps(g), τὸ ἐφόλκιον. καὶ οἱ ἐφολκίοις πλέοντες <λέμβαρχοι> s.v. λέμβος

(648) **Lemb**: a small skiff, *epholkion* (small boat towed after a ship). *Lembarchoi*: those who sail on *epholkion*.

Greek: *Hesychii Alexandrini lexicon*, 2 vols., ed. K. Latte. Munksgaard, Copenhagen 1953-66.

In his *Lexicon* Hesychius gathered a collection of unusual and obscure Greek words. With more than 50,000 entries and many *hapax legomena*, it is one of the most valuable sources of our knowledge of Ancient Greek language and dialectical variations. However, due to the fact that it exists in only one copy (*Codex Marcianus Graecus* 622 from the 15th century), in which there are many errors, interpolations, corrections and inconsistences, it is hard to say what was the original version and what are later additions. For our purposes there are two important points in this entry: first, that in the 5th or 6th century the term *lembos* was considered a curiosity

5. Written sources on lembs and Liburnians from the 4th c. BC to Late Antiquity

with an unclear meaning. Second, that it is found synonymously with the term *epholkion*, which we have already seen in Posidonius and Pausanias of Halicarnassus.

A-XXX Procopius (*c.* 507 - after 555)

The Wars

A-XXX.1

ἐπεὶ δὲ οἱ στασιῶται πάντες ἐκάθευδον, ἐξελθόντες ἐκ τοῦ ἱεροῦ ἐς τὴν Θεοδώρου τοῦ ἐκ Καππαδοκίας οἰκίαν ἦλθον, ὅς αὐτοὺς δειπνῆσαί τε οὔ τι προθυμουμένους ἠνάγκασε, καὶ ἐς τὸν λιμένα διακομίσας **ἐς λέμβον νεώς δή τινος** ἐσεκόμισεν, ὅς δὴ ἐνταῦθα Μαρτίνῳ παρεσκευασμένος ἐτύγχανεν. εἵποντο δὲ Προκόπιός τε, ὃς τάδε ξυνέγραψε, καὶ τῆς Σολόμωνος οἰκίας ἄνδρες πέντε μάλιστα
4.14.38–39

And when all the mutineers were sleeping they went out from the sanctuary and entered the house of Theodorus, the Cappadocian who compelled them to dine although they had no desire to do so, and conveyed them to the harbour and put them on **the skiff of a certain ship**, which happened to have been made ready there by Martinus. And Procopius also, who wrote this history, was with them, and about five men of the house of Solomon.
p. 339

A-XXX.2

οὗ δὴ τῶν νηῶν ὁρμισαμένων τοὺς ἱστοὺς ξυνέβαινε τῶν ἐπάλξεων καθυπερτέρους εἶναι. αὐτίκα οὖν **τοὺς λέμβους τῶν νηῶν** ἅπαντας τοξοτῶν ἐμπλησάμενος ἀπεκρέμασεν ἄκρων ἱστῶν. ὅθεν δὴ κατὰ κορυφὴν βαλλόμενοι οἱ πολέμιοι ἐς δέος τι ἄμαχον ἦλθον καὶ Πάνορμον εὐθὺς ὁμολογίᾳ Βελισαρίῳ παρέδοσαν.
5.5.14-16

Now when the ships had anchored there, it was seen that the masts were higher than the parapet. Straightway, therefore, he filled all **the small boats of the ships** with bowmen and hoisted them to the tops of the masts. And when from these boats the enemy were shot at from above, they fell into such an irresistible fear that they immediately delivered Panormus to Belisarius by surrender.
p. 47

A-XXX.3

[...] Βελισάριος ἐξεῦρε τόδε. ἔμπροσθεν τῆς γεφύρας, ἧς ἄρτι πρὸς τῷ περιβόλῳ οὔσης ἐμνήσθην, σχοίνους ἀρτήσας ἐξ ἑκατέρας τοῦ ποταμοῦ ὄχθης ὡς ἄριστα ἐντεταμένας, **ταύταις τε λέμβους δύο** παρ' ἀλλήλους ξυνδήσας, πόδας δύο ἀπ' ἀλλήλων διέχοντας, ᾗ μάλιστα ἡ τῶν ὑδάτων ἐπιρροὴ ἐκ τοῦ τῆς γεφύρας κυρτώματος ἀκμάζουσα κατῄει, μύλας τε **δύο ἐν λέμβῳ ἑκατέρῳ** ἐνθέμενος ἐς τὸ μεταξὺ τὴν μηχανὴν ἀπεκρέμασεν, ᾗ τὰς μύλας στρέφειν εἰώθει. ἐπέκεινα δὲ ἄλλας τε ἀκάτους ἐχομένας τῶν ἀεὶ ὄπισθεν κατὰ λόγον ἐδέσμευε, καὶ τὰς μηχανὰς τρόπῳ τῷ αὐτῷ ἐπὶ πλεῖστον ἐνέβαλε.
5.19.20-21

And so Belisarius hit upon the following device Just below the budge which I lately mentioned as being connected with the circuit-wall, he fastened ropes from the two banks of the river and stretched them as tight as he could, and then attached to them **two boats** side by side and two feet apart, where the flow of the water comes down from the arch of the bridge with the greatest force, and placing two mills **on either boat**, he hung between them the mechanism by which mills are customarily turned And below these he fastened other boats, each attached to the one next behind in order, and he set the water-wheels between them in the same manner for a great distance.
p. 191

A-XXX.4

ἡ μὲν γὰρ ὁδὸς, ἡ τοῦ ποταμοῦ ἐν ἀριστερᾷ ἐστιν, ὥσπερ μοι ἐν τοῖς ἔμπροσθεν λόγοις ἐρρήθη, πρὸς τῶν πολεμίων ἐχομένη Ῥωμαίοις τηνικαῦτα ἀπόρευτος ἦν, ἡ δὲ αὐτοῦ ἐπὶ θάτερα, ὅσα γε παρ' ὄχθην, ἀστίβητος παντάπασι τυγχάνει οὖσα. διὸ δὴ **τοὺς λέμβους νηῶν τῶν μειζόνων** ἀπολεξάμενοι, σανίσι τε αὐτοὺς ὑψηλαῖς κύκλῳ τειχίσαντες, ὅπως οἱ πλέοντες πρὸς τῶν πολεμίων ἥκιστα βάλλωνται, τοξότας τε καὶ ναύτας ἐσεβίβασαν κατὰ λόγον ἑκάστου. τῶν τε φορτίων ἐν αὐτοῖς ὅσα φέρειν οἷοί τε ἦσαν ἐνθέμενοι, διὰ τοῦ Τιβέριδος ἐς Ῥώμην πνεῦμα τηρήσαντες σφίσιν ἐπίφορον ἀνέπλεον, καὶ τοῦ στρατοῦ μέρος ἐν δεξιᾷ τοῦ ποταμοῦ παρεβεβοηθήκει.
6.7.6–8

For the road which is on the left of the river was held by the enemy, as stated by me in the previous narrative, and not available for the use of the Romans at that time, while the road on the other side of it is altogether unused, at least that part of it which follows the river-bank. They therefore selected **the small boats belonging to the larger ships**, put a fence of high planks around them on all sides, in order that the men on board might not be exposed to the enemy's shots, and embarked archers and sailors on them in numbers suitable for each boat And after they had loaded the boats with all the height they could carry, they waited for a favouring wind and set sail toward Rome by the Tiber, and a portion of the army followed them along the right bank of the river to support them.
p. 349

A-XXX.5

πλεύσαντες οὖν ἐκ τοῦ Ῥωμαίων λιμένος Γενούᾳ προσέσχον, ἣ Τουσκίας μέν ἐστιν ἐσχάτη, παράπλου δὲ καλῶς Γάλλων τε καὶ Ἰσπανῶν κεῖται. ἔνθα δὴ τάς τε ναῦς ἀπολιπόντες καὶ ὁδῷ πορευόμενοι πρόσω ἐχώρουν, **τοὺς λέμβους τῶν νηῶν** ἐν ταῖς ἁμάξαις ἐνθέμενοι, ὅπως ἂν Πάδον τὸν ποταμὸν διαβαίνουσι μηδὲν σφίσιν ἐμπόδιον εἴη.
6.12.29–30.

They set sail, accordingly, from the harbour of Rome and put in at Genoa, which is the last city in Tuscany and well situated as a port of call for the voyage to Gaul and to Spain There they left their ships and travelling by land moved forward, placing **the boats of the ships** on their wagons, in order that nothing might prevent their crossing the river Po.
p. 393

A-XXX.6

προϊόντος δὲ τοῦ κακοῦ καὶ τῆς ἀπωλείας τοῖς πολιορκουμένοις ἐπὶ μέγα χωρούσης γνώμῃ Κόνωνος **ἐς λέμβον τινὰ** λάθρα ἐσβὰς ἐτόλμησε παρὰ τὸν στρατηγὸν Δημήτριον ἰέναι μόνος. ἐκ δὲ τοῦ παραδόξου σωθείς τε καὶ ξυγγενόμενος τῷ Δημητρίῳ ἐθάρσυνέ τε μάλιστα καὶ ἐς ταύτην δὴ τὴν πρᾶξιν ἐνῆγε. Τουτίλας δὲ τὸν πάντα λόγον ἀμφὶ τῷ στόλῳ τούτῳ ἀκούσας δρόμωνας μὲν πολλοὺς ἄριστα πλέοντας ἐν παρασκευῇ εἶχεν, ἐπειδὴ δὲ κατῆραν ἐς τὴν ἐκείνῃ ἀκτὴν οἱ πολέμιοι Νεαπόλεως οὐ μακρὰν ἄποθεν, ἐλθὼν ἐκ τοῦ ἀπροσδοκήτου κατέπληξέ τε καὶ ἐς φυγὴν ἅπαντας ἔτρεψε. καὶ αὐτῶν πολλοὺς μὲν ἔκτεινεν, ἐζώγρησε δὲ πλείστους, διέφυγον δὲ ὅσοι **ἐς τῶν νεῶν τοὺς λέμβους** ἐσπηδῆσαι κατ' ἀρχὰς ἴσχυσαν, ἐν τοῖς καὶ Δημήτριος ὁ στρατηγὸς ἦν. τὰς γὰρ ναῦς ἁπάσας σὺν αὐτοῖς

As the situation became worse and the loss of life among the besieged was becoming serious, this man, acting on the advice of Conon, had the daring to embark secretly on **a skiff** and go alone to the general Demetrius. And having, to everybody's surprise, made the voyage in safety and coming before Demetrius, he endeavoured with all his power to stir him to boldness, and urged him on to undertake the task before him. But Totila had heard the whole truth about this fleet and was holding many ships of the swiftest sort in readiness ; and when the enemy put in at that part of the coast, not far from Naples, he came upon them unexpectedly, and filling them with consternation turned the whole force to flight. And although he killed many of them, he captured a very large number, and there escaped

5. Written sources on lembs and Liburnians from the 4th c. BC to Late Antiquity

φορτίοις, αὐτοῖς ἀνδράσιν, οἱ βάρβαροι εἷλον. οὗ δὴ καὶ Δημήτριον τὸν Νεαπόλεως ἐπίτροπον εὗρον. γλῶσσάν τε καὶ χεῖρας ἄμφω ἀποτεμόντες οὐκ ἔκτειναν μέν, οὕτω δὲ λωβησάμενοι ὅπη βούλοιτο ἀφῆκαν ἰέναι. ταύτην τε Τουτίλᾳ τὴν δίκην Δημήτριος γλώσσης ἀκολάστου ἐξέτισεν.
7.6.22-26.

only as many as succeeded at the first in leaping **into the small boats of the ships**, among whom was Demetrius the general. For the barbarians captured all the ships with their cargoes, crews and all, among whom they found Demetrius, the governor of Naples. And cutting off his tongue and both his hands, they did not indeed kill him, but released him thus mutilated to go where he would. This then was the penalty which Demetrius paid to Totila for an unbridled tongue.
p. 205

A-XXX.7
(9) αὐτὸς δὲ ἐς ἕνα τῶν δρομώνων ἐσβὰς τοῦ τε στόλου ἡγεῖτο καὶ τὰς ἀκάτους ἐφέλκειν ἐκέλευεν, οὗ δὴ τὸν πύργον ἐτύγχανε τεκτηνάμενος. (10) τοῦ δὲ πύργου ὕπερθεν **λέμβον** τινὰ ἔθετο, πίσσης τε καὶ θείου καὶ ῥητίνης αὐτὸν ἐμπλησάμενος καὶ τῶν ἄλλων ἁπάντων ὅσα δὴ τοῦ πυρὸς βρῶσις ὀξυτάτη γίνεσθαι πέφυκε.
[...]
(18) τότε δὴ Βελισάριος τὰς ἀκάτους, ἐφ᾽ ὧν οἱ ὁ πύργος πεποίηται, ὡς ἀγχοτάτω ἀγαγὼν θατέρου τῶν πολεμίων πύργου, ὃς δὴ ἐπὶ τῆς κατὰ τὸν Πόρτον ὁδοῦ ἐπ᾽ αὐτὸ τοῦ ποταμοῦ τὸ ὕδωρ εἱστήκει, ἐκέλευε **τὸν λέμβον** ὑφάψαντας ὕπερθεν τῶν πολεμίων (19) τοῦ πύργου ῥίπτειν. καὶ Ῥωμαῖοι μὲν κατὰ ταῦτα ἐποίουν. ἐμπεσὼν δὲ τῷ πύργῳ **ὁ λέμβος** αὐτόν τε αὐτίκα μάλα ἐνέπρησε καὶ ξὺν αὐτῷ Γότθους ἅπαντας, ἐς διακοσίους μάλιστα ὄντας.
7.19.9-19

Then he himself embarked on one of the swift oats and led on the fleet, giving orders to tow the oats on which he had constructed the tower. Now he had placed on the top of the tower **a little boat** which he had caused to be filled with pitch, sulphur, resin, and all the other substances on which fire naturally feeds most fiercely.
p. 314
[...] Just at that moment Belisarius brought the skiffs on which the tower had been built as close as possible to one of the towers of the enemy— the one which stood on the road to Portus at the very edge of the water—and gave orders to set fire to **the little boat** and throw it on top of the enemy's tower. And the Romans carried out this order. Now when **this little boat** fell upon the tower, it very quickly set fire to it, and not only was the tower itself consumed, but also all the Goths in it, to the number of about two hundred.
pp. 317-319

Greek: *Procopii Caesariensis opera omnia* 1-2, eds J. Haury, G. Wirth. Teubner, Leipzig 1962-63.

English: *Procopius. History of the Wars, Volume II: Books 3-4. (Vandalic War); Volume III: Books 5-6.15. (Gothic War); Volume IV: Books 6.16-7.35 (Gothic War); Volume V: Books 7.36-8. (Gothic War)*, transl. H. B. Dewing. LCL 81, 107, 173, 217 (1916-28).

On Buildings

A-XXX.8

(1) Ταῦτα μὲν Ἰουστινιανῷ βασιλεῖ τῇδε πεποίηται. ἐν δὲ Ἀλεξανδρείᾳ ἐξείργασται τάδε. Νεῖλος ποταμὸς οὐκ ἄχρι ἐς τὴν Ἀλεξάνδρειαν φέρεται, ἀλλ' ἐπὶ πόλισμα ἐπιρρεύσας, ὃ δὴ Χαιρέου ἐπονομάζεται, ἐπ' ἀριστερὰ τὸ λοιπὸν ἵεται, ὅρια τά γε Ἀλεξανδρέων ἀπολιπών. (2) διὸ δὴ οἱ πάλαι ἄνθρωποι, ὡς μὴ ἀμοιροίη τὸ παράπαν ἡ πόλις, διώρυχα ἐκ τῆς Χαιρέου κατορύξαντες βαθεῖάν τινα βραχεῖα τοῦ ποταμοῦ ἐς αὐτὴν ἐκροῇ διεπράξαντο ἐσιτητὰ εἶναι. οὗ δὴ καὶ ἄλλας τινὰς ἐκροὰς ἐκ λίμνης Μαρίας ἐσβάλλειν ξυμβαίνει. (3) ἐπὶ ταύτης δὲ τῆς διώρυχος μεγάλαις μὲν ναυσὶ πλώϊμα οὐδαμῇ γίνεται, **ἐς λέμβους** δὲ τὸν Αἰγύπτιον σῖτον ἐκ τῆς Χαιρέου μεταβιβάσαντες, οὕσπερ καλεῖν διαρήματα νενομίκασιν, ἔς τε τὴν πόλιν διακομίζουσιν, ἵνα δὴ ἐξικνεῖσθαι δυνατά ἐστι τῷ κατὰ τὴν διώρυχα ποταμῷ, καὶ κατατίθενται ἐν χώρῳ ὅνπερ Ἀλεξανδρεῖς καλοῦσι Φιάλην.
6.1.1–3

Thus were these things done by the Emperor Justinian. And at Alexandria he did the following. The Nile River does not flow all the way to Alexandria, but after flowing to the town which is named from Chaereüs, it then turns to the left, leaving aside the confines of Alexandria. Consequently the men of former times, in order that the city might not be entirely cut off from the river, dug a very deep canal from Chaereüs and thus by means of a short branch made the river accessible to it. There too, as it chances, are the mouths of certain streams flowing in from Lake Maria [today Mariout]. In this canal it is by no means possible for large vessels to sail, so at Chaereüs they transfer the Egyptian grain to **boats** which they are wont to call diaremata, and thus convey it to the city, which they are enabled to reach by way of the canal-route, and they deposit it in the quarter of the city which the Alexandrians call Phialê.
pp. 361-363

A-XXX.9

(5) ἐπειδάν τε ναῦς ἀνέμῳ ἢ κλύδωνι βιαζομένη τοῦ διάπλου ἐντὸς ὑπὲρ τοῦ μηνοειδοῦς τὴν ἀρχὴν γένηται, τὸ ἐνθένδε αὐτῇ ἐπανιέναι ἀμήχανά ἐστιν, ἀλλὰ συρομένη τὸ λοιπὸν (6) ἔοικε, καὶ διαφανῶς ἐπίπροσθεν ἀεὶ ἑλκομένη. καὶ ἀπ' αὐτοῦ, οἶμαι, τὸν χῶρον οἱ πάλαι ἄνθρωποι τοῦ πάθους (7) τῶν νεῶν ἕνεκα Σύρτεις ὠνόμασαν. οὐ μὴν οὐδὲ διανεῦσαι τοῖς πλοίοις ἄχρι ἐς τὴν ἠϊόνα δυνατὰ γεγένηται. πέτραι γὰρ ὕφαλοι διακεκληρωμέναι τὰ πλεῖστα τοῦ κόλπου πλώϊμα οὐ ξυγχωροῦσιν ἐνταῦθα εἶναι, ἀλλ' ἐν τοῖς (8) βράχεσι τὰς ναῦς διαχρῶνται. **μόνοις δὲ τοῖς λέμβοις** οἱ πλωτῆρες τούτων δὴ τῶν νηῶν οἷοί τέ εἰσι διασῴζεσθαι, ἂν οὕτω τύχοι, μετὰ κινδύνων τὰς διεξόδους ποιούμενοι.
6.3.5–8

When a ship driven by wind or wave gets inside the opening and beyond the chord of the crescent, it is then impossible for it to return, but from that moment it seems 'to be drawn' (suresthai) and appears distinctly to be dragged steadily forward. From this fact, I suppose, the men of ancient times named the place Syrtes because of the fate of the ships. On the other hand, it is not possible for the ships to make their way to the shore, for submerged rocks scattered over the greater part of the gulf do not permit sailing there, since they destroy the ships in the shoals. **Only in small boats** are the sailors of such ships able to save themselves, with good luck, by picking their way amid perils through the outlets.
pp. 371-373

Greek: *Procopii Caesariensis opera omnia* 4, eds J. Haury, G. Wirth. Teubner, Leipzig 1964.

English: *Procopius. On Buildings*, transl. H. B. Dewing, G. Downey. LCL 343 (1940).

For Procopius the term *λέμβος* clearly denotes a small boat, small enough to be able to be lifted by hoisting (**A-XXX.2**) or even thrown into the air (**A-XXX.7**). An interesting point is that in the text, lembs are called *diaremata* (διαρήματα), the word which appears to be a *hapax* *legomenon* with an unclear meaning. However, there is this accompanying footnote by the translator G. Downey (p. 363, n.2):

> *Trans-shipment of grain in Egypt is mentioned in several papyri, which call it the διέρασις (or διαίρασις) τοῦ δημοσίου πυροῦ; in the documents in which reference is made to the vessels into which the grain was transferred, they are called διεράματα [...] The spelling in the present passage may be an error of the author or of a copyist. Procopius' evidence, which has not been used in connection with that of the papyri, confirms [...] that a διέραμα was a hopper for lading grain into a vessel.*

A-XXXI Agathias Scholasticus (*c.* 532 - after 580)

The Histories

20. (4) μόλις δὲ περὶ πλήθουσαν ἀγορὰν οἱ Ῥωμαῖοι τὴν διάβασιν ἐγνωκότες διεταράχθησάν τε καὶ περὶ πλείστου ἐποιοῦντο προτερῆσαι ἀνὰ τὸ ἄστυ τοὺς πολεμίους. τοιγάρτοι τάς τε τριήρεις καὶ ὅσαι τριακόντοροι παρώρμουν πληρώσαντες ὀξύτατα τοῦ ποταμοῦ τῷ ῥῷ κατεφέροντο. (5) ἀλλὰ φθάσας ὁ Ναχοραγὰν ἀμφὶ τὸ μεσαίτατον τῆς τε Νήσου καὶ τοῦ ἄστεος ἐτύγχανεν ἤδη, ξύλοις τε καὶ **λέμβοις** τὸ εὖρος ἅπαν τῆς δίνης διαφραξάμενος τό τε τῶν ἐλεφάντων στῖφος ὄπισθεν ἐπιστήσας, ἐς ὅσον οἷόν τε ἦν αὐτοῖς τῆς βάσεως ἐφικνεῖσθαι. (6) ὁ δὲ τῶν Ῥωμαίων στόλος ταῦτα πόρρωθεν ἰδόντες αὐτίκα πρύμναν ἐκρούοντο πολλῇ τε χρώμενοι ἐς τοὔμπαλιν τῇ εἰρεσίᾳ, χαλεπώτατα πρὸς τὸ ἀντίπρωρόν τε καὶ ῥοῶδες τῆς φορᾶς ἀνταναγόμενοι, ἀπεχώρησαν. (7) δύο δὲ ὅμως ναῦς κενὰς τῶν ἐμπλεόντων εἷλον οἱ Πέρσαι.
3.20.4-7

20. (4) It was not until late in the morning that the Romans realized, to their alarm, that the Persians had crossed over. Consequently they were most anxious to reach the town before the enemy and manned all the triremes and thirty-oared ships which they had moored nearby. The boats were propelled downstream at a very great speed. (5) But Nachoragan had had a very good start and was in fact already half way between Nesos and the town. At this point he laid a barrier of timber and **small boats** right across the river, massing his elephants behind it in lines which extended as far as they could wade. (6) Seeing this from a distance, the Roman fleet immediately began to back water. They had a hard job rowing in reverse with the current against them, but they pulled manfully at the oars and managed to back away. (7) Even so the Persians captured two empty boats which their crews had abandoned.
p. 90

Greek: *Agathiae Myrinaei historiarum libri quinque*, ed. R. Keydell [Corpus Fontium Historiae Byzantinae. Series Berolinensis 2]. De Gruyter, Berlin 1967.

English: *Agathias, The Histories*, transl. and intr. J. D. Frendo. De Gruyter, Berlin–New York 1975.

5.2.2. Latin sources

A-1 Titus Macc(i)us Plautus (*c.* 250 – *c.* 184 BC)

The Two Bacchises

A-1.1

278. CH. Postquam aurum apstulimus, in navem conscendimus / domi cupiente. Forte ut adsedi in stega, / dum circumspecto, atque ego **lembum** conspicor / longum, strigorem maleficum exornarier.
NI. Perii hercle: **lembus** ille mihi laedit latus.
CH. Is erat communis cum hospite et praedonibus.
NI. Adeo n' me fuisse fungum, ut qui illi crederem, / quom mi ipsum nomen eius Archidemides / clamaret dempturum esse, si quid crederem?
CH. Is **lembus** nostrae navi insidias dabat. / Occepi ego observare eos quam rem gerant. Interea e portu nostra nauis soluitur. / Ubi portu eximus, homines remigio sequi, / neque aues nec venti citius. quoniam sentio / quae res gereretur, navem extemplo statuimus. / quoniam vident nos stare, occeperunt ratem / tardare in ponto.
NI. Edepol mortalis malos! / quid denique agitis?
CH. Rursum in portum recipimus.
NI. Sapienter factum a vobis. Quid illi postea?
CH. Revorsionem ad terram faciunt vesperi.
NI. Aurum hercle auferre voluere: ei rei operam dabant.
CH. Non me fefellit, sensi, eo exanimatus fui. / Quoniam videmus auro insidias fieri, / quoniam videmus auro insidias fieri, / capimus consilium continuo: pos[t]tridie / auferimus aurum omne illis praesentibus / palam atque aperte, ut illi id factum sciscerent.
NI. Scite hercle: cedo quid illi?
CH. Tristes ilico, / quom extemplo a portu ire nos cum auro vident, / subducunt **lembum** capitibus quassantibus. / Nos apud Theotimum omne aurum depos<i>uimus, Qui illic sacerdos est Dianae Ephesiae.
Act 2: 278-307

278. CH. After we took the gold away, we went onto the ship, wishing to go home. As I sat down on the deck by chance, I saw, while I was looking around, **a long fast-sailer**, solid and evil, being prepared.
NI. I'm done for! That **fast-sailer** is ramming me amidships.
CH. It was shared between your friend and pirates.
NI. How can I have been so weak in the head as to trust him, when his very name Archidemides was shouting at me that if I entrusted him with anything, he'd dematerialize it?
CH. That **fast-sailer** was lying in wait for our ship. I began to observe what they were doing. Meanwhile our ship set sail from the harbor. As we were leaving the harbor, these people were rowing after us: neither birds nor winds are faster. When I realized what was going on, we immediately brought the ship to a standstill. When they saw us halted, they began to slow down their boat on the open sea.
NI. Bad people they are! What did you do in the end?
CH. We returned to the harbor.
NI. Wise of you. What did they do after this?
CH. They returned to the shore in the evening.
NI. They wanted to steal the gold; that's what they were after.
CH. It didn't take me in, I saw through it, that's why I was beside myself. Since we saw that a trap was being set for the gold, we made a plan at once. The next day we took all the gold ashore in their presence, openly and
publicly, to let them know that this had been done.
NI. Clever indeed. Tell me, what did they do?
CH. They were cast down as soon as they saw us coming with the gold from the harbor, and they put their **swift-boat** on shore with shaking heads. We left all the gold with Theotimus, who is a priest of Diana of Ephesus there...
pp. 395-399

5. Written sources on lembs and Liburnians from the 4th c. BC to Late Antiquity

A-1.2

953. CH. ... Paria item tria eis tribus sunt fata nostro huic Ilio. / Nam dudum primo ut dixeram nostro seni mendacium / et de hospite et de auro et de **lembo**, ibi signum ex arce iam abstuli. Iam duo restabant fata tunc, nec magis id ceperam oppidum.
Act 4: 953-959

953. ... For this Ilium of ours there are also three fates, parallel to the other three: first, when a while ago I told our old man a lie about his friend, the gold, and **the boat**, I took away the statue from the citadel.
p. 467

Latin: *T. Macci Plauti comoediae 2*, eds G. Goetz, F. Schoell. Teubner, Leipzig, 1904.

English: *Plautus. Amphitryon. The Comedy of Asses. The Pot of Gold. The Two Bacchises. The Captives*, ed. and transl. W. de Melo. LCL 60 (2011).

A-1.3 *Menaechmuses*

MES. Non tu istas meretrices nouisti, ere.
SOS. Tace, inquam --- / Mihi dolebit, non tibi, siquid ego stulte fecero. / Mulier haec stulta atque ins<c>itast: quantum perspexi modo, Est hic praeda nobis.
MES. Peri<i>. / iam ne abis? periit probe: / Ducit **lembum** dierectum nauis praedatoria. / Sed ego ins<c>itus, qui + drome postulem moderarier: / Dicto me emit audientem, haud imperatorem sibi.
Act 2: 437-44

MES. You don't know those prostitutes, master.
SOS. Be quiet, I tell you, and go away. I will feel the pain, not you, if I do something stupid. This woman is stupid and naive; as far as I could see just now, there's booty in here for us.
Exit MENAECHMUS into Erotium's house.
MES. (calling after him) I'm done for! Are you going away already? (to the audience) He's perished properly. **The pirate ship** is leading our sailing boat to its ruin. But it's stupid of me to expect to restrain my master. He bought me as his servant, not as his commander.
p. 471

Latin: *T. Macci Plauti comoediae 4*, eds G. Goetz, F. Schoell. Teubner, Leipzig 1895.

English: *Plautus. Casina. The Casket Comedy. Curculio. Epidicus. The Two Menaechmuses*, ed. and transl. W. de Melo. LCL 61 (2011).

The Merchant:

A-1.4
CHA. Perdidisti me, pater. / Eho tu, eho tu, quin cauisti, ne eam uideret, uerbero? / Quin, sceleste, <eam> abstrudebas, ne eam conspiceret pater? / AC. Quia negotiosi eramus nos nostris negotiis: / Armamentis complicandis, [et] componendis studuimus. / Dum haec aguntur, **lembo** aduehitur tuos pater **pauxillulo**, / Neque quisquam hominem conspicatust, donec in + naui super.
Act 1: 188-194

CHAR. You've ruined me, father. (to Acanthio) Hey you, hey you, why didn't you guard against him seeing her, you thug? You criminal, why didn't you hide her away so that my father wouldn't spot her?
ACAN. Because we were busy with our own business: we were concentrated on folding and packing up the tackle. While this was going on, your father came round **in a tiny boat**, and no one spotted him until he went onto the ship.
p. 35

A-1.5
DEM. [...] Ad portum hinc abii mane cum luci semul. / Postquam id quod uolui transegi, atque ego conspicor / Nauem ex Rhodo quast heri | aduectus filius. / Conlubitumst illuc mihi nescioqui uisere. / Inscendo in **lembum** | atque ad nauem deuehor: / Atque ibi ego aspicio forma eximia mulierem, / Filius quam aduexit meus matri ancillam suae.
Act 2: 255-261

I went to the harbor in the morning, at sunrise. After sorting out what I wanted, I spot the ship from Rhodes on which my son arrived yesterday. Somehow, I fancy seeing that. I go on board **a boat** and go to the ship. There I set eyes on a woman of outstanding beauty, someone my son brought here as a maid for his mother.
p. 40

Latin: *T. Macci Plauti comoediae* 4, eds G. Goetz, F. Schoell. Teubner, Leipzig 1895.

English: *Plautus. The Merchant. The Braggart Soldier. The Ghost. The Persian*, ed. and transl. W. de Melo. LCL 163 (2011).

For Plautus, the word used in connection with the term *lembus* in *The Two Bacchises* is *strigor* (**A-1.1** v. 282). It is an exotic word, not well testified in classical Latin literature.[190] For Plautus, a lemb is a pirate boat, '*faster than birds and winds*' (A-1.1). It is a tiny boat and it is used to reach a bigger ship. In the last fragment (**A-1.5**) Plautus is obviously referring to the comic meaning of the lemb as it is found in Anaxandrides, (**A-IV**) accentuating it in its diminutive form. The translation is rather far-fetched, since the '*lemb*' is not a pimp, but some sort of a flatterer, coaxer, sycophant, etc. However, much of the themes and motives Plautus chose are adaptations of previous Greek comedies, and in this respect it is hard to say how much first-hand knowledge he must have had about the term *lembus* and how much he took from earlier Greek sources. However, it is interesting to note that he does not use the term *lembus* in a specifically comic sense as we have seen in later Greek sources mentioning that it was a favourite motif in comedies.

---

[190] The epitome of Verrius Flacus, *De Verborum Significatione* s.v. *strigores* describes *strigores* as *densarum virium homines* – '*men of great strength*'.

5. Written sources on lembs and Liburnians from the 4th c. BC to Late Antiquity

A-2 Lucius Accius (170 - *c.* 90 BC)

Fragments from Deiphobo

'**Lembus**', navicula brevis piscatoria. Accius Deiphobo Eo ante noctem extenta retia ut proueherem et statuerem, / Forte aliquanto solito **lembo** sum progressus longius. Frag. 248-249 (=Nonius, *De Comp. Doct.* 13.534.1)	**Lembus** is a fishermen's small and short boat. Accius in *Deiphobus* That thither I might bring / Before the night my nets of yesterday, / And spread them there, it chanced that in my **wherry** I sailed a little farther than my wont. p. 411

Latin: *Scaenicae Romanorum poesis fragmenta, Vol. I: Tragicorum fragmenta*, ed. O. Ribbeck. Teubner, Leipzig 1871.

English: *Remains of Old Latin, Volume II: Livius Andronicus. Naevius. Pacuvius. Accius*, transl. E. H. Warmington. LCL 314 (1936).

This quote is in the vein of Theocritus' (**A-VII**) depiction of *lemb* as a fishermen's boat. Accius was educated in Greece and Asia Minor and there is no reason to think that he was not familiar with Theocritus' *Idylls*, more so since his tragedies rely heavily on Euripides (W.-L. Liebermann, *NP*).

A-3 Sextus Turpilius (2nd century BC)

Fragments of comedies

Lemniae V **Lembi** redeuntes domum duae ad nostram adcelerarunt ratem. Frag. 5 (=Nonius, *De Comp. Doct.* 13.534.7)	*Lemnae* The **lembs**, returning home, together hastened our skiff.
Leucadia XIV --- hortari coepi nostros ilico, Vt celerent **lembum**. Frag. 14 (=Nonius, *De Comp. Doct.* 13.534.4)	I began to encourage our people on the spot so that they rush to the **lemb**.

Latin: *Scaenicae Romanorum poesis fragmenta, Vol. II: Comicorum fragmenta*, ed. O. Ribbeck. Teubner, Leipzig, 1873.

These fragments were probably taken from Greek comedies, Turpilus' main source was the Greek poet Menander.

A-4 Lucius Cornelius Sisenna (c. 118 - 67 BC)

Histories

Otacilium legatum cum scaphis ac **lembis** ---   The legate Otacilius with skiffs and **lembs** [...].
Frag. 38 (=Nonius Marcellus, *De Comp. Doct.*
13.534.11-12)

Latin: *Historicorum Romanorum reliquiae* 1, ed. H. Peter, 2nd ed. Teubner, Leipzig 1914.

This fragment consists only of a direct object in accusative and a propositional phrase, probably *ablativus comitativus*. Cornell comments on the fragment in the following way:

> *It is, in fact, very probable that Otacilius is the correct reading in the passage of Suetonius. What is more, the historian was a freedman, and though freedmen served aboard ships ..., it is incredible that a freedman should have been the commander. There are two Otacilii in Pompeius Strabo's consilium in 89 ... and one of the Ilerdenses there enfranchised is an Otacilius. We may note that there was a long tradition of naval service among the Otacilii, going back to M. Otacilius Crassus (cos. 263, 246) and his brother Titus (cos. 261) in the First Punic War, and T. Otacilius Crassus (pr. 217) in the Hannibalic War. In the civil war of the 40s we find another Otacilius Crassus in a naval context.*[191]

Sisenna's *Historiae*, a historical work on the Social War, on the rule of the Marius and his supporters and the dictatorship of Sulla was held in high regard. However, only tfragments survive (W. Kierdorf, *NP*). The fragment testifies that lembs might have been part of the Roman military fleet around 90 BC.

A-5 Marcus Terentius Varro (Reatinus) (116-27 BC)

On agriculture

non multo aliter tuendum hoc pecus in pastu atque ouillum, [quod] tamen habent sua propria quaedam, quod potius siluestribus saltibus delectantur quam pratis. studiose enim de agrestibus fruticibus pascuntur atque in locis cultis uirgulta carpunt. itaque a carpendo caprae nominatae. ab hoc in lege locationis fundi excipi solet, ne colonus capra[m] natum in fundo pascat. harum enim dentes inimici sationi[s], quas etiam astrologi ita receperunt in caelum, ut extra **lembum** duodecim signorum excluserint: sunt duo haedi et capra non longe a tauro.
2.3.6–8

The care of this animal in the matter of feeding is about the same as that of the sheep, though each has certain peculiarities; thus, the goat prefers wooded glades to meadows, as it eats eagerly the field bushes and crops the undergrowth on cultivated land. Indeed, their name *capra* is derived from *carpere*, to crop. It is because of this fact that in a contract for the lease of a farm the exception is usually made that the renter may not pasture the offspring of a goat on the place. For their teeth are injurious to all forms of growth; and though the astronomers have placed them in the sky, they have put them **outside the circle** of the twelve signs—there are two kids and a she-goat not far from Taurus.
pp. 347-349

---

[191] Cornell 2013: 388. For the correct reading of Otacilius, see Kaster 1992: 120-24.

5. Written sources on lembs and Liburnians from the 4th c. BC to Late Antiquity

Latin: *Marcus Terentius Uarro, Res rusticae*, ed. G. Goetz. Teubner, Leipzig 1929.

English: *Cato, Varro. On Agriculture.* transl. W. D. Hooper, H. Boyd Ash. LCL 283 (1934).

This is the only place in which the term *lembus* is used as a synonym for the zodiac, an area of the sky that extends approximately 8° north or south of the ecliptic, and is divided into twelve signs, each occupying 30° of celestial longitude. Moreover, Varro never uses the term *lembus* in any other context. This *hapax* makes it very hard to decide on which basis Varro, otherwise very careful in using his language, used the term *lembus* as a symbol for the zodiac. We can only guess whether he refers to the shape of lembs, which reminds one of the portion of the sky called the zodiac or, maybe, refers to the carrying powers of lembs.

A-6 Gaius Iulius Caesar (100-44 BC)

Commentarii belli civilis

43. His rebus cognitis Marcius Rufus quaestor in castris relictus a Curione cohortatur suos ne animo deficiant. Illi orant atque obsecrant ut in Siciliam navibus reportentur. Pollicetur magistrisque imperat navium ut primo vespere omnes scaphas ad litus appulsas habeant. Sed tantus fuit omnium terror ut alii adesse copias Iubae dicerent alii cum legionibus instare Varum iamque se pulverem venientium cernere—quarum rerum nihil omnino acciderat—alii classem hostium celeriter advolaturam suspicarentur. Itaque perterritis omnibus sibi quisque consulebat. Qui in classe erant proficisci properabant. Horum fuga navium onerariarum magistros incitabat. Pauci **lenunculi** ad officium imperiumque conveniebant. Sed tanta erat completis litoribus contentio qui potissimum ex magno numero conscenderent ut multitudine atque onere nonnulli deprimerentur, reliqui hoc timore propius adire tardarentur. 2.43

43. Learning of this, Marcius Rufus, the quaestor left by Curio at the camp, urged his men not to lose heart. They begged and pleaded to be ferried back to Sicily. He promised, and ordered the ship captains to have all of their longboats drawn up on shore at sunset. But so great was the universal terror that some were saying that Juba's forces were at hand, others that Varus and his legions were imminent and that the dust cloud of their approach was already visible to them—not one of these things had happened—and others suspected that the enemy fleet was going to sail against them soon. Everyone was terrified, so each acted for himself. Those in the fleet set out in a hurry. Their flight spurred on the transport captains. A few **small vessels** mustered as per their obligations and orders, but such was the struggle on the packed beaches over who of the huge number was going to embark, that some boats were sunk by the weight of the crowd and the rest hesitated to come closer for fear of this. pp. 189-191

Latin and English: *Caesar. Civil War*, ed. and transl. C. Damon. LCL 39 (2016).

A-7 Publius Vergilius Maro (Vergil/Virgil) (70-19 BC)

Georgics:

uidi lecta diu et multo spectata labore / degenerare tamen, ni uis humana quotannis / maxima quaeque manu legeret: sic omnia fatis / in peius ruere ac retro sublapsa referri, / non aliter quam qui aduerso uix flumine **lembum** / remigiis subigit, si bracchia forte remisit, /atque illum in praeceps prono rapit alueus amni. 1.197–203	I have seen seeds, though picked long and tested with much pains, yet degenerate, if human toil, year after year, culled not the largest by hand. Thus by law of fate all things speed towards the worse and slipping away fall back; even as if one, whose oars can scarce force his **skiff** against the stream, should by chance slacken his arms, and lo! headlong down the current the channel sweeps it away. p. 113

Latin: *P. Vergili Maronis Opera*, ed. R. A. B. Mynors. Oxford Classical Texts, Oxford 1969.

English: *Virgil. Eclogues. Georgics. Aeneid: Books 1-6.* Trans. H. Rushton Fairclough, Rev. G. P. Goold. LCL 63 (1916).

In these verses Vergil uses the term *lembus* as a part of a simile. The point of the simile is the following: only by investing a lot of effort and labour can we, humans, fight the detrimental forces of nature that tend to deteriorate the best seed if not properly attended to. To make this point clear, Vergil compares it to a man in a lemb: he can use his force to guide the boat upstream, but at the moment he stops rowing, the boat will inevitably start going downstream. This simile is in the same vein as the general approach of the whole *Georgica*. Vergil's didactic poem on agriculture is today widely understood not merely as a practical manual for farmers, but, the agricultural motifs are understood to have been used as: *'a vehicle for wider-reaching statements of religious, philosophical (more precisely: Epicurean), cultural-historical or political relevance.'* (W. Suerbaum, *NP*). For us, the especially interesting moment is that Vergil's usage of the term *lembus* in this simile implies that the it must be a understood as rather small boat, a skiff, of such a size that one person can push it against the stream by rowing.

A-8 Titus Livius (Livy) (59 BC - AD 17)

Ab Urbe condita:

A-8.1 legati ab Orico ad M. Valerium praetorem venerunt, praesidentem classi Brundisio Calabriae que circa litoribus, nuntiantes Philippum primum Apolloniam temptasse **lembis biremibus** centum viginti flumine adverso subvectum [...]. 24.40.2-3	Legates came from Oricum to Marcus Valerius, the praetor, who with his fleet was guarding Brundisium and the neighbouring coast of Calabria. They reported that Philip had first sailed up the river with a hundred and twenty **small vessels having two banks of oars** and attacked Apollonia [...]. p. 303

5. Written sources on lembs and Liburnians from the 4th c. BC to Late Antiquity

A-8.2
7 inde nauibus acceptis ab Achaeis – errant autem tres quadriremes et biremes totidem – Anticyram traiecit. 8 inde quinqueremibus septem et **lembis** uiginti amplius, quos ut adiungeret Carthaginiensium classi miserat in Corinthium sinum, profectus ad Erythras Aetolorum, quae prope Eupalium sunt, escensionem fecit.
28.8.6-8.

Then on receiving ships—they were three quadriremes and as many biremes—from the Achaeans, he [Philip] sailed over to Anticyra. From there he set sail with seven quinqueremes and more than twenty **light vessels** previously sent by him into the Gulf of Corinth to be added to the Carthaginian fleet, and made a landing at Erythrae, in Aetolia and near Eupalium.
pp. 33-35

A-8.3
9 Ab Andro Cythnum traiecerunt; ibi dies aliquot oppugnanda urbe nequiquam absumpti, et quia uix operae pretium erat abscessere. 10 ad Prasias – continentis Atticae is locus est – Issaeorum uiginti **lembi** classi Romanorum adiuncti sunt. ii missi ad populandos Carystiorum agros; cetera classis Geraestum, nobilem Euboeae portum, dum ab Carysto Issaei redirent, tenuit.
31.45.9-10

From Andros the allies crossed to Cythnus. Here several days were wasted on an assault on the city, and they eventually abandoned the attempt because it was scarcely worth the trouble. At Prasiae—this is a place on the Attic mainland—twenty **light craft** from Issa joined the Roman fleet. These were dispatched to conduct raids on the territory of the people of Carystus, and the rest of the fleet docked at the famous Euboean port of Geraestus to await the return of the Issaei from Carystus.
pp. 135-137

A-8.4
26 paeneinsula est Peloponnesus, angustis Isthmi faucibus continenti adhaerens, nulli apertior neque opportunior quam nauali bello. 27 si centum tectae naues et quinquaginta leuiores apertae et triginta Issaei **lembi** maritimam oram uastare, et expositas prope in ipsis litoribus urbes coeperint oppugnare, in mediterraneas scilicet nos urbes recipiemus, tamquam non intestino et haerente in ipsis uisceribus uramur bello?
32.21.26–27

[The praetor Aristaenus' speech] The Peloponnese is a peninsula, joined to the mainland by the narrow strip of the Isthmus, and it is above all exposed and vulnerable to naval attack. Suppose a hundred decked ships, fifty lighter open-decked vessels, and thirty Issaean **cutters** begin to conduct raids on our seacoast and attack the cities that are exposed to them, practically right on the shoreline. We shall retreat into our inland cities, shall we—as though we were not burning with an internal war, one clinging to our vitals?
p. 227

A-8.5
11 eo Philocles regius et ipse praefectus mille et quingentos milites per Boeotiam duxit; praesto fuere ab Corintho **lembi** qui praesidium id acceptum Lechaeum traicerent.
32.23.11

Philocles, who was also an officer of the king, brought 1,500 men through Boeotia to this spot; there **boats** from Corinth were waiting to receive the troops and take them across to Lechaeum.
p. 235

A-8.6
9 in sinu Maliaco prope Nicaeam litus elegere. eo rex ab Demetriade cum quinque **lembis** et una naue rostrata uenit: 10 erant cum eo principes Macedonum et Achaeorum exsul, uir insignis, Cycliadas.
32.32.9–10

They chose as a site a beach in the Malian Gulf near Nicaea, and the king came there from Demetrias with five **cutters** and a warship. With him were some leading Macedonians and a distinguished Achaean exile, Cycliadas.
p. 261

A-8.7
10 ipse cum classe centum tectarum nauium, ad hoc leuioribus nauigiis cercuris que ac **lembis** ducentis proficiscitur, 11 simul per omnem oram Ciliciae Lyciae que et Cariae temptaturus urbes quae in dicione Ptolomaei essent, simul Philippum - necdum enim debellatum erat - exercitu nauibus que adiuturus.
33.19.10–11

He [Antiochus] then set out himself with a fleet of 100 ships with decks, plus 200 lighter vessels, Cyprian **cutters** and pinnaces. His aim was to strike at the cities under Ptolemy's control all along the coast of Cilicia, Lycia, and Caria, and at the same time to assist Philip (for the war against Philip was not yet finished) with his army and ships.
p. 341

A-8.8
11 naualines quoque magnae copiae conueniebant: iam ab Leucade L. Quinctius quadraginta nauibus uenerat, iam Rhodiae duodeuiginti tectae naues, iam Eumenes rex circa Cycladas insulas erat cum decem tectis nauibus, triginta **lembis** mixtis que aliis minoris formae nauigiis.
34.26.11

In addition, large naval forces were beginning to assemble there: Lucius Quinctius had already arrived from Leucas with forty ships; there were eighteen decked vessels from Rhodes; and King Eumenes was now off the Cyclades with ten decked ships, thirty **cutters**, and an assortment of other craft of smaller dimensions.
pp. 495-497

A-8.9
[...] 5 naues quas ciuitatibus maritimis ademisset redderet, neue ipse nauem ullam praeter duos **lembos**, qui non plus quam sedecim remis agerentur, haberet; [...]
34.35.5

Nabis was to return the vessels that he had appropriated from the coastal communities and himself retain no ship apart from two **cutters** that were propelled by no more than sixteen oars.
p. 525

A-8.10
1 Comparauerat et tyrannus modicam classem ad prohibenda si qua obsessis mari submitterentur praesidia, tres tectas naues et **lembos** pristesque, tradita uetere classe ex foedere Romanis.
35.26.1

To prevent any assistance being sent by sea to the beleaguered populace, the tyrant [Nabis] had also put together a small fleet comprising three decked ships and some pinnaces and **cutters** (his old fleet having been surrendered to the Romans under the terms of the treaty).
p. 69

5. Written sources on lembs and Liburnians from the 4th c. BC to Late Antiquity

A-8.11
4 cum derexissent ad terram proras, quindecim ferme eis naues circa Myonnesum apparuerunt, quas primo ex classe regia praetor esse ratus institit sequi; apparuit deinde piraticos celoces et **lembos** esse. 5 Chiorum maritimam oram depopulati, cum omnis generis praeda reuertentes postquam uidere ex alto classem, in fugam uerterunt. et celeritate superabant leuioribus et ad id fabrefactis nauigiis, et propiores terrae erant; 6 itaque priusquam adpropinquaret classis, Myonnesum perfugerunt, unde se e portu ratus abstracturum naues, ignarus loci sequebatur praetor.
37.27.4–5

After they [the Romans] turned their prows landward, some fifteen ships appeared before them off Myonnesus and the praetor, initially believing them to be from the king's fleet, proceeded to give chase. But it then became clear that they were the **cutters** and skiffs of buccaneers. They had been raiding the coastline of Chios and were returning with all manner of plunder; and when they saw the Roman fleet from the open sea they turned to flight. They had the advantage of speed with their lighter vessels (expressly built for that) and they were also closer to land. As a result, they made good their escape to Myonnesus before the fleet could get near them.
p. 369

A-8.12
2 [...] Pleuratus, Illyriorum rex, cum sexaginta **lembis** Corinthium sinum inuectus, adiunctis Achaeorum quae Patris erant nauibus, maritima Aetoliae uastabat. 3 aduersus quos mille Aetoli missi, quacumque se classis circumegerat per litorum anfractus, breuioribus semitis occurrebant.
38.7.2–3

The Illyrian king Pleuratus had sailed with sixty **light vessels** into the Corinthian Gulf and, joining the Achaean ships anchored at Patrae, was ravaging the coastal areas of Aetolia. A thousand Aetolians were sent to combat them and, wherever the fleet followed the twisting shoreline, these would confront them by taking shorter routes.
p. 21

A-8.13
6 [...] Lutarius Macedonibus, per speciem legationes ab Antipatro ad speculandum missis, duas tectas naues et tres **lembos** adimit. iis alios atque alios dies noctes que trauehendo intra paucos dies omnes copias traiecit.
38.16.6

Lutarius commandeered two decked ships and three **light vessels** from some Macedonians who, though ostensibly on a diplomatic mission, had really been sent by Antipater to spy on them; and with these, by making repeated crossings day and night, he shipped all the troops across in a matter of days.
pp. 51-53

A-8.14
9 proficiscuntur ab Thessalonica Aeneam ad statum sacrificium, quod Aeneae conditori cum magna caerimonia quotannis faciunt. 10 ibi die per sollemnes epulas consumpto, nauem praeparatam a Poride, sopitis omnibus, de tertia uigilia conscendunt tamquam redituri in Thessalonicam; sed traicere in Euboeam erat propositum. 11 ceterum in aduersum uentum nequicquam eos tendentes prope terram lux oppressit, et regii qui praeerant custodiae portus **lembum**

They [Theoxena and Poris and their children] left Thessalonica for Aenea to attend a traditional sacrifice that the citizens of the town celebrate each year with great ceremony in honor of their founder Aeneas. At Aenea they spent a day over the ritual feasts and then at about the third watch, when all were asleep, boarded a ship that had been made ready in advance by Poris, ostensibly to return to Thessalonica, though in fact their intention was to cross to Euboea. However, struggling vainly against a headwind, they were

115

armatum ad pertrahendam eam nauem miserunt, cum graui edicto, ne reuerterentur sine ea.
40.4.9–11

still close to land when dawn came upon them and the king's men responsible for guarding the port sent out an armed **clipper** to take the ship in tow, with strict orders not to return without it.
p. 395

A-8.15
6 praemissus a praetore est frater <M>. Lucretius cum quinquereme una, iussus que ab sociis ex foedere acceptis nauibus ad Cephallaniam classi occurrere. 7 ab Reginis triremi una <sumpta>, ab Locris duabus, ab Vritibus quattuor, praeter oram Italiae superuectus Calabriae extremum promunturium, Ionio mari Dyrrachium traicit. 8 ibi decem ipsorum Dyrrachinorum, duodecim Issaeorum, quinquaginta quattuor Genti regis **lembos** nanctus, simulans se credere eos in usum Romanorum comparatos esse, omnibus abductis die tertio Corcyram, inde protinus in Cephallaniam traicit. 9 C. Lucretius praetor, ab Neapoli profectus, superato freto die quinto in Cephallaniam transmisit.
42.48.6–9

The praetor sent his brother Marcus Lucretius on in advance with one quinquereme, with orders to meet the fleet at Cephallania with the ships received from the allies according to treaty. After taking on one trireme from Rhegium, two from Locri, four from the district of Uria, he coasted along Italy past the farthest cape of Calabria and crossed the Ionian sea to Dyrrhachium. There he came upon ten **cutters** of the Dyrrhachians themselves, twelve of the Issaeans, and fifty-four of King Gentius, and pretending that he believed they had been collected for the use of the Romans, took them all with him on the third day to Corcyra and thence crossed at once to Cephallania.
p. 439

A-8.16
6 haec parantibus his, decem regii **lembi** ab Thessalonica cum delectis Gallorum auxiliaribus missi, cum in salo stantes hostium naues conspexissent, ipsi obscura nocte, simplici ordine, quam poterant proxime litus tenentes, intrarunt urbem.
44.12.6

During the preparation of these works, ten Macedonian **scout-ships**, sent from Thessalonica with picked Gallic auxiliaries, saw that the besieging ships were anchored out to sea, and in the dead of night, in single column, hugging the shore as closely as possible, entered the city.
p. 129

A-8.17
28 1 Perseus, post reditum ab Eumene Herophontis spe deiectus, Antenorem et Callippum praefectos classis cum quadraginta **lembis** - adiectae ad hunc numerum quinque pristes erant - Tenedum mittit, 2 ut inde sparsas per Cycladas insulas naues, Macedoniam cum frumento petentes, tutarentur. 3 Cassandreae deductae naues in portus primum qui sub Atho monte sunt, <in>de de Tenedum placido mari cum traiecissent, stantes in portu Rhodias apertas naues Eudamum que praefectum earum inuiolatos atque etiam benigne appellatos dimiserunt. 4 cognito deinde in latere altero quinquaginta onerarias suarum stantibus in ostio portus Eumenis rostratis, quibus + Diamius + praeerat, inclusas esse, 5 circumuectus propere ac summotis terrore hostium nauibus, onerarias

XXVIII After the disappointment of Herophon's return from the court of Eumenes, Perseus sent Antenor and Callippus, his naval commanders, with forty scout-ships—there were five **cutters** added to this number—to Tenedos, in order to protect from that base the ships scattered through the Cyclades islands on their way to Macedonia with grain. The ships were launched at Cassandrea, and after crossing first to the harbours under Mount Athos and thence on a calm sea to Tenedos, they sent away unharmed and even with kind addresses some undecked Rhodian ships which were at the port with Eudamus their commander. Then Antenor discovered that on the other side of the island fifty of their freight ships were blockaded by warships of Eumenes, under command of

datis qui prosequerentur decem **lembis** in Macedoniam mittit, ita ut in tutum prosecuti redirent Tenedum. 6 nono post die ad classem iam ad Sigeum stantem redierunt. inde Subota - insula est interiecta Elaeae et Chio - traiciunt. 7 forte postero die quam Subota classis tenuit quinque et triginta naues, quas hippagogus uocant, ab Elaea profectae cum equitibus Gallis equis que, Phanas, promunturium Chiorum, petebant, unde transmittere in Macedoniam possent; Attalo ab Eumene mittebantur. 8 has naues per altum ferri cum ex specula signum datum Antenori esset, profectus ab Subotis inter Erythrarum promunturium Chium que, qua artissimum fretum est, iis occurrit. 9 nihil minus credere praefecti Eumenis quam Macedonum classem in illo uagari mari: nunc Romanos esse, nunc Attalum aut remissos aliquos ab Attalo ex castris Romanis Pergamum petere. 10 sed cum iam adpropinquantium forma **lemborum** haud dubia esset, et concitatio remorum derectaeque in se prorae hostes adpropinquare aperuissent, tunc iniecta trepidatio est. 11 cum resistendi spes nulla esset inhabili que nauium genere et Gallis uix quietem ferentibus in mari, 12 pars eorum, qui propiores continenti litori erant, in Erythraeam enarunt, pars uelis datis ad Chium naues eiecere relictis que equis effusa fuga urbem petebant. 13 sed propius urbem **lembi** accessu que commodiore cum exposuissent armatos, partim in uia fugientes Gallos adepti Macedones ceciderunt, partim ante portam exclusos. clauserant enim Chii portas, ignari qui fugerent aut sequerentur. 14 octingenti ferme Gallorum occisi, ducenti uiui capti; equi pars in mari fractis nauibus absumpti, parti neruos succiderunt in litore Macedones. 15 uiginti eximiae equos formae cum captiuis eosdem decem **lembos** quos ante miserat Antenor deuehere Thessalonicam iussit, et primo quoque tempore ad classem reuerti; Phanis se eos exspectaturum. 16 triduum ferme classis ad urbem <stetit>. Phanas inde progressi sunt, et spe celerius regressis decem **lembis** euecti Aegaeo mari Delum traiecerunt. 29 1 Dum haec geruntur, legati Romani C. Popillius et C. Decimius et C. Hostilius, a Chalcide profecti tribus quinqueremibus Delum cum uenissent, **lembos** ibi Macedonum quadraginta et quinque regis Eumenis quinqueremes inuenerunt.

Damius, which were stationed at the mouth of the harbour. Antenor promptly sailed around and by threat caused the enemy ships to retreat: the freight ships were sent to Macedonia under convoy of ten **scout-ships**, which were instructed to return to Tenedos after seeing them safe. On the ninth day thereafter they returned to the fleet, which was now anchored at Sigeum. Thence they crossed to Subota, an island lying between Elaea and Chios. On the day after the arrival of the fleet at Subota, thirty-five of the ships called horse-transports, setting out from Elaea with Galatian cavalry and their mounts, were making for Phanae, a cape of Chios, from which they could cross to Macedonia. They were being sent to Attalus by Eumenes. When a signal reached Antenor from a lookout post that these ships were at sea, he started from Subota and met them between the cape of Erythrae and Chios, where the strait is narrowest. Eumenes' officers least of all suspected that a Macedonian fleet was at large in that sea; now they thought them to be Romans, now Attalus, or some men sent back by Attalus from the Roman camp and on their way to Pergamum. But when the shape of the approaching **scout-ships** was unmistakable and the rapid motion of the oars and the pointing of the prows head-on revealed that enemies were approaching, then panic fell upon them. Since there was no hope of resistance, both because of the unwieldy type of vessel and because the Galatians could hardly withstand an undisturbed voyage, some of them, who were nearer to the mainland, swam ashore at Erythrae and some set sail for Chios, ran their ships aground, and abandoning their horses fled in rout to the city. But as the **scout-ships** put troops ashore nearer the city at a more convenient landing-place, the Macedonians overtook the Galatians and cut them down, partly as they fled along the road and partly when they were shut out of the city gates. For the Chians had closed their gates, not knowing who were fleeing or who pursuing. About eight hundred of the Galatians were killed and two hundred taken alive; some of the horses were destroyed in the sea as the ships were wrecked, some were hamstrung on shore by the Macedonians. Antenor ordered twenty horses of exceptional beauty, along with

2 sanctitas templi insulae que inuiolatos praestabat omnes. itaque permixti Romani que et Macedones et Eumenis nauales socii in templo, indutias religione loci praebente, uersabantur. 3 Antenor Persei praefectus, cum aliquas alto praeferri onerarias naues ex speculis significatum foret, 4 parte **lemborum** ipse insequens, parte per Cycladas disposita, praeterquam si quae Macedoniam peterent omnes aut supprimebat aut spoliabat naues. quibus poterat Popillius <aut suis> aut Eumenis nauibus succurrebat; 5 sed <e>uecti nocte binis aut ternis plerumque **lembis** Macedones fallebant. 6 per id fere tempus legati Macedones Illyrii que simul Rhodum uenerunt, quibus auctoritatem addidit non **lemborum** modo aduentus passim per Cycladas atque Aegaeum uagantium mare, sed etiam coniunctio ipsa regum Persei Genti que et fama cum magno numero peditum equitum que uenientium Gallorum.

44.28–29.6

the prisoners, to be taken to Thessalonica by the same ten **scout-ships** he had sent before, which he ordered to return as soon as possible to the fleet; he said he would await them at Phanae. For about three days the fleet anchored before the city. Then they moved on to Phanae, and setting sail on the unexpectedly early arrival of the ten **scout-ships**, crossed the Aegean Sea to Delos.

XXIX. While these events were taking place, the Roman envoys, Gaius Popilius, Gaius Decimius, and Gaius Hostilius, set out from Chalcis in three five-banked ships and on arriving at Delos found there forty Macedonian **scout-ships** and five five-bankers of King Eumenes. The holiness of the temple and the island kept them all from harm. And so the Roman, the Macedonian, and Eumenes' sailors mingled in the temple under the truce provided by the sacredness of the place. Whenever signals came from the lookouts that any freight-ships were passing out at sea, Antenor, Perseus' officer, would himself pursue with some **ships**, while others of his ships were distributed among the Cyclades, and either sank or plundered every ship not sailing for Macedonia. Popilius would come to the rescue with what ships he had either of his own or Eumenes'; but the Macedonians would evade him by sailing at night mostly in groups of two or three **ships**.

At about this same time, the envoys from Macedonia and Illyria arrived together at Rhodes, and weight was lent their words not only by the arrival of the **scout-ships** roaming all around the Cyclades and the Aegean Sea, but also by the very fact of the combination between the kings Perseus and Gentius, and the rumour of the arrival of the Gauls in great numbers both of infantry and cavalry.

pp. 179-185

5. Written sources on lembs and Liburnians from the 4th c. BC to Late Antiquity

A-8.18
12 Anicius praetor eo tempore Apolloniae auditis quae in Illyrico gererentur, praemissis que ad Appium litteris ut se ad Genusum opperiretur, triduo et ipse in castra uenit, 13 et ad ea quae habebat auxilia Parthinorum iuuenta duobus milibus peditum et equitibus ducentis - peditibus Epicadus equitibus Agalsus praeerat -, parabat ducere in Illyricum, maxime ut Bassanitas solueret obsidione. tenuit impetum eius fama **lemborum** uastantium maritimam oram. 14 octoginta erant **lembi**, auctore Pantaucho missi a Gentio ad Dyrrachinorum et Apolloniatium agros populandos. 15 tum classis ad [...] to eo tradiderunt se.
44.30.12–15

He [the praetor Anicius] added to the auxiliaries he had two thousand infantry and two hundred cavalry of the young men of the Parthini—Epicadus commanded the infantry, Algalsus the cavalry—and made ready to march into the region of Illyria, especially in order to relieve the siege of Bassania. His urgency was restrained by a rumour of **scout-ships** ravaging the coast. There were eighty of these **ships**, sent by Gentius at the suggestion of Pantauchus to plunder the territory of Dyrrachium and Apollonia. Then fleet at ... they surrendered.
pp. 189-191

A-8.19
12 [...] rex in domum se recepit, pecunia que et auro argento que in **lembos**, qui in Strymone stabant, delatis, et ipse ad flumen descendit. 13 Thraces nauibus se committere non ausi domos dilapsi, et alia militaris generis turba; Cretenses spe pecuniae secuti. et quoniam in diuidendo plus offensionum quam gratiae erat, quinquaginta talenta iis posita sunt in ripa diripienda. 14 ab hac direptione cum per tumultum naues conscenderent, **lembum** unum in ostio amnis multitudine grauatum merserunt. Galepsum eo die, postero Samothracam, quam petebant, perueniunt; 15 ad duo milia talentum peruecta eo dicuntur.
44.45.12–15

Thereafter the king [Evander] went home, put his money, his gold, and his silver into the **scout-ships** which were moored in the Strymon and himself went down to the river. The Thracians and the rest of the mob of soldiers, not daring to trust themselves to ships, slipped away to their homes; the Cretans followed Perseus in hopes of cash. Because apportionment would create more hard feelings than gratitude, fifty talents were set out on the river-bank for them to scramble for. When after this scramble they were boarding the ships in riotous fashion, they sank one **scout-ship** at the mouth of the river by overcrowding it. On that day the party reached Galepsus, on the next Samothrace, their destination; it is said that two thousand talents were brought there.
p. 239

A-8.20
6 1 Ceterum tanto facinore in unicum relictum amicum admisso, per tot casus expertum, proditum que quia non prodiderat, omnium ab se abalienauit animos. 2 pro se quisque transire ad Romanos, fugae que consilium capere solum prope relictum coegerunt; Oroandem +que+ Cretensem, cui nota Thraciae ora <erat> quia mercaturas in ea regione fecerat, appellat ut se sublatum <in> **lembum** ad Cotym deueheret. 3 Demetrium est portus in promunturio quodam Samothracae; ibi **lembus** stabat. sub occasum solis deferuntur quae ad usum necessaria erant; defertur et pecunia quanta clam deferri poterat. 4 rex

By fleeing they forced the king [Evander] too, left almost alone, to adopt some plan of flight; finally he appealed to Oroandes the Cretan, who knew the coast of Thrace from having made trading voyages there, to take him aboard **a scout-ship** and carry him to Cotys. There is a harbour Demetrium on a certain headland of Samothrace; **the scout-ship** was anchored there. At sun-down the needful equipment was carried to the shore; all the money that could be brought secretly was also carried down. In the middle of the night, the king himself with three companions of his flight went by a back door of the house into

ipse nocte media cum tribus consciis fugae <per> posticum aedium in propinquum cubiculo hortum, atque inde, maceriam aegre transgressus, ad mare peruenit. 5 Oroandes tantum <moratus> dum pecunia deferretur, primis tenebris soluerat nauem ac per altum Cretam petebat. 6 postquam in portu nauis non inuenta est, uagatus Perseus aliquamdiu in litore, postremo timens lucem iam adpropinquantem, in hospitium redire non ausus, in latere templi prope angulum obscurum delituit.
45.6.1–6

a garden next to his bedroom and thence, after scrambling with difficulty over a wall, reached the sea. Oroandes had waited just long enough for the money to be brought down, and in the early darkness had weighed anchor and was sailing directly for Crete. When the ship was not found at the harbour, Perseus wandered about for some time on the shore and, finally, fearing the imminent approach of dawn, did not dare to return to the house where he was entertained, but took cover near an out-of-the-way corner on one side of the temple.
pp. 263-265

A-8.21
1 Victoriae Romanae fama cum peruasisset in Asiam, Antenor, qui cum classe **lemborum** ad Phanas stabat, Cassandriam inde traiecit. 2 C. Popillius, qui Deli in praesidio nauibus Macedoniam petentibus erat, postquam debellatum in Macedonia et statione summotos hostium **lembos** audiuit, dimissis et ipse + adticis + nauibus ad susceptam legationem peragendam nauigare Aegyptum pergit, 3 ut prius occurrere Antiocho posset quam ad Alexandreae moenia accederet.
45.10.1-3

When a report of the Roman victory penetrated to Asia, Antenor, who was lying off Phanae with his fleet of **scout-ships**, crossed from there to Cassandria. When Gaius Popillius, who was at Delos to protect ships making for Macedonia, heard that the war had been brought to an end in Macedonia and that the enemy s**cout-ships** had left their post, he for his part dismissed the ships of Attalus and proceeded to sail for Egypt to complete the mission on which he had started, so that he might be able to meet Antiochus before he reached the walls of Alexandria.
p. 273

A-8.22
12 Aetolis dimissis, Acarnanum citata gens. in his nihil nouatum nisi quod Leucas exempta est Acarnanum concilio. 13 quaerendo deinde latius qui publice aut priuatim partium regis fuissent in Asiam quoque cognitionem extendere; 14 et ad Antissam in Lesbo insula diruendam, traducendos Methymnam Antissaeos Labeonem miserunt, quod Antenorem regium praefectum, quo tempore cum **lembis** circa Lesbum est uagatus, portu receptum commeatibus iuuissent. 15 duo securi percussi uiri insignes, Andronicus Andronici filius Aetolus, quod patrem secutus arma contra populum Romanum tulisset, et Neo Thebanus, quo auctore societatem cum Perseo iunxerant.
45.31.12–15

When the Aetolians had been dismissed, the Acarnanian League was called up. No changes were made concerning them, except that Leucas was removed from the Acarnanian federation. In the course of more sweeping inquiries as to support of the king, either individual or by political units, the investigation was extended to Asia, and Labeo was sent to destroy Antissa on the island of Lesbos and to move its inhabitants to Methymna, because when Antenor, the king's admiral, had been roaming about Lesbos with **his scout-ships**, the people of Antissa had received him into their harbour and aided him with provisions. Two men of distinction were beheaded by the consul, Andronicus, son of Andronicus, an Aetolian, because following his father he had borne arms against the Roman People, and Neon of Thebes, who had induced his people to make an alliance with Perseus.
p. 357

5. Written sources on lembs and Liburnians from the 4th c. BC to Late Antiquity

A-8.23

9 rex Gentius cum liberis et coniuge et fratre Spoletium in custodiam ex senatus consulto ductus, ceteri captiui Romae in carcerem coniecti; recusantibus que custodiam Spoletinis, Iguuium reges traducti. 10 reliquum ex Illyrico praedae ducenti uiginti **lembi** erant; de Gentio rege captos eos Corcyraeis et Apolloniatibus et Dyrrachinis Q. Cassius ex senatus consulto tribuit.
45.43.9-10

King Gentius with his wife, his children, and his brother, was taken to Spoletium for safe-keeping, in accordance with a decree of the senate. The other prisoners were thrown into the prison at Rome. When the people of Spoletium refused to take custody, the royal family was transferred to Iguvium. There remained from the Illyrian booty two hundred and twenty **scout-ships**; as prizes taken from King Gentius they were presented to the people of Corcyra, Apollonia, and Dyrrhachium by Quintus Cassius, in accordance with a decree of the senate.
p. 403

Latin: *Oxford Classical Texts: Titi Livi: Ab Urbe Condita, Vol. 3: Libri XXI-XXV*, ed. J. Briscoe. OUP, Oxford 2016.

Oxford Classical Texts: Titi Livi: Ab Urbe Condita, Vol. 4: Libri XXVI-XXX, eds

R. S. Conway, S. K. Johnson. OUP, Oxford 1935.

Oxford Classical Texts: Titi Livi: Ab Urbe Condita, Vol. 5: Libri XXXI-XXXV, ed. A. H. McDonald. OUP, Oxford 1965.

Oxford Classical Texts: Titi Livi: Ab Urbe Condita, Vol. 6: Libri XXXVI-XL, ed. P. G. Walsh. OUP, Oxford 1999.

English: Livy, *History of Rome, Volume VI: Books 23-25; Volume VIII: Books 28-30*, transl. F. Gardner Moore. LCL 355, 381 (1940-49).

Livy, *History of Rome, Volume IX: Books 31-34; Volume X: Books 35-37; Volume XI: Books 38-40*, ed. and transl. J. C. Yardley. LCL 295, 301, 313 (2017-18).

Livy, *History of Rome, Volume XII: Books 40-42*, transl. E. T. Sage, A. C. Schlesinger. LCL 332 (1938).

Livy, *History of Rome, Volume XIII: Books 43-45*, transl. A. C. Schlesinger. LCL 396 (1951).

As we have seen in the above citations, Livy uses the term *lembus* abundantly and in different context. Thus, lembs belonged to:

- the Macedonians **A-8.1-2, A-8.5-6, A-8.17, A-8.21-22**
- the Issaeans **A-8.3-4**
- the Illyrian kingdom **A-8.12, A-8.16, A-8.18-19, A-8.23**
- the Dyrrhachians **A-8.15**
- the Gauls **A-8.13**
- the Thessalians (Theoxena) **A-8.14**
- the Cyprians **A-8.7**

- the Pergamonians **A-8.8**
- the Spartans **A-8.9-10**
- the pirates from Myonnessus **A-8.11**
- the Cretans **A-8.20**

Moreover, from the description in the text, we learn that lembs had or could have two rows of oars (*biremes*, **A-8.1**), but not more than sixteen oars (**A-8.9**). They were light and speedy (**A-8.11**, **A-8.17**), and could be armed (**A-8.14**). Lembs had a recognisable shape by pointing their prows head-on (**A-8.17**). As for their freight capacity, ten lembs could carry twenty horses and two hundred prisoners (**A-8.17**). Lembs could also come in rather large numbers: the booty acquired after the defeat of Illyrian king Genthius consisted of two hundred and twenty lembs (**A-8.23**).

A-9 Quintus Curtius Rufus (1st century).

History of Alexander

A-9.1
16.1 Praefecti Alexandri in obsidione urbis perseverabant non tam suis viribus <fisi> quam ipsorum, qui ob sidebantur, voluntate. Nec fefellit opinio: namque inter Apolloniden et duces militum orta seditio inrumpendi in urbem occasionem dedit, 17.1 cumque porta effracta cohors Macedonum intrasset, oppidani olim consilio proditionis agitato adgregant se Amphotero et Hegelocho, Persarumque praesidio caeso Pharnabazus cum Apollonide et Athenagora vincti traduntur, XII triremes cum suo milite ac remige, praeter eas XXX inanes et L piratici **lembi** Graecorumque III milia a Persis mercede conducta. His in supplementum copiarum suarum distributis piratisque supplicio adfectis captivos remiges adiecere classi suae.
4.5.16.1–18.6

Alexander's generals persisted in the siege of the city [Chios], relying not so much on their own strength as on the inclination of the besieged. And they were not mistaken; for a disagreement which arose between Apollonides and the leaders of the soldiers gave an opportunity for forcing their way into the city, and after a gate had been broken down and a cohort of Macedonians had entered, the townsmen, who had previously planned to betray the city, attached themselves to Amphoterus and Hegelochus, the Persian garrison was slain, and Pharnabazus as well as Apollonides and Athenagoras were bound and surrendered to the Macedonians, also twelve triremes with their soldiers and oarsmen, and besides these, thirty ships without crews, and **fifty piratical boats** and 3000 Greeks serving as mercenaries with the Persians. These last were distributed as a reinforcement of the Macedonian forces, the pirates were put to death, and the captured oarsmen were enrolled in the fleet.
pp. 211-213

A-9.2

21.2 Nec dubitavit Aristonicus primus intrare, secuti sunt ducem piratici **lembi**, at, dum applicant navigia crepidini portus, obicitur a vigilibus claustrum, et qui proximi excubabant ab eisdem excitantur.
4.5.21.2–4

Aristonicus did not hesitate to enter first and [ten][192] **pirate vessels** followed their leader; and while they were bringing the ships up to the quay of the port, the guards put the barrier in place and summoned those who were on watch near by.
p. 213

Latin: *Quintus Curtius Rufus, Geschichte Alexanders des Grossen*, eds K. Müller, H. Schönfeld. De Gruyter, Berlin 1954.

English: *Quintus Curtius. History of Alexander, Volume I: Books 1-5*, transl. J. C. Rolfe. LCL 368 (1946).

In this passage Curtius describes the battle between the Persians and the Macedonians in 332 BC. Amphoterus and Hegelochus were commissioned to drive the Persian garrisons from the islands in the Aegean Sea. They were fully successful, since the inhabitants of the Aegean islands were anxious to get rid of the Persian yoke. Curtius asserts that *lembs* were part of the Persian fleet in the Aegean.

A-10 Gaius Plinius Secundus (Pliny the Elder) (23/24-79)

Historia naturalis

206.1 Nave primus in Graeciam ex Aegypto Danaus advenit; antea ratibus navigabatur inventis in mari Rubro inter in-sulas a rege Erythra. reperiuntur qui Mysos et Troianos priores excogitasse in Hellesponto putent, cum transirent adversus Thracas. etiam nunc in Britannico oceano v<i>tiles corio circumsutae fiunt, in Nilo ex papyro ac scirpo et harundine. 207.1 longa nave Iasonem primum navigasse Philostephanus auctor est, Hegesias Parhalum, Ctesias Samiramin, Archemachus Aegaeonem, biremem Damastes Erythraeos fecisse, triremem Thucydides Aminoclen Corinthium, quadriremem Aristoteles Carthaginienses, 208.1.

Danaus first came from Egypt to Greece by ship; before that time rafts were used for navigation, having been invented by King Erythras for use between the islands in the Red Sea. Persons are found who think that vessels were devised earlier on the Hellespont by the Mysians and Trojans when they crossed to war against the Thracians. Even now in the British ocean coracles are made of wicker with hide sown round it, and on the Nile canoes are made of papyrus, rushes and reeds The first voyage made in a long ship is attributed by Philostephanus to Jason, by Hegesias to Parhalus, by Ctesias to Samiramis, and by Archemachus to Aegaeo. Further advances were as follows:—

---

[192] I do not see the reason why this number is put here. LB.

quinqueremem <M>nesi<gi>ton Salaminios, sex ordinum <X>enagoras Syracusios, ab ea ad decemremem <M>nesigiton Alexandrum Magnum, ad duodecim ordines Philostephanus Ptolemaeum Soterem, ad quindecim Demetrium Antigoni, ad XXX Ptolemaeum Philadelphum, ad XL Ptolemaeum Philopatorem, qui Tryphon cognominatus est. onerariam Hippus Tyrius invenit, **lembum** Cyrenenses, cumbam Phoenices, celetem Rhodii, 209.1 cercyrum Cyprii. siderum observationem in navigando Phoenices, remum Copae, latitudinem eius Platae<ae>, vela Icarus, malum et antennam Daedalus, hippegum Samii aut Pericles Atheniensis, tectas longas Thasii; antea ex prora tantum et puppi pugnabatur. rostra addidit Pis<a>eus Tyrreni, ancoram Eupalamus, eandem bidentem Anacharsis, harpagones et manus Pericles Atheniensis, adminicula gubernandi Tiphys. classe princeps depugnavit Minos.
7.206-209

Vessel	Inventor	Authority
double-banked galley trireme	the Erythraeans Aminocles of Corinth	Damastes Thucydides

Vessel	Inventor	Authority
Quadrireme	the Carthaginians	Aristotle
Quinquereme	the Salaminians	Mnesigiton
galleys of six banks	the Syracusans	Xenagoras
up to ten banks	Alexander the Great	Mnesigiton
up to twelve	Ptolemy Soter	Philostephanus
up to fifteen	Demetrius son of Antigonus	Ditto
up to thirty	Ptolemy Philadelphus	Ditto
up to forty	Ptolemy Philopator surnamed Tryphon.	Ditto

The freight-ship was invented by Hippus of Tyre, **the cutter** by the Cyrenians, the skiff by the Phoenicians, the yacht by the Rhodians, the yawl by the Cyprians; the Phoenicians invented observing the stars in sailing, the town of Copae invented the oar, the city of Plataea the oar-blade, Icarus sails, Daedalus mast and yard, the Samians or Pericles of Athens the cavalry transport, the Thasians decked longships—previously the marines had fought from the bows and stern only. Pisaeus son of Tyrrenus added beaks, Eupalamus the anchor, Anacharsis the double-fluked anchor, Pericles of Athens grappling-irons and claws, Tiphys the tiller. Minos was the first who fought a battle with a fleet.
pp. 645-647

Latin: *Plinius maior (Caius Plinius Secundus), Naturalis historia*, eds L. Ian, C. Mayhoff. Teubner, Leipzig 1892-1909.

English: *Pliny. Natural History, Volume II: Books 3-7*, transl. H. Rackham. LCL 352 (1942).

The *Natural History* was most likely composed between AD 77-79, and it remained unfinished due to Pliny the Elder's death in the eruption of Mount Vesuvius. The material comes almost exclusively from Greek textbooks (Aristoteles, Theophrastus, Hippocratics), mostly directly, or from epitomes, and to a lesser extent from handbooks, as had been presumed earlier. In this passage referring to the origin of ships, Pliny attributes the origin of lemb to the Cyrenians with no reference as to the origin of this claim.

A-11 Sextus Iulius Frontinus (second half of 1st century AD)

Strategems

14. M. Cato, cum Ambraciam eo tempore quo sociae naues ab Aetolis oppugnabantur inprudens uno **lembo** appulisset, quamquam nihil se cum praesidii haberet, coepit signum uoce gestu que dare, quo uideretur subsequentis suorum nauis uocare; ea que adseueratione hostem terruit, tamquam plane adpropinquarent, qui quasi ex proximo citabantur. Aetoli, ne aduentu Romanae classis opprimerentur, reliquerunt oppugnationem.
2.7.14

14. Marcus Cato, having inadvertently landed with a single **galley** in Ambracia at a time when the allied fleet was blockaded by the Aetolians, although he had no troops with him, began nevertheless to make signals by voice and gesture, in order to give the impression that he was summoning the approaching ships of his own forces. By this earnestness he alarmed the enemy, just as though the troops, whom he pretended to be summoning from near at hand, were visibly approaching. The Aetolians, accordingly, fearing that they would be crushed by the arrival of the Roman fleet, abandoned the blockade.
p. 177

Latin: *Frontinus (Sextus Iulius Frontinus), Strategemata, libri I-III*, ed. R. I. Ireland. Teubner, Leipzig 1990.

English: *Frontinus. Stratagems. Aqueducts of Rome*, transl. C. E. Bennett, M. B. McElwain. LCL 174 (1925).

A-12 Marcus Cornelius Fronto (2nd century)

Ad Antoninum imp. 1.2

3. [...] quod nunc vides provenisse et, quamquam non semper ex summis opibus ad eloquentiam velificaris, tamen sipharis et remis te tenuisse iter, atque ut primum vela pandere necessitas impulit, omnes eloquentiae studiosos ut **lembos** et celoces facile praeteruehi.
1.2.3

3. [...] This you see has now come to pass, and although you have not always set every sail in pursuit of eloquence, yet you have held on your course with topsails and with oars, and as soon as ever necessity has forced you to spread all your canvas, you are easily distancing all devotees of eloquence like so many **pinnaces** and yachts.
p. 39

Latin and English: *Fronto. Correspondence, Volume II*, transl. C. R. Haines. LCL 113 (1920).

A-13 Aulus Gellius (2nd century)

Attic Nights:

13.1
Navium autem, quas reminisci tunc potuimus, appellationes hae sunt: gauli, corbitae, caudicae, longae, hippagines, cercuri, celoces vel, ut Graeci dicunt, κέλητες, **lembi**, oriae, **lenunculi**, actuariae, quas Graeci ἱστιοκώποθς vocant vel ἐπακτρίδας, prosumiae vel geseoretae vel oriolae, stlattae, scaphae, pontones, vetutiae moediae, phaseli, parones, myoparones, lintres, caupuli, camarae, placidae, cydarum, ratariae, catascopium.
10.25.5

The names of ships which I recalled at the time are these: merchant-ships, cargo-carriers, skiffs, warships, cavalry-transports, cutters, fast cruisers, or, as the Greeks call them, κέλητες, **barques**, smacks, **sailing-skiffs**, light galleys, which the Greeks call ἱστιοκόποι or ἐπακτρίδες, scouting-boats, galliots, tenders, flat-boats, vetutiae moediae, yachts, pinnaces, long-galliots, scullers' boats, caupuls, arks, fair-weather craft, pinks, lighters, spy-boats.
p. 287

13.2
Et praeterea pro alio quoque adverbio dicitur, id est "statim" factum, quod in his Vergili versibus existimatur, ubi obscure et insequenter particula ista posita est:
sic omnia fatis / In peius ruere ac retro sublapsa referri; / Non aliter quam qui adverso vix flumine **lembum** / Remigiis subigit, si brachia forte remisit, / Atque illum in praeceps prono rapit alveus amni.
10.29.4

Atque is said to have been used besides for another adverb also, namely statim, as is thought to be the case in these lines of Virgil,[193] where that particle is employed obscurely and irregularly:
Thus, by Fate's law, all speeds towards the worse, / And giving way, falls back; e'en as if one / Whose oars can barely force his **skiff** upstream / Should chance to slack his arms and cease to drive; / Then straightway (*atque*) down the flood he's / swept away.
p. 295

Latin and English: *Gellius. Attic Nights, Volume II: Books 6-13*, transl. J. C. Rolfe. LCL 200 (1927).

Gellius's only work is *Attic Nights*, comprising 20 books. *'The work [...] is aimed at his averagely educated contemporaries. The declared pedagogical goal is to convey correct conduct within the contemporary culture of conversation. [...]'* (H. Krasser, *NP*). In **A-13.1** Gellius mentions the lemb as a separate category of ships. The list itself does not seem to be ordered in any way (e.g. by size, function or anything else) so it lets us conclude nothing beyond the fact that it was considered a sort of ship. However, it is interesting that Gellius mentions both lembs and *lenunculi* as two separate and different ships despite the fact that the word *lenunculus* is a diminutive form of *lembus*. This indicates that in Gellius' times the terms *lembus* and *lenunclus* were so widely and differently used that their common etymological origin was lost. In **A-13.2**, Gellius, cites Virgil (**A-7**), where lemb is taken as an example of a boat with oars which one person can force upstream.

---

[193] Verg. *Georg.* 1.199-203 (**A-7**).

5. Written sources on lembs and Liburnians from the 4th c. BC to Late Antiquity

A-14 Gaius Iulius Solinus (mid 3rd century)

The wonders of the world:

§ 52 46 Indorum nemora in tam proceram sublimantur excelsitatem, ut transiaci ne sagittis quidem possint. 47 Pomaria ficus habent, quarum codices in orbem spatio sexaginta passuum extuberantur; ramorum umbrae ambitu bina stadia consumunt; foliorum latitudo formae Amazonicae peltae conparatur; pomum eximiae suauitatis. 48 Quae palustria sunt, harundinem creant ita crassam, ut fissis internodiis **lembi** uice uectitet nauigantes. E radicibus eius umor dulcis exprimitur ad melleam suauitatem. 49 Tylos Indiae insula est; ea fert palmas, oleam creat, uineis abundat. Terras omnes hoc miraculo sola uincit, quod quaecumque in ea arbos nascitur, numquam caret folio.
52.46-49

§ 52. 46 The trees of the Indians are raised to such a lofty height that they cannot shoot arrows over them. 47 They have orchards of figs. The trunks measure 60 paces around, and the shadows of the branches consume two stadia in every direction. The size of their leaves is comparable to that of Amazonian shields. Their fruit is of an excellent sweetness. 48 There are marshes which grow reeds of such thickness, that cloven in half, they can serve sailors as **boats**. From their roots a pleasant liquid is expressed, as sweet as honey. 49 Tylos is an Indian island. It bears date-palms, produces oil, and abounds in vines. It conquers all lands with this sole marvel: whatsoever tree grows there is never without leaves.

Latin: *C. Ivlii Solini Collectanea rervm memorabilivm*, ed. Th. Mommsen. Weidmann, Berlin 1845.

English: *Gaius Iulius Solinus, the Polyhistor*, transl. A. Apps, PhD diss. Macquarie University 2011.

Solinus is the author of the text entitled *De mirabilibus mundi* (*The wonders of the world*), also known as *Collectanea rerum memorabilium* (*Collection of Curiosities*), and *Polyhistor*. The text is a combination of geographical, historical, mythological, economical and fictional descriptions of the known world. Solinus relies heavily on Pliny and Pomponius Mela. In the above passage, Solinus claims that in India the reed can grow so big as to be cut in half and turned into a *lemb*. Regardless of the truthfulness of such a claim (the size of the reed), the implication of this is that Solinus understands the term lembus to be something of a canoe-like boat.

A-15 Ammianus Marcellinus (c. 330-400)

Res Gestae

A-15.1

nam sole orto magnitudine angusti gurgitis sed profundi a transitu arcebantur et, dum piscatoris quaerunt **lenunculos** uel innare temere contextis ratibus parant, effusae legiones, quae hiemabant tunc apud Siden, isdem impetu occurrere ueloci. et signis prope ripam locatis ad manus comminus conserendas denseta scutorum compage semet scientissime praestruebant, | ausos quoque aliquos fiducia nandi uel cauatis arborum truncis | amnem permeare latenter facillime trucidarunt.
14.2.10

10. For when the sun rose, they were prevented from crossing by the size of the stream, which was narrow but deep. And while they were hunting for **fishermen's boats** or preparing to cross on hastily woven hurdles, the legions that were then wintering at Side poured out and fell upon them in swift attack. And having set up their standards near the river-bank, the legions drew themselves up most skillfully for fighting hand to hand with a close formation of shields; and with perfect ease they slew some, who had even dared to cross the river secretly, trusting to swimming, or in hollowed out tree trunks.
p. 19.

A-15.2

nec enim gentem ullam bella cientem per se superauit aut uictam fortitudine suorum comperit ducum | uel addidit quaedam imperio aut usquam in necessitatibus summis | primus uel inter primos est uisus, sed ut pompam nimis extentam | rigentia que auro uexilla et pulchritudinem stipatorum ostenderet | agenti tranquillius populo haec uel simile quidquam uidere nec speranti umquam nec optanti: 3 ignorans fortasse quosdam ueterum principum in pace quidem lictoribus fuisse contentos, ubi uero proeliorum ardor nihil perpeti poterat segne, alium anhelante rabido flatu uentorum **lenunculo** se commisisse piscantis, alium ad Deciorum exempla | uouisse pro re publica spiritum, alium hostilia castra per semet ipsum cum militibus infimis explorasse, diuersos denique actibus inclaruisse magnificis, ut glorias suas posteritatis celebri memoriae commendarent.
16.10.2-3

2. For neither in person did he [Constantinus Augustus] vanquish any nation that made war upon him, nor learn of any conquered by the valour of his generals; nor did he add anything to his empire; nor at critical moments was he ever seen to be foremost, or among the foremost; but he desired to display an inordinately long procession, banners stiff with goldwork, and the splendour of his retinue, to a populace living in perfect peace and neither expecting nor desiring to see this or anything like it. 3. Perhaps he did not know that some of our ancient commanders in time of peace were satisfied with the attendance of their lictors; but when the heat of battle could tolerate no inaction, one, with the mad blast of the winds shrieking, entrusted himself to a **fisherman's skiff**,[194] another, after the example of the Decii, vowed his life for the commonwealth; a third in his own person together with common soldiers explored the enemy's camp; in short, various among them became famous through splendid deeds, so that they commended their glories to the frequent remembrance of posterity.
pp. 243-245

---

[194] Julius Caesar; see Luc. v. 533 ff.

5. Written sources on lembs and Liburnians from the 4th c. BC to Late Antiquity

A-15.3
17 statimque, ne alacritas intepesceret pugnatorum, impositi lintribus per abdita ducti uelites expediti occuparunt latibula Sarmatarum, quos repentinus fefellit aspectus gentiles **lembos** et nota remigia conspicantes.
17.13.17

17. And at once, for fear that the ardour of the warriors might cool, light-armed troops were put into skiffs, and taking the course which offered the greatest secrecy, came upon the lurking-places of the Sarmatians; and the enemy were deceived as they suddenly came in sight, seeing their native **boats** and the manner of rowing of their own country.
p. 391

A-15.4
3 atque ut lateret stationarios milites, fundum in Hiaspide, qui locus Tigridis fluentis alluitur, pretio non magno mercatur. hoc que commento cum nullus causam ueniendi ad extremas Romani limitis partes iam possessorem cum plurimis auderet exigere, per familiares fidos peritos que nandi occultis saepe colloquiis cum Tamsapore habitis, qui tractus omnes aduersos ducis potestate tunc tuebatur, et antea cognitus misso a Persicis castris auxilio uirorum pernicium **lembis** impositus cum omni penatium dulcedine nocte concubia transfretatur ex contraria specie Zopyri illius similis Babylonii proditoris.
18.5.3

3. And to the end that he might elude the sentinels, he bought at no great price a farm in Iaspis, a place washed by the waters of the Tigris. And since because of this device no one ventured to ask one who was now a landholder with many attendants his reason for coming to the utmost frontier of the Roman empire, through friends who were loyal and skilled in swimming he held many secret conferences with Tamsapor, then acting as governor of all the lands across the river, whom he already knew; and when active men had been sent to his aid from the Persian camp, he embarked in **fishing boats** and ferried over all his beloved household in the dead of night, like Zopyrus, that famous betrayer of Babylon, but with the opposite intention.
p. 429

A-15.5
2 cum que ad locum uenisset, unde nauigari posse didicit flumen, **lembis** escensis, quos opportune fors dederat plurimos, per alueum, quantum fieri potuit, ferebatur occulte [...].
21.9.2.

2. And when he [Julianus Augustus] came to the place where he learned that the river was navigable, embarking in **boats**, of which by a fortunate chance there was a good supply, he was carried down the channel of the river as secretly as possible [...].
p. 129

A-15.6
3 imperator ipse breuibus **lembis**, quos post exustam classem docuimus remansisse, cum paucis transuectus eadem nauigia ultro citro que discurrere statuit, dum omnes conueheremur, tandemque uniuersi praeter mersos ad ulteriores uenimus margines fauore superi numinis discrimine per difficiles casus extracti.
25.8.3

3. The emperor himself with a few others crossed in **the small boats**, which, as I have said, survived the burning of the fleet, and ordered the same craft to go back and forth, until we were all transported. At last all of us (except those who were drowned) reached the opposite bank, saved from danger by the favour of the supreme deity after many difficulties.
p. 539

A-15.7

18 Pars eorum, si agros uisuri processerunt longius aut alienis laboribus uenaturi, Alexandri Magni itinera se putant aequiperasse uel Caesaris, aut si a lacu Auerni **lembis** inuecti sunt pictis Puteolos, Duili certamen, maxime cum id uaporato audeant tempore. ubi si inter aurata flabella laciniis sericis insiderint muscae uel per foramen umbraculi pensilis radiolus irruperit solis, queruntur, quod non sunt apud Cimmerios nati.
28.4.18

18. Some of them, if they make a longish journey to visit their estates, or to hunt by the labours of others, think that they have equalled the marches of Alexander the Great or of Caesar; or if they have sailed in their **gaily-painted boats** from the Lake of Avernus to Puteoli, it is the adventure of the golden fleece, especially if they should dare it in the hot season. And if amid the gilded fans flies have lighted on the silken fringes, or through a rent in the hanging curtain a little ray of sun has broken in, they lament that they were not born in the land of the Cimmerians.
pp. 147-149

A-15.8

18 et stagnantibus ciuitatis residuis membris, quae tenduntur in planitiem molliorem, montes soli et, quidquid insularum celsius eminebat, a praesenti metu defendebatur et, ne multi inedia contabescerent undarum magnitudine nusquam progredi permittente, **lembis** et scaphis copia suggerebatur abunde ciborum. at uero ubi tempestas molliuit et flumen retinaculis ruptis redit ad solitum cursum, absterso metu nihil postea molestum exspectabatur.
29.6.17

18. While all the remaining quarters of the city, which extend down to a gentler level, were under water, the mountains alone, and such buildings as were especially high, were protected from present danger. And since the height of the waters prevented movement anywhere on foot, a supply of food was furnished in abundance by **boats** and skiffs, for fear that many people might starve to death. But, in fact, when the stormy weather moderated, and the river, which had broken its bonds, returned to its usual course, all fear was dispelled and no further trouble was looked for.
p. 293

A-15.9

5 contra Augustus escensis amnicis **lembis** saeptus ipse quoque multitudine castrensium ordinum tutius prope ripas accessit signorum fulgentium nitore conspicuus et immodestis gestibus murmure que barbarico tandem sedato post dicta et audita ultro citro que uersus amicitia media sacramenti fide firmatur.
30.3.5

5. On the other side the Augustus embarked on some **river-boats**, himself also hedged by a throng of military officers and conspicuous amid the brilliance of flashing standards, and cautiously approached the shore. Finally, the savages ceased their immoderate gesticulation and barbaric tumult, and after much had been said and heard on both sides, friendship was confirmed between them by the sanctity of an oath.
p. 317

Latin: *Ammianus Marcellinus, Rerum gestarum libri qui supersunt, Vol. 1: Libri XIV-XXV; Vol. 2: Libri XXVI-XXXI*, eds W. Seyfarth, L. Jacob-Karau, I. Ulmann. Teubner, Munich 1999-2011.

English: *Ammianus Marcellinus. History, Volume I: Books 14-19; Volume II: Books 20-26; Volume III: Books 27-31. Excerpta Valesiana*, transl. J. C. Rolfe. LCL 300, 315, 331 (1939-50).

5. Written sources on lembs and Liburnians from the 4th c. BC to Late Antiquity

As for our topic, Ammianus, interestingly, claims that lembs were *'native Sarmatian boats'* (**A-15.3**). Ammianus also mentions lembs as predominantly river boats (**A-15.1, A-15.5-6**) which could also be used as a boat for the shortest of distances, a sort of pleasure, *'gaily-painted'* boat for amusement (**A-15.7**).

A-16 Postumius (?) Rufius Festus Avienus (mid 4th century)

De ora maritima

siquis dehinc Ab insulis Oestrymnicis **lembum** audeat Vrgere in undas, axe qua Lycaonis Rigescit aethra, caespitem Ligur[g]um subit Cassum incolarum: namque Celtarum manu, Crebris que dudum praeliis uacuata sunt.	If anyone then dared to drive his **ship** from the Extreme West [Portugese?] Islands to the seas, where the air frozen with cold by Lycaon's chariot, he would come to the lands of the Ligurians, empty of inhabitants; because they are depopulated by the hands of the Celts and frequent battles that took place in the past.

128–133

Latin: *Rufi Festi Avieni Carmina*, ed. A. Holder. Wagner, Innsbruck 1887.

In the poem Avienus mentions lembs as traveling in the vicinity of the *'Oestryminian islands'*. The term *Oestriminis* refers to the 'Far West', probably Portugal or some other part of the Iberian Peninsula.

A-17 Decimus Magnus Ausonius (*c.* 310 - *c.* 394)

The Moselle

194 tota natant crispis iuga motibus et tremit absens / pampinus et uitreis uindemia turget in undis. / adnumerat uirides derisus nauita uites, / nauita caudiceo fluitans super aequora **lembo** / per medium, qua sese amni confundit imago / collis et umbrarum confinia conserit amnis. / 200 Haec quoque tam dulces celebrant spectacula pompas, / remipedes medio certant cum flumine **lembi** / et uarios ineunt flexus uirides que per oras / stringunt attonsis pubentia germina pratis.	194 Whole hills float on the shivering ripples: here quivers the far-off tendril of the vine, here in the glassy flood swells the full cluster. The deluded boatman tells o'er the green vines—the boatman whose **skiff** of bark floats on the watery floor out in mid-stream, where the pictured hill blends with the river and where the river joins with the edges of the shadows. 200 And when oared **skiffs** join in mimic battle in mid-stream, how pleasing is the pageant which this sight affords! They circle in and out and graze the sprouting blades of the cropped turf along the green banks. pp. 239-241

Latin: *Decimi Ausonii Burdigalensis Opuscula*, ed. S. Prete. Teubner, Leipzig 1978.

English: *Ausonius. Volume I: Books 1-17*, transl. H. G. Evelyn-White. LCL 96 (1919).

A-18 Claudius Mamertinus (fl. c. AD 362)

Gratiarum actio de consulatu suo Iuliano imp.

8.1. O facundia potens Graecia! omnium tuorum principum gesta in maius extollere sola potuisti, sola factorum glorias ad uerborum copiam tetendisti. 2. Tu nauem unam propter aurati uelleris furtum et uirginis raptum in caelum usque sublatam sideribus consecrasti. Tu puerum, inuentorem serendi, draconum alitum curru uolantem semina in terras sparsisse iactasti. 3. Quid tu, si ad scribendas celebrandas que res principis nostri animum adieceris, de Iuliani **lembis liburnisque** factura es? quae non modo nihil cuiquam adimunt neque urbes hospitas populant, sed ultro omnibus populis immunitates priuilegia pecunias largiuntur. 4. Qua dignitate describes classem per maximi fluminis tractum remis uentis que uolitantem, tum principem nostrum alta puppe sublimem non per cuiuscemodi agros frumenta spargentem, sed Romanis oppidis bonas spes libertatem diuitias diuidentem, tum ex parte altera in barbaricum solum terrorem bellicum trepidationes fugas formidines obserentem?

8.1 O Greece renowned for eloquence! You alone have been successful in exalting to the very limits of credibility the deeds of all your princes, you alone have matched in fluency of speech the glory of their exploits. 2. You it was who, for the theft of a golden fleece and the stealing away of a maiden, raised a certain ship to the heavens and consecrated it amongst the stars. You it was who published abroad how that youth, inventor of sowing? borne in a chariot drawn by winged dragons, scattered seeds broadcast over the land. 3. If you were to undertake to recount and celebrate our prince's career, what would you make of Julian's **fast cutters and brigantines**, since not only did they not pillage anything from a single person nor devastate any of the towns which gave them hospitality but issued ,what is more, to all the peoples, a largesse of exemptions, privileges and gifts of money? 4. With what majesty would you describe the fleet gliding down this mightiest of rivers, propelled by oars and the winds alike, and our prince presiding aloft on the poop, not scattering grain here and there across the fields, but distributing amongst the Roman towns great optimism, liberty and wealth, whilst on the other hand casting over the barbarian lands the fear of war, confusion, panic and terror?
pp. 19-20

Latin: *Panégyriques latins*, Vol. 3, ed. É. Galletier. Les Belles Lettres, Paris 1955.

English: *The Emperor Julian Panegyric and Polemic: Claudius Mamertinus, John Chrysostom, Ephrem the Syrian*, 2nd ed., transl. M. M. Morgan, ed. S. N. C. Lieu. TTH 2 (1989).

Not much is known about Claudius Mamertinus, other than the fact that a propagandistic speech is attributed to him. The speech praises the emperor Julian and contrasts his successes with the failures of the emperors before him, especially Julian's rival Constantius. For us, this text is of interest because it is the only known text that mentions both lembs and liburnians in one sentence, one term next to the other. The context of the usage of both terms is self-explanatory: they are meant to represent Julian's fleet that was present in Greece. From this it may be suggested that the fleet consisted primarily of lembs and liburnians or that lembs and liburnians were the most representative type of vessels in the fleet. However, against this suggestion it may be added that Mamertinus: *'made significant use of words which are seldom found outside Latin poetry'*.[195] This would in turn suggest that the usage of both terms here had more of a poetic ring to it, as relatively antiquated terms, rather than a precise technical meaning, describing certain types of vessels.

A-19 St Jerome (347-420)

Dialogues against the Pelagians

A-19.1
hoc et nos dicimus, posse hominem non peccare, si uelit, pro tempore, pro loco, pro imbecillitate corporea, quamdiu intentus est animus, quamdiu chorda nullo uitio laxatur in cithara. quod si paululum remiserit, quomodo qui aduerso flumine **lembum** trahit, si remiserit manus, statim retrolabitur et fluentibus aquis, quo non uult, ducitur; sic humana condicio, si paululum se remiserit, discit fragilitatem suam et multa se non posse cognoscit.
3.4

We also say that it is possible for man not to sin for a while, depending on the circumstances and the weakness of the body, so long as his spirit is attentive, so long as the strings of the lyre are not slackened by any defect. But, if he relaxes for even a moment, even as a man rowing his **boat** upstream immediately slips back if he relaxes his hands and is carried by the current of the river in a direction he does not want to go, so also the sate of man, if it relaxes for even a moment, get to know its weakness and realises that it cannot do much of itself.
p. 352.

Latin: *PL*, ser. 1 Vol. 23. Vrayet, Paris 1845.

English: *Saint Jerome, Dogmatic and Polemical Works*, transl. J. N. Hritzu [The Fathers of the Church: a new translation 53]. Catholic University of America Press, Washington DC 1965.

Life of Hilarion

A-19.2
mirabatur omnis ciuitas et magnitudo signi Salonis quoque percrebuerat. quod intelligens senex in breui **lembo** clam nocte fugit, et inuenta post biduum oneraria naui perrexit cyprum. cum que inter maleam et cytheram piratae, derelicta classe in littore, quae non antemna sed conto regitur, duobus haud paruis myoparonibus occurrissent, denuo hinc inde fluctus uerrente remige, omnes qui in naui erant trepidare, flere, discurrere, praeparare contos, et quasi non sufficeret unus nuntius, certatim seni piratas adesse dicebant. quos ille procul intuens subrisit et conuersus ad discipulos dixit: "modicae", inquit, "fidei quare trepidatis?

The whole city was astounded and new of this extraordinary miracle spread as far as Salona. When the old man learned about this, he stole away by night in a little **boat**; after two das he found a cargo ship and went on to reach Cyprus. Between Malea and Cythera the pirates who had abandoned their fleet (propelled not by sails but by oars) on the shore, attacked them in two large vessels. As the waves beat them on all sides, all the oarsmen whowere on the ship were terrified: they wept and rushed about and prepared the poles. As if it were not sufficient for one person to report it, they all crowded round to give the old man the news that pirates were

---
[195] Lieu 1989: 7.

numquid plures hi quam pharaonis exercitus? tamen omnes deo uolente submersi sunt". loquebatur his et nihilominus spumantibus rostris hostiles carinae imminebant iactu tantum lapidis medio. stetit ergo in prora nauis et porrecta contra uenientes manu: "hucusque", ait, "uenisse sufficiat". o mira rerum fides: statim resiluere nauiculae et impellentibus contra remis ad puppim impetus redit. mirabantur piratae post tergum se redire nolentes toto que corporis nisu, ut ad nauigium peruenirent, laborantes uelocius multo quam uenerant ad littus ferebantur.
§41

coming. Hilarion looked at the pirates in the distance, smiled, and turning to the disciples, said, 'You of little faith, why are you afraid? I suppose these are more numerous than Pharaoh's army? And yet they were all drowned by the will of God.' Even as he spoke the enemy ships were drawing very close, their prows foaming, now only a stone's throw away. So Hilarion went and stood in the prow of the ship, and holding his hand up to them as they approached, he said, 'No need to come any further.' It is hard to believe but at once the ships withdrew and though the oars impelled them in the opposite direction, their attack moved back. The pirates, still unwilling to retreat, were amazed. Using all their physical strength they tried to reach the ship but were driven back to the shore much faster than they had come.
p. 112

Latin: *PL,* ser. 1 Vol. 22. Vrayet, Paris 1845.

English: *Early Christian Lives,* transl. and ed. C. White. Penguin Classics 1998.

In **A-19.1** we have the same trope from Vergil (see **A-7**): a single man can row upstream, but as long as he stops rowing, the stream will drag him downstream again. Since the word for the boat used by Vergil in this context was the *lembus,* it has become a standard word used in this metaphor.

A-20 Aurelius Prudentius Clemens (348/349 - after 405)

Crowns of Martyrdom

ecquis virorum strenue
cumbam peritus pellere remo,
rudente et carbaso,
secare qui pontum queas,
rapias palustri e caespite
corpus, quod intactum iacet,
levique vectum **lembulo**
amplum per aequor auferas?
5.449–456

Some man of you who knows how to drive a boat briskly on with oar and rope and canvas and can plough the sea, take the body from the swampy grass where it lies untouched, and in a **swift wherry** carry it away over the wide waters!
pp. 195-197

Latin and English: *Prudentius, Against Symmachus 2. Crowns of Martyrdom. Scenes from History. Epilogue,* transl. H. J. Thomson. LCL 398 (1953).

5. Written sources on lembs and Liburnians from the 4th c. BC to Late Antiquity

The poet, probably for the sake of fitting the words into the metric scheme, used a somehow pleonastic *levis lembulus*, 'light, little lemb' phrase. From the context of the poem, it is used synonymously with *cumba*. The *cumba* or *cymba*, etymologically stemming from κύμβη, was a vessel considered Phoenician and generally understood to be a small boat, a skiff. A typical usage of the boat was by Charon to ferry the dead across the rivers of Hades.

A-21 Historia Augusta (*c.* 400)

The Deified Aurelian

XXXVIII [...] Aurelianus Augustus Ulpio patri. quasi fatale quiddam mihi sit, ut omnia bella quaecumque gessero, omnes motus ingravescant, ita etiam seditio intramurana bellum mihi gravissimum peperit. monetarii auctore Felicissimo, ultimo servorum, cui procurationem fisci mandaveram, rebelles spiritus extulerunt. hi compressi sunt septem milibus **lembariorum** et ripariensium et castrianorum et Daciscorum interemptis. unde apparet nullam mihi a dis inmortalibus datam sine difficultate victoriam.
38.3-4

From Aurelian Augustus to Ulpius his father. Just as though it were ordained for me by Fate that all the wars that I wage, and all commotions only become more difficult, so also a revolt within the city has stirred up for me a most grievous struggle. For under the leadership of Felicissimus, the lowest of all my slaves, to whom I had committed the care of the privy-purse, the mint-workers have shown the spirit of rebellion. They have indeed been crushed, but with the loss of seven thousand men, **boatmen**, bank-troops, camp-troops and Dacians. Hence it is clear that the immortal gods have granted me no victory without some hardship.
p. 271

Latin and English: *Historia Augusta, Volume III: The Two Valerians. The Two Gallieni. The Thirty Pretenders. The Deified Claudius. The Deified Aurelian. Tacitus. Probus. Firmus, Saturninus, Proculus and Bonosus. Carus, Carinus and Numerian*, transl. D. Magie. LCL 263 (1932).

For our purposes it is interesting that this is the only surviving source with the word *lembarius*, the boatman of lemb. The context is an army along the river Danube.

A-22 Claudius Claudianus (about 400)

A-22.1 On Stilicho's Consulship

[...] nec spicula supplex / iam torquet Garamans; repetunt deserta fugaces / Autololes; pauidus proiecit missile Mazax. / cornipedem Maurus nequiquam hortatur anhelum; / praedonem **lembo** profugum uentis que repulsum / suscepit merito fatalis Thabraca portu / expertum quod nulla tuis elementa paterent / hostibus, et laetae passurum iurgia plebis / fracturum que reos humili sub iudice uultus.
21.1.352–362

[...] the Garamantian hurls not his spears but begs for mercy, the swift-footed Autololes fly to the desert, the terror-stricken Mazacian flings away his arms, in vain the Moor urges on his flagging steed. The brigand flees **in a small boat** and driven back by the winds met with his just fate in the harbour of Tabraca, discovering that no element offered refuge, Stilicho, to thine enemies.
p. 391

Latin and English: *Claudian. Panegyric on Probinus and Olybrius. Against Rufinus 1 and 2. War against Gildo. Against Eutropius 1 and 2. Fescennine Verses on the Marriage of Honorius. Epithalamium of Honorius and Maria. Panegyrics on the Third and Fourth Consulships of Honorius. Panegyric on the Consulship of Manlius. On Stilicho's Consulship 1*, transl. M. Platnauer. LCL 135 (1922).

A-22.2 Panegyric on the consulship of Fl. Manlius Theodorus

mobile ponderibus descendat pegma reductis / inque chori speciem parcentes ardua flammas / scaena rotet: uarios effingat Mulciber orbes / per tabulas inpune uagus pictae que citato / ludant igne trabes et non permissa morari / fida per innocuas errent incendia turres. / lasciui subito confligant aequore **lembi** / stagnaque remigibus spument inmissa canoris.
17.325–335

Let the counterweights be removed and the mobile crane descend, lowering on to the lofty stage men who, wheeling chorus-wise, scatter flames; let Vulcan forge balls of fire to roll innocuously across the boards, let the flames appear to play about the sham beams of the scenery and a tame conflagration, never allowed to rest, wander among the untouched towers. Let **ships** meet in mimic warfare on an improvised ocean and the flooded waters be lashed to foam by singing oarsmen.
p. 363

Latin and English: *Claudian. On Stilicho's Consulship 2-3. Panegyric on the Sixth Consulship of Honorius. The Gothic War. Shorter Poems. Rape of Proserpina*, transl. M. Platnauer. LCL 136 (1922).

A-22.2 offers a new usage of the term *lembus*. Here we read that lembs were used in the theatre to enact naval battles.

5. Written sources on lembs and Liburnians from the 4th c. BC to Late Antiquity

A-23 Gaius Sollius Apollinaris Sidonius (c. 430 - 489)

Letters

A-23.1

[...] a Zephyro plebeius et tumultuarius frutex frequenterque **lemborum** superlabentum ponderibus inflexus; hunc circa lubrici scirporum cirri plicantur [...]
2.2.17

[...] On the west is a vulgar and disorderly growth of weeds, which is often bent under the weight of **the yachts** that speed over it; round this growth slippery tufts of bulrushes wrap themselves; [...].
p. 433

A-23.2

SIDONIVS AGRICOLAE SVO SALVTEM
1. Misisti tu quidem **lembum** mobilem solidum lecti capacem iamque cum piscibus; tum praeterea gubernatorem longe peritum, remiges etiam robustos expeditosque, qui scilicet ea rapiditate praetervolant amnis adversi terga qua defluit. sed dabis veniam quod invitanti tibi in piscationem comes venire dissimulo; namque me multo decumbentibus nostris validiora maeroris retia tenent, quae sunt amicis quaeque et externis indolescenda. unde te quoque puto, si rite germano moveris adfectu, quo temporis puncto paginam hanc sumpseris, de reditu potius cogitaturum.
2.12.1

SIDONIUS TO HIS DEAR AGRICOLA, GREETING
1. You have sent me a **boat** which is swift and substantial, big enough to hold a couch and a load of fish too; also a boatman of wide experience and oarsmen so strong and brisk that they fly over the surface of the water as swiftly up-stream as down-stream. But you must excuse me for not availing myself of your invitation to join you in a fishing excursion; for with illness in our family I am held here by a much stronger kind of net, a net of affliction, which must needs bring grief to friends and strangers alike: so I think that if you feel a genuine brotherly affection for me, as soon as ever you take up this sheet you will think rather of returning here.
p. 471

A-23.3 *Panegyric on Anthemius*

quin et Aremoricus piratam Saxona tractus sperabat, cui pelle salum sulcare Britannum ludus et assuto glaucum mare findere **lembo**.
369-371

The Aremorican region too expected the Saxon pirate, who deems it but sport to furrow the British waters with hides, cleaving the blue sea in **a stitched boat**.
p. 151

Latin and English: *Sidonius. Poems. Letters: Books 1-2*, transl. W. B. Anderson. LCL 296 (1936).

In the passages above, there are two points to be mentioned. First, the lemb is described as being a fast and swift boat of such a size as to carry a couch (*lectum*), a load of fish, a boatman and a number of oarsmen (**A-23.2**). It would roughly correspond to our understanding of a 'yacht', as translated in **A-23.1**. Second, in **A-23.3** is it mentioned as '*a stitched boat*' (*lembus assutus*). This is the only time that the *lembus* is described as a stitched boat.

A-24 Fabius Planciades Fulgentius (*c.* 500)

Expositio sermonum antiquorum ad grammaticum Calcidium

30. [Quid sit lembum.] **Lembum** est genus nauicellae uelocissimae, quos dromones dicimus, sicut Uirgilius ait: 'Quam qui auerso uix flumine **lembum** remigiis subigit'	30. What a *lembus* is. A ***lembus*** is a kind of very fast small boat, what we call cutters (dromones), as Vergil says: 'Who with his oars scarcely moved his **little craft** against the current'.

Latin: *Fabii Planciadis Fulgentii V. C. Opera*, ed. R. Helm. Teubner, Leipzig 1848.

English: *Fulgentius the Mythographer*, transl. L. G. Whitbread. Ohio State University Press, Columbus 1971.

From this text we can understand that the term *lembus* was antiquated in Fulgentius' times, for he inserted it in his *Expositio sermonum antiquorum*. With this information in mind, we may also approach the previous quotes of Sidonius (**A-22**), since Sidonius was almost a contemporary of Fulgentius. Fulgentius connects lemb with a new type of warship, the *dromon*, showing that the term liburna was already antiquated in his time (see also **B-VIII** below).

A-25 Isidorus Hispalensis (*c.* 560-636)

Etymologiarum siue Originum libri XX

Lembus nauicula breuis, qui alia appellatione dicitur et cumba et caupulus, sicut et lintris, id est carabus, quem in Pado paludibusque utuntur. 19.1.25	**Lemb** is a sort of boat, which is otherwise called *cumba* and *caupulus*, as well as *lintris*, that is, *carabus*, which is used in the river Po and in marshes.

Latin: *Isidori Hispalensis Episcopi Etymologiarum Sive Originum Libri XX*, ed. W. M. Lindsay. Clarendon Press, Oxford 1911.

In this passage we find the same definition of the lemb as a short boat that we have already seen in Nonius's transmission of Accius (**A-2**). Moreover, Isidorus offers several synonyms for it: *cumba, caupulus, lintris* (which as an alternative for *linter* is found in Nonius and Isidorus) and *carabus*. Two of them – *lintris* and *caupulus* – we have also found in Aulus Gellius (**A-13**), but as names of different types of vessel. It is noteworthy that Isidorus understands the *lembus* as a type of vessel which has local equivalents (versions) in northern Italy – the river Po and the marshes in its vicinity.

5. Written sources on lembs and Liburnians from the 4th c. BC to Late Antiquity

5.3. Liburnian

5.3.1. Ancient Greek sources

B-I Hecataeus of Miletus (c. 560-480 BC)

Fragmenta

Λιβυρνοί· ἔθνος προσεχὲς τῶι ἐνδοτέρωι μέρει τοῦ Ἀδριατικοῦ κόλπου. Ἑκαταῖος Εὐρώπηι. τὸ θηλυκὸν Λιβυρνίς. καὶ Λιβυρναῖοι. ὠνομάσθησαν δὲ ἀπό τινος Λιβυρνοῦ † Ἀττικοῦ. εὕρηται τὰ Λιβυρνικὰ σκάφη. καὶ Λιβυρνικὴ μανδύη εἶδος ἐσθῆτος.
BNJ 1 F93
(=Steph. Byz., *Ethnica* 415.7-10, s.v. Λιβυρνοί)

Liburnoi. A people next to the most interior part of the Adriatic Gulf. Hekataios mentions them in his *Europe*. The feminine form of the adjective is *Liburnis*. There is also the variant form *Liburnaioi*. They are named after a certain Liburnos from Attica(?). The adjective Liburnian is found in reference to ships. And a Liburnian cloak is an article of clothing.

Greek and English: F. Pownall, 'Hekataios of Miletos (1)', *BNJ*.

Hecataeus is one of the earliest Greek prose authors of whom we have many fragments preserved (c. 370), and the earliest author to mention the Liburni. The above fragment comes from Stephanus of Byzantium's geographical dictionary entitled *Ethnica* (Ἐθνικά), written around a thousand years after Hecataeus' time. Hecataeus is not directly referred to in this passage as directly mentioning the Liburnian vessels τὰ Λιβυρνικὰ σκάφη, and it seems to be Stephanus' own observation (see p. 176 below). However, the reference to: '*Libyrnos is otherwise unknown, and the text is thought to be corrupt because the reference to Attica makes little sense in this context.*' (Pownall, BNJ).[196] The reference to the '*Liburnian cloak*' (Λιβυρνικὴ μανδύη) may be a reference to Aeschylus' fragment 711f, referred again by Stephanus: Λιβυρνικῆς μίμημα μανδύης χιτών ('*The garment is an imitation of a Liburnian woollen cloak*').

B-II Philoxenus Alexandrinus (1st century BC)

Περὶ μονοσυλλάβων ῥημάτων:

ΓΑΥΛΟΣ· λέγεται γὰρ καὶ τριήρης γαῦλος διὰ τὸ πλεῖστα δέχεσθαι, ὥς φησιν Ἡρόδοτος ἐν τῇ τρίτῃ τῶν Ἱστοριῶν. ἔστι δὲ εἶδος πλοίου λῃστρικοῦ, ἥτις καὶ Λίβυρνος καλεῖται.
Frag. 79b
(=*Etymologicum Magnum*, 222.25-32)

GAULOS [a galleon, round-built Phoenician merchant vessel] is also called a trireme *gaulos*, because it can carry very many things, as Herodotus says in the third book of his *Histories* [3.136,1]. And it is a sort of a pirate vessel, which is also called *liburnos*.

Greek: *Die Fragmente des Grammatikers Philoxenos*, ed. C. Theodoridis [Sammlung griechischer und lateinischer Grammatiker 2]. De Gruyter, Berlin 1976.

---

[196] The reference to Liburnus could have from a Liburnian myth of origins recorded by Hecataeus, Džino 2017: 71.

In this passage Philoxenus calls the Liburnian ship in its masculine form, *liburnos*, rather than most common feminine names. The lexicographical fragment is not completely clear, and it can mean two things, depending on the interpretation of the particle δὲ. First, if the particle is understood as narrative transition, liburnians (*libyrnoi*) are another name for *gaulos* in context of piracy: when the name *gaulos* refers to a pirate ship it can also be called a *libyrnos*, probably suggesting that a liburnian was of a similar shape as *gaulos*; therefore *gaulos* is a generic name of which *libyrnos* is a species. Second, if the particle is understood with a conjunctive meaning, then *gaulos*, also called triereme *gaulos*, is the same as *libyrnos*, with the difference that the term *libyrnos* is usually associated with piracy. *Gaulos* was a type of Phoenician merchant ship with a beamy and rounded hull.[197]

B-III Plutarch of Chaeronea (*c.* 45 - before 125)

Life of Antonius

B-III.1

Ἐκείνη δὲ γνωρίσασα σημεῖον ἀπὸ τῆς νεὼς ἀνέσχε· καὶ προσενεχθεὶς οὕτω καὶ ἀναληφθεὶς ἐκείνην μὲν οὔτε εἶδεν οὔτε ὤφθη, παρελθὼν δὲ μόνος εἰς πρῷραν ἐφ' ἑαυτοῦ καθῆστο σιωπῇ, ταῖς χερσὶν ἀμφοτέραις ἐχόμενος τῆς κεφαλῆς. ἐν τούτῳ δὲ **λιβυρνίδες** ὤφθησαν διώκουσαι παρὰ Καίσαρος· ὁ δὲ ἀντίπρωρον ἐπιστρέφειν τὴν ναῦν κελεύσας τὰς μὲν ἄλλας ἀνέστειλεν, Εὐρυκλῆς δ' ὁ Λάκων ἐνέκειτο σοβαρῶς, λόγχην τινὰ κραδαίνων ἀπὸ τοῦ καταστρώματος ὡς ἀφήσων ἐπ' αὐτόν. ἐπιστάντος δὲ τῇ πρῴρᾳ τοῦ Ἀντωνίου καί "Τίς οὗτος," εἰπόντος, "ὁ διώκων Ἀντώνιον;" "Ἐγώ," εἶπεν, "Εὐρυκλῆς ὁ Λαχάρους, τῇ Καίσαρος τύχῃ τὸν τοῦ πατρὸς ἐκδικῶν θάνατον." ὁ δὲ Λαχάρης ὑπ' Ἀντωνίου λῃστείας αἰτίᾳ περιπεσὼν ἐπελεκίσθη. πλὴν οὐκ ἐνέβαλεν ὁ Εὐρυκλῆς εἰς τὴν Ἀντωνίου ναῦν, ἀλλὰ τὴν ἑτέραν τῶν ναυαρχίδων (δύο γὰρ ἦσαν) τῷ χαλκώματι πατάξας περιερρόμβησε, καὶ ταύτην τε πλαγίαν περιπεσοῦσαν εἷλε καὶ τῶν ἄλλων μίαν, ἐν ᾗ πολυτελεῖς σκευαὶ τῶν περὶ δίαιταν ἦσαν.
67.1-4

LXVII. Cleopatra recognized him and raised a signal on her ship; so Antony came up and was taken on board, but he neither saw her nor was seen by her. Instead, he went forward alone to the prow and sat down by himself in silence, holding his head in both hands. At this point, **Liburnian ships** were seen pursuing them from Caesar's fleet; but Antony ordered the ship's prow turned to face them, and so kept them all off, except the ship of Eurycles the Laconian, who attacked vigorously, and brandished a spear on the deck as though he would cast it at Antony. And when Antony, standing at the prow, asked, "Who is this that pursues Antony?" the answer was, "I am Eurycles the son of Lachares, whom the fortune of Caesar enables to avenge the death of his father." Now, Lachares had been beheaded by Antony because he was involved in a charge of robbery. However, Eurycles did not hit Antony's ship, but smote the other admiral's ship (for there were two of them) with his bronze beak and whirled her round, and as she swung round sideways he captured her, and one of the other ships also, which contained costly equipment for household use.
pp. 288-289

Greek: *Plutarchi vitae parallelae* 3.1, ed. K. Ziegler, 2nd ed. Teubner, Leipzig 1971.
English: *Plutarch. Lives, Volume IX: Demetrius and Antony. Pyrrhus and Gaius Marius*, transl. B. Perrin. LCL 101 (1920).

---

[197] Casson 1971: 66, see also below p. 176-77.

5. Written sources on lembs and Liburnians from the 4th c. BC to Late Antiquity

Life of Pompeius

B-III.2

Ἐν δὲ τῷ χρόνῳ τούτῳ μεγάλη συνέστη Πομπηΐῳ δύναμις, ἡ μὲν ναυτικὴ καὶ παντελῶς ἀνανταγώνιστος (ἦσαν γὰρ αἱ μάχιμοι πεντακόσιαι, **λιβυρνίδων** δὲ καὶ κατασκόπων ὑπερβάλλων ἀριθμός), ἱππεῖς δέ, Ῥωμαίων καὶ Ἰταλῶν τὸ ἀνθοῦν, ἑπτακισχίλιοι, γένεσι καὶ πλούτῳ καὶ φρονήμασι διαφέροντες· τὴν δὲ πεζὴν σύμμικτον οὖσαν καὶ μελέτης δεομένην ἐγύμναζεν ἐν Βεροίᾳ καθήμενος οὐκ ἀργός, ἀλλ᾽ ὥσπερ ἀκμάζοντι χρώμενος αὐτῷ πρὸς τὰ γυμνάσια.
64.1-2

LXIV. In the meantime a great force was gathered by Pompey. His navy was simply irresistible, since he had five hundred ships of war, while the number of his **light galleys** and fast cruisers was immense; his cavalry numbered seven thousand, the flower of Rome and Italy, preeminent in lineage, wealth, and courage; and his infantry, which was a mixed multitude and in need of training, he exercised at Beroea, not sitting idly by, but taking part in their exercises himself, as if he had been in the flower of his age.
pp. 280-281

Greek and English: *Plutarch. Lives, Volume V: Agesilaus and Pompey. Pelopidas and Marcellus*, transl. B. Perrin. LCL 87 (1917).

Life of Cato Minor

B-III.3

ἔνθα δὴ καὶ μάλιστα τῆς γνώμης κατάφωρος ἔδοξε γεγονέναι Πομπήϊος. ὥρμησε μὲν γὰρ ἐγχειρίσαι τῷ Κάτωνι τὴν τῶν νεῶν ἡγεμονίαν· ἦσαν δὲ πεντακοσίων μὲν οὐκ ἐλάττους αἱ μάχιμοι, **λιβυρνικὰ** δὲ καὶ κατασκοπικὰ καὶ ἄφρακτα παμπληθῆ· ταχὺ δὲ ἐννοήσας ἢ διδαχθεὶς ὑπὸ τῶν φίλων ὡς ἕν ἐστι κεφάλαιον Κάτωνι πάσης πολιτείας ἐλευθερῶσαι τὴν πατρίδα, κἂν γένηται κύριος τηλικαύτης δυνάμεως, ἧς ἂν ἡμέρας καταγωνίσωνται Καίσαρα, τῆς αὐτῆς ἐκείνης ἀξιώσει καὶ Πομπήϊον τὰ ὅπλα καταθέσθαι καὶ τοῖς νόμοις ἕπεσθαι, μετέγνω, καίπερ ἤδη διειλεγμένος αὐτῷ, καὶ Βύβλον ἀπέδειξε ναύαρχον.
54.4-6

Pompey was thought to have made his opinion of Cato manifest. For he determined to put the command of his fleet into the hands of Cato, and there were no less than five hundred fighting ships, besides **Liburnian craft**, look-out ships, and open boats in great numbers. But he soon perceived, or was shown by his friends, that the one chief object of Cato's public services was the liberty of his country, and that if he should be made master of so large a force, the very day of Caesar's defeat would find Cato demanding that Pompey also lay down his arms and obey the laws. Pompey therefore changed his mind, although he had already conferred with Cato about the matter, and appointed Bibulus admiral.
pp. 366-367

Greek: *Plutarchi vitae parallelae* 2.1, ed. K. Ziegler 2nd ed. Teubner, Leipzig 1964.

English: Plutarch. *Lives, Volume VIII: Sertorius and Eumenes. Phocion and Cato the Younger*, transl. B. Perrin. LCL 100 (1919).

The first of these three passages (**B-III.1**) describes an event from the Battle of Actium in 31 BC, while **B-III.2** and **B-III.3** describe Pompey's fleet from the year 49–48 BC. From the description we learn that:

- Liburnians were a part of Caesar's fleet.[198]
- Liburnians had a deck, from which Eurycles threw a spear at Antony.
- Liburnians had or could have a bronze beak (ram). Eurycles hit one of the admiral's ships with it and swung it around.

Plutarch mentions that, on the one hand, there are '*ships of war*' (μάχιμοι) and, on the other hand (δὲ) liburnians and scout-ships (or fast cruisers, κατάσκοποι), which implies that liburnians were not considered to be ships of war properly speaking, and were in the same group as scout-ships, but also different from them. Plutarch also repeats this distinction between ships of war, or, as it is differently translated by the same translator, '*fighting ships*' (μάχιμοι), on the one hand, and, on the other hand, liburnians, scout-ships ('*look-out ships*', κατασκοπικὰ) and '*open boats*' (unprotected, unarmoured ships, ἄφρακτα). Strictly speaking, this would imply that liburnians could have been armoured or protected, since they are distinguished from the unarmed or open boats by Plutarch.[199]

B-IV Appian of Alexandria (*c.* 90-160)

Civil wars

B-IV.1

χρόνῳ δὲ τῆς τε χώρας καὶ πόλεως κατασχεῖν Βρίγας ἐκ Φρυγῶν ἐπανελθόντας καὶ Ταυλαντίους ἐπ' ἐκείνοις, Ἰλλυρικὸν ἔθνος, ἐπὶ δὲ τοῖς Ταυλαντίοις ἕτερον γένος Ἰλλυριῶν Λιβυρνούς, οἳ τὰ περίοικα νηυσὶ ταχείαις ἐλῄζοντο· **καὶ Λιβυρνίδας ἐντεῦθεν ἡγοῦνται Ῥωμαῖοι τὰς ναῦς τὰς ταχείας, ὧν ἄρα πρώτων ἐς πεῖραν ἦλθον.** οἱ δ' ἐκ τῶν Λιβυρνῶν ἐξελαθέντες ἀπὸ τοῦ Δυρραχίου Κερκυραίους ἐπαγόμενοι θαλασσοκρατοῦντας ἐξέβαλον τοὺς Λιβυρνούς· καὶ αὐτοῖς οἱ Κερκυραῖοι σφετέρους ἐγκατέμιξαν οἰκήτορας, ὅθεν Ἑλληνικὸν εἶναι δοκεῖ τὸ ἐπίνειον. τὴν δ' ἐπίκλησιν ὡς οὐκ αἴσιον ἐναλλάξαντες οἱ Κερκυραῖοι καὶ τήνδε ἀπὸ τῆς ἄνω πόλεως Ἐπίδαμνον ἐκάλουν, καὶ Θουκυδίδης οὕτως ὠνόμαζεν· ἐκνικᾷ δ' ὅμως τὸ ὄνομα, καὶ Δυρράχιον κληίζεται.
2.6.39

At a later period the Briges, returning from Phrygia, took possession of the city and the surrounding country. They were supplanted by the Taulantii, an Illyrian tribe, who were displaced in their turn by the Liburnians, another Illyrian tribe, who were in the habit of making piratical expeditions against their neighbours with very swift ships. **Hence the Romans call swift ships Liburnians because these were the first ones they came in conflict with.** The people who had been expelled from Dyrrachium by the Liburnians procured the aid of the Corcyreans, who then ruled the sea, and drove out the Liburnians. The Coreyreans mingled their own colonists with them and thus it came to be considered a Greek port; but the Corcyreans changed its name, because they considered it unpropitious, and called it Epidamnus from the town just above it, and Thucydides gives it that name also. Nevertheless, the former name prevailed finally and it is now called Dyrrachium.
pp. 298-301

---

[198] See also **B-2.2** (Caesar) and p. 177 below.
[199] This point is accentuated by Morrison 1996: 259.

5. Written sources on lembs and Liburnians from the 4th c. BC to Late Antiquity

B-IV.2

Ἑτοίμου δὲ τοῦ στόλου γενομένου, αὖθις ὁ Καῖσαρ ἀνήγετο καὶ ἐς Ἱππώνειον παραπλεύσας δύο μὲν τέλη πεζῶν Μεσσάλαν ἔχοντα περᾶν ἐκέλευσεν ἐς Σικελίαν ἐπὶ τὸ Λεπίδου στρατόπεδον καὶ σταθμεύειν ἐς τὸν κόλπον διελθόντα τὸν εὐθὺ Ταυρομενίου, τρία δ' ἔπεμπεν ἐπὶ Στυλίδα καὶ πορθμὸν ἄκρον ἐφεδρεύειν τοῖς ἐσομένοις·Ταῦρον δ' ἐς τὸ Σκυλάκιον ὄρος, ὃ πέραν ἐστὶ Ταυρομενίου, περιπλεῖν ἐκ Τάραντος ἐκέλευε. καὶ ὁ μὲν περιέπλει διεσκευασμένος ἐς μάχην ὁμοῦ καὶ εἰρεσίαν· καὶ τὰ πεζὰ αὐτῷ παρωμάρτει, καὶ εἰρεσίαν· καὶ τὰ πεζὰ αὐτῷ παρωμάρτει, προερευνώντων τήν τε γῆν ἱππέων καὶ **λιβυρνίδων** τὴν θάλασσαν. καὶ ὁ Καῖσαρ ὧδε ἔχοντι ἐξ Ἱππωνείου ἐπιδραμὼν ἐπιφαίνεται κατὰ τὸ Σκυλάκιον, καὶ τὴν εὐταξίαν ἀποδεξάμενος ἐπανῆλθεν ἐς τὸ Ἱππώνειον. ὁ δὲ Πομπήιος, ὥς μοι προείρηται, τάς τε ἐς τὴν νῆσον ἀποβάσεις ἐφύλασσεν ἁπάσας καὶ τὰς ναῦς ἐν Μεσσήνῃ συνεῖχεν ὡς βοηθήσων, ὅποι δεήσειεν.
5.11.103

When the fleet was ready Octavian set sail again. He landed at Vibo and ordered Messala, who had two legions of infantry, to cross over to Sicily, join the army of Lepidus, pass through to the bay in front of Tauromenium, and station himself there, and three legions he sent to Stylis and the extremity of the straits, to await events. He ordered Taurus to sail round from Tarentum to Mount Scylacium, which is opposite Tauromenium. Taurus did so, having prepared himself for fighting as well as for rowing. His infantry kept even pace with him, cavalry reconnoitering by land and **liburnians** by sea. While he was making this movement Octavian, who had advanced from Vibo, made his appearance near Scylacium, and, after giving his approval to the good order of the forces, returned to Vibo. Pompeius, as I have already said, guarded all the landing places on the island and retained his fleet at Messana, in order to send aid where it might be needed.
pp. 550-551

B-IV.3

111. Καίσαρα δ' ἐν τοῖς ὑπηρετικοῖς ἐς πολὺ τῆς νυκτὸς ἀνακωχεύοντα καὶ βουλευόμενον, εἴτε ἐς Κορνιφίκιον ἐπανέλθοι διὰ μέσων τοσῶνδε ναυαγίων εἴτε ἐς Μεσσάλαν διαφύγοι, θεὸς ἐς τὸν Ἀβάλαν λιμένα παρήνεγκε μεθ' ἑνὸς ὁπλοφόρου, χωρὶς φίλων τε καὶ ὑπασπιστῶν καὶ θεραπόντων. καί τινες ἐκ τῶν ὀρῶν ἐς πύστιν τῶν γεγονότων καταθέοντες εὗρον αὐτὸν τό τε σῶμα καὶ τὴν ψυχὴν ἐσταλμένον, καὶ ἐς ἀκάτιον ἐξ ἀκατίου μεταφέροντες, ἵνα διαλάθοι, μετεκόμισαν ἐς Μεσσάλαν οὐ μακρὰν ὄντα. ὁ δ' εὐθύς, ἔτι ἀθεράπευτος, ἔς τε Κορνιφίκιον ἔστελλε **λιβυρνίδα** καὶ πανταχοῦ διὰ τῶν ὀρῶν περιέπεμπεν, ὅτι σῴζοιτο, Κορνιφικίῳ τε πάντα ἐπικουρεῖν ἐκέλευε καὶ αὐτὸς ἔγραφεν αὐτίκα πέμψειν βοήθειαν. θεραπεύσας δὲ τὸ σῶμα καὶ ἀναπαυσάμενος ὀλίγον ἐς Στυλίδα νυκτὸς ἐξῄει, παραπεμπόμενος ὑπὸ τοῦ Μεσσάλα, πρὸς Καρρίναν τρία ἔχοντα ἐπὶ τοῦ πρόπλου τέλη· καὶ τῷδε μὲν ἐκέλευσε διαπλεῖν ἐς τὸ πέραν, ἔνθα καὶ αὐτὸς ἔμελλε διαπλευσεῖσθαι, Ἀγρίππαν δὲ γράφων ἠξίου κινδυνεύοντι Κορνιφικίῳ πέμπειν Λαρώνιον μετὰ στρατιᾶς ὀξέως. Μαικήναν δ' αὖθις ἐς Ῥώμην ἔπεμπε διὰ τοὺς νεωτερίζοντας· καί τινες παρακινοῦντες ἐκολάσθησαν. καὶ Μεσσάλαν ἐς Δικαιάρχειαν ἔπεμπεν, ἄγειν τὸ πρῶτον καλούμενον τέλος ἐς τὸ Ἱππώνειον.

111. Octavian spent the greater part of the night among his small boats, in doubt whether he should go back to Cornificius through the scattered remains of his fleet, or take refuge with Messala. Providence brought him to the harbour of Abala with a single armour-bearer, without friends, attendants, or slaves. Certain persons, who had come down from the mountain to learn the news, found him shattered in body and mind and brought him in rowboats (changing from one to another for the purpose of concealment) to the camp of Messala, which was not far distant. Straightway, and before he had attended to his bodily wants, he dispatched a **liburnian** to Cornificius, and set word throughout the mountains that he was safe, and ordered all his forces to help Cornificius, and wrote to him that he would send him aid forthwith. After attending to his own person and taking a little rest, he set forth by night, accompanied by Messala, to Stylis, where Carinas was stationed with three legions ready to embark, and ordered him to set sail to the other side, whither he would shortly follow. He wrote to Agrippa and urged him to send Laronius with an army to the rescue of Cornificius with all speed. He sent Maecenas again to Rome on account of the revolutionists;

112. Τὰ μὲν οὖν πεζὰ πάντα Κορνιφικίῳ παραδοὺς ὁ Καῖσαρ ἐκέλευσε τοὺς κατὰ τὴν γῆν πολεμίους ἀπομάχεσθαι καὶ πράσσειν, ὅ τι ἐπείγοι· αὐτὸς δὲ ταῖς ναυσὶν ἔτι πρὸ ἡμέρας ἀνήγετο ἐς τὸ πέλαγος, μὴ καὶ τοῦδε αὐτὸν ἀποκλείσαιεν οἱ πολέμιοι. καὶ τὸ μὲν δεξιὸν ἐπέτρεπε Τιτινίῳ, τὸ δὲ λαιὸν Καρισίῳ, **λιβυρνίδος** δὲ αὐτὸς ἐπέβαινε καὶ περιέπλει πάντας παρακαλῶν· ἐπὶ δὲ τῇ παρακλήσει τὰ στρατηγικὰ σημεῖα, ὡς ἐν κινδύνῳ μάλιστα ὤν, ἀπέθετο. ἐπαναχθέντος δὲ τοῦ Πομπηίου δὶς μὲν ἐπεχείρησαν ἀλλήλοις, καὶ τὸ ἔργον ἐς νύκτα ἐτελεύτησεν. ἁλισκομένων δὲ καὶ πιμπραμένων τῶν Καίσαρος νεῶν, αἱ μὲν ἀράμεναι τὰ βραχέα τῶν ἱστίων ἀπέπλεον ἐς τὴν Ἰταλίαν, τῶν παραγγελμάτων καταφρονοῦσαι· καὶ αὐτὰς ἐπ᾽ ὀλίγον οἱ τοῦ Πομπηίου διώξαντες ἐπὶ τὰς ὑπολοίπους ἀνέστρεψαν, καὶ τῶνδε τὰς μὲν ᾕρουν ὁμοίως, τὰς δὲ ἐνεπίμπρασαν. ὅσοι δ᾽ ἐξ αὐτῶν ἐς τὴν γῆν ἐσενήχοντο, τοὺς μὲν οἱ ἱππέες οἱ τοῦ Πομπηίου διέφθειρον ἢ συνελάμβανον, οἱ δ᾽ ἐς τὸ τοῦ Κορνιφικίου στρατόπεδον ἀνεπήδων, καὶ αὐτοῖς ὁ Κορνιφίκιος ἐπιθέουσιν ἐπεχείρει, τοὺς κούφους ἐκπέμπων μόνους· οὐ γὰρ εὔκαιρον ἐδόκει κινεῖν φάλαγγα δύσθυμον ἀντικαθημένων πεζῶν μεγαλοφρονουμένων, ὡς εἰκὸς ἦν, ἐπὶ νίκῃ.

5.12.111-12

and some of these, who were stirring up disorder, were punished. He also sent Messala to Puteoli to bring the first legion to Vibo.

112. Octavian placed all of his infantry under charge of Cornificius, and ordered him to drive back the enemy and do whatever the exigency required. He himself took ship before daylight and went seaward lest the enemy should enclose him on this side also, giving the right wing of the fleet to Titinius and the left to Carisius, and embarking himself on a **liburnian**, with which he sailed around the whole fleet, exhorting them to have courage. Having done this he lowered the general's ensign, as is customary in times of extreme danger. Pompeius put to sea against him, and they encountered each other twice, the battle ending with the night. Some of Octavian's ships were captured and burned; others spread their small sails and made for the Italian coast, contrary to orders. Those of Pompeius followed them a short distance and then turned against the remainder, capturing some and burning others. Some of the crews swam ashore, most of whom were slaughtered or taken prisoners by Pompeius' cavalry. Some of them set out to reach the camp of Cornificius, who sent only his light-armed troops to assist them as they came near, because he did not consider it prudent to move his disheartened legionaries against the enemy's infantry, who were naturally much encouraged by their victory.

pp. 562-565

Greek: *Appiani Historia Romana* 2, eds L. Mendelssohn, P. Viereck. Teubner, Leipzig 1905.

English: Appian. *Roman History, Volume III: The Civil Wars, Books 1-3.26*, transl. H. White. LCL 4 (1913).

Illyrian wars:

B-IV.4
[...] καὶ ναυτικοὶ μὲν ἐπὶ τοῖς Ἀρδιαίοις ἐγένοντο Λιβυρνοί, γένος ἕτερον Ἰλλυριῶν, οἳ τὸν Ἰόνιον καὶ τὰς νήσους ἐλῄστευον ναυσὶν ὠκείαις τε καὶ κούφαις, ὅθεν ἔτι νῦν Ῥωμαῖοι τὰ κοῦφα καὶ ὀξέα δίκροτα **Λιβυρνίδας** προσαγορεύουσιν.
1.3

[...] The Liburni, another Illyrian tribe, were next to the Ardiaei as a nautical people. These practised piracy in the Adriatic Sea and islands, with their light, fast-sailing pinnaces, from which circumstance the Romans to this day call their own light, swift biremes '**Liburnians**'.

p. 57

5. Written sources on lembs and Liburnians from the 4th c. BC to Late Antiquity

B-IV.5

οἱ ὅμοροι προσέθεντο αὐτῷ καταπλαγέντες, Ἱππασῖνοί τε καὶ Βεσσοί. ἑτέρους δὲ αὐτῶν ἀποστάντας, Μελιτηνοὺς καὶ Κορκυρηνούς, οἳ νήσους ᾤκουν, ἀνέστησεν ἄρδην, ὅτι ἐλῄστευον τὴν θάλασσαν· καὶ τοὺς μὲν ἡβῶντας αὐτῶν ἔκτεινε, τοὺς δ᾽ ἄλλους ἀπέδοτο. **Λιβυρνῶν δὲ τὰς ναῦς** ἀφείλετο, ὅτι καὶ οἵδε ἐλῄστευον. Ἰαπόδων δὲ τῶν ἐντὸς Ἄλπεων Μοεντῖνοι μὲν καὶ Αὐενδεᾶται προσέθεντο αὐτῷ προσιόντι, Ἀρουπῖνοι δ᾽, οἳ πλεῖστοι καὶ μαχιμώτατοι τῶνδε τῶν Ἰαπόδων εἰσίν, ἐκ τῶν κωμῶν ἐς τὸ ἄστυ ἀνῳκίσαντο, καὶ προσιόντος αὐτοῦ ἐς τὰς ὕλας συνέφυγον. ὁ δὲ Καῖσαρ τὸ ἄστυ ἑλὼν οὐκ ἐνέπρησεν, ἐλπίσας ἐνδώσειν αὐτούς· καὶ ἐνδοῦσιν οἰκεῖν ἔδωκεν.
4.16

When these were conquered, the Hippasini and the Bessi, nei3ghbouring tribes, were overcome by fear and surrendered themselves to him [Augustus]. Others which had revolted, the Maltese and the Corcyreans, who inhabited islands, he destroyed utterly, because they practised piracy, putting the young men to death and selling the rest as slaves. He deprived **the Liburnians of their ships** because they also practised piracy. The Moentini and the Avendeatae, two tribes of the Japydes, dwelling within the Alps, surrendered themselves to him at his approach. The Arupini, who are the most numerous and warlike of these Iapydes, betook themselves from their villages to their city, and when he arrived there they fled to the woods. Augustus took the city, but did not burn it, hoping that they would deliver themselves up, and when they did so he allowed them to occupy it.
p. 81

Greek: *Appiani Historia Romana* 1, eds E. Gabba, A.G. Roos, P. Viereck. Teubner, Leipzig 1939.

English: Appian. *Roman History, Volume II*, ed. and transl. B. McGing. LCL 3 (1912).

Appian explains the origin of the name *liburnian* attributed to the ships. He claims in **B-IV.1** that the Liburnians were the first people using fast boats with whom the Romans came into contact. The Romans then metonymically transferred the name of the people to the sort of fast ships they saw them using. From this account, it does not appear that the Romans took the ship names from the Liburnians. This is reconfirmed in **B-IV.4**, where Appian gives a more detailed description of the original liburnians (*light, fast-sailing pinnaces*) and accentuates that the Romans gave this name to *their own* light, swift biremes *liburnians*. Here it is interesting that McGing in the Loeb edition added to English translation *their own* – though there are no such words in the Greek original text. More recent translation of Šašel Kos provides a much better solution here: *They* [Liburni] *were active in piracy in the Ionian Sea and the islands with fast and light vessels, after which the Romans even today call their light and swift double-banked galleys 'liburnians'.*[200]

Secondly, liburnians were supposedly used for reconnoitering (**B-IV.2** προερευνώντων [...] λιβυρνίδων τὴν θάλασσαν), which gives them a certain purpose within a battle fleet. They were used for scouting, as well as being employed in direct contact with other ships (**B-III.1**, above). It was similar to the function lembs performed in early Roman fleets, as pointed out by Morrison:

> *Polybios (1.53.9), as has been seen, describes the Roman fleet moving along the coast of Sicily towards Lilybaion in 249 BC preceded by* lemboi *'which are accustomed to move ahead of* (proplein) *fleets'; and later Appian (CW 5.103) speaks of liburnians performing the same function when Octavian's*

---
[200] Šašel Kos 2005: 55.

commander Taurus moved his army and fleet from Tarentum to join Octavian opposite Tauromenium. Appian says that reconnaissance was effected on land by cavalry and at sea by liburnians. Liburnians are also spoken of as used for fleet communication by Octavian in the same campaign and later at Aktion.[201]

Finally, in **B-IV.3**, we have a description of a liburnian in Octavian's fleet as a sort of a sailing messenger ship used for quick manoeuvering.

B-V (Pseudo)Lucian (early 4th century)

Amores

ΛΥΚΙΝΟΣ: Ἐπ᾽ Ἰταλίαν μοι διανοουμένῳ ταχυναυτοῦν σκάφος εὐτρέπιστο τούτων τῶν δικρότων, οἷς μάλιστα χρῆσθαι Λιβυρνοὶ δοκοῦσιν ἔθνος Ἰονίῳ κόλπῳ παρῳκισμένον. 6	Lycinus: I had in mind going to Italy and a swift ship had been made ready for me. It was one of the double-banked vessels which seem particularly to be used by the Liburnians, a race who live along the Ionian Gulf.

Greek and English: *Lucian. Soloecista. Lucius or The Ass. Amores. Halcyon. Demosthenes. Podagra. Ocypus. Cyniscus. Philopatris. Charidemus. Nero*, transl. M. D. MacLeod. LCL 432 (1967).

The authorship of the text Ἔρωτες (*Loves*, or *The two kinds of love*, known in Latin as *Amores* or alternatively in English as *Affairs of the Heart*) is disputed, the most probable date is the early 4th century AD. Another reason to date this text later than Lucian is also the geographical inaccuracy in the text. The text is written in a very pure Attic style and with a lot of linguistic versatility which resembles Lucian's style.

B-VI Pseudo-Callisthenes (*c.* 338)

Historia Alexandri Magni

B-VI.1

καὶ τούτους καθοπλίσας μεθ᾽ ὧν εἶχεν ἀπὸ τοῦ πατρὸς στρατιωτῶν λαμβάνει παρὰ τῶν τῆς Μακεδονίας <θησαυρῶν> (3) χρυσοῦ νενομισμένου τάλαντα ο΄. καὶ ναυπηγήσας τριήρεις, ἔτι δὲ **λίβερνα**, διαπεράσας ἀπὸ Μακεδονίας διὰ τοῦ Θερμώδοντος ποταμοῦ εἰς τὴν ὑπερκειμένην Θρᾴκην ὑπήκοον φύσει τυγχάνουσαν διὰ τὴν τοῦ πατρὸς Φιλίππου δύναμιν, ἐκεῖθεν παραλαβὼν ἀργυρίου αἱρετοὺς καὶ ἀργυρίου τάλαντα φ΄ ᾤχετο (4) ἐπὶ Λυκαονίαν. 1.26.2-4	He took from the treasury of the Macedonians seventy talents of current gold and built triremes, also **galleys**, and journeyed from Macedonia through the Thermodon river to Thrace lying to the north, which happened to be subject to the rule of his father Philip. From the Thracians he received silver presents and 500 talents of silver, then went to Lycaonia.

---

[201] Morrison 1996: 264.

5. Written sources on lembs and Liburnians from the 4th c. BC to Late Antiquity

B-VI.2

(1) Ὁ δὲ Ἀλέξανδρος παραλαβὼν τὰ στρατεύματα ἐπείγετο εἰς τὴν Αἴγυπτον, πέμψας **τὰ λίβερνα** περιμένειν αὐτὸν εἰς Τρίπολιν. ἔκαμνε δὲ (2) τὸ στράτευμα τῆς ὁδοιπορίας δυσχεροῦς οὔσης. ὑπαντῶντες <δὲ> τῷ Ἀλεξάνδρῳ κατὰ πᾶσαν πόλιν οἱ προφῆται τοὺς ἰδίους θεοὺς κομίζοντες ἀνηγόρευον αὐτὸν νέον Σεσόγχωσιν κοσμοκράτορα.
1.34.1-4

Then Alexander heading the army proceeded into Egypt after sending **the ships** to Tripolis to await him. The army grew weary for the journey was hard. Now the prophets, the caretakers of the country's gods, on meeting Alexander in every city, proclaimed him a new Sesonchosis, ruler of the world.

Greek: *Historia Alexandri Magni (Pseudo-Callisthenes)* 1, ed. W. Kroll. Weidmann, Berlin 1926.

In his recent commentary on the text, Nawotka commented on the mention of liburnian vessels as follows:

... τριήρεις, ἔτι δὲ λίβερνα: *triremes, powered by a sail and three rows of oarsmen on each side, were the most common combat vessels of classical antiquity. The liburnians (Liburna), spelled also* λιβυρνή, λιβυρνίς, *were the most typical Roman ships, powered by a sail and two rows of oarsmen on each side, being smaller and more agile than triremes ... Naming a* λίβερνα *in this context is anachronistic, reflecting the late date of composition of the* Alexander Romance.[202]

B-VII Eunapius (c. 347-414)

Fragmenta historica

Λίβερνα, εἶδος πλοίου· καράβια. "πηξάμενος δρομάδας τριακοντήρεις Λιβερνίδων τύπῳ."
Frag. 81

Liberna, a type of vessel: a small κάραβος. '[He] having built thirty-oared fast ships (runners) by liburnian type.'

Greek: *Historici Graeci minores* 1, ed. L. Dindorf. Teubner, Leipzig 1870.

The word κάραβος is of an unknown origin; with Pre-Greek, Ancient Macedonian or Semitic loan all having been suggested. It was linked with the Latin *scarabaeus, carabus*. LSJ offers three meanings of the term: a kind of beetle, probably a longhorn beetle; a kind of crustacean, probably a crayfish; and a small boat. The reference to δρομάδας τριακοντήρεις could have been the first mention of the *dromon*, the aforementioned Late Antique ship, which replaced the liburnian.[203]

---

[202] Nawotka 2017: 91.
[203] Pryor, Jeffreys 2006: 123.

B-VIII Zosimus (*c.* 500)

Historia nova

πλοῖα γὰρ ἦν αὐτῷ πρὸς ναυμαχίαν ἀρκοῦντα, **Λίβερνα** ταῦτα καλούμενα, ἀπό τινος πόλεως ἐν Ἰταλίᾳ κειμένης ὀνομασθέντα, καθ' ἣν ἐξ ἀρχῆς τούτων τῶν πλοίων τὸ εἶδος ἐναυπηγήθη. 5.30.3	He gathered enough vessels for the naval battle, so called **Liberna**, named thus after a certain town in Italy, where the form of these vessels was initially built.

Greek: *Zosime. Histoire nouvelle*, vols. 1-3.2, ed. F. Paschoud. Les Belles Lettres, Paris 1971-89.

For our purpose it is interesting note the inaccuracy with which Zosimus describes the origin of the name of liburnians. This is probably evidence for liburnians already being replaced by *dromons* at that time. A similar error is made a century later by Isidorus, who thought that the liburnians originated from Libya.[204]

5.3.2. Latin sources

B-1 Gaius Iulius Caesar (100-44 BC)

Commentarii belli civilis

B-1.1
Frumenti vim maximam ex Thessalia Asia Aegypto Creta Cyrenis reliquisque regionibus comparaverat. Hiemare Dyrrachii Apolloniae omnibusque oppidis maritimis constituerat ut mare transire Caesarem prohiberet eiusque rei causa omni ora maritima classem disposuerat. praeerat Aegyptiis navibus Pompeius filius Asiaticis D. Laelius et C. Triarius Syriacis C. Cassius Rhodiis C. Marcellus cum C. Coponio **Liburnicae** atque Achaicae classi Scribonius Libo et M. Octavius. toti tamen officio maritimo M. Bibulus praepositus cuncta administrabat ad hunc summa imperii respiciebat.
3.5

Pompey had procured a very large quantity of provisions from Thessaly, Asia, Egypt, Crete, Cyrene, and the other regions. He had decided to winter at Dyrrachium and Apollonia and all the coastal towns in order to prevent Caesar from crossing the Adriatic, and for this purpose he had distributed his fleet all along the coast. The commander of the ships from Egypt was his son Pompeius, of the ships from Asia Minor, Decimus Laelius and Gaius Triarius, of those from Syria, Gaius Cassius, of those from Rhodes, Gaius Marcellus with Gaius Coponius, of the **Liburnians** and the fleet from Achaia, Scribonius Libo and Marcus Octavius. Marcus Bibulus had been placed in overall charge of the maritime operation and was running everything; supreme authority fell to him.
p. 199.

---

[204] Isid. *Etym.* 19.1.12: *Liburnae dictae a Libyis; naves enim sunt negotiatorum.* Similar connection of liburnians and Libya is seen in Scholia on Horace – Panciera 1956: 132 n.3.

5. Written sources on lembs and Liburnians from the 4th c. BC to Late Antiquity

B-1.2

Discessu **Liburnarum** ex Illyrico M. Octavius cum iis quas habebat **navibus** Salonas pervenit. Ibi concitatis Dalmatis reliquisque barbaris Issam a Caesaris amicitia avertit. Conventum Salonis cum neque pollicitationibus neque denuntiatione periculi permovere posset oppidum oppugnare instituit. Est autem oppidum et loci natura et colle munitum, sed celeriter cives Romani, ligneis effectis turribus, his sese munierunt. Et cum essent infirmi ad resistendum propter paucitatem hominum, crebris confecti vulneribus ad extremum auxilium descenderunt servosque omnes puberes liberaverunt. Et praesectis omnium mulierum crinibus tormenta effecerunt. Quorum cognita sententia Octavius quinis castris oppidum circumdedit atque uno tempore obsidione et oppugnationibus eos premere coepit. Illi omnia perpeti parati maxime a re frumentaria laborabant. Quare missis ad Caesarem legatis auxilium ab eo petebant. Reliqua, ut poterant, incommoda per se sustinebant. Et longo interposito spatio cum diuturnitas oppugnationis neglegentiores Octavianos effecisset nacti occasionem meridiani temporis discessu eorum, pueris mulieribusque in muro dispositis ne quid cotidianae consuetudinis desideraretur, ipsi manu facta cum iis quos nuper maxime liberaverant in proxima Octavi castra irruperunt. His expugnatis eodem impetu altera sunt adorti, inde tertia et quarta et deinceps reliqua. Omnibusque eos castris expulerunt et magno numero interfecto reliquos atque ipsum Octavium in naves confugere coegerunt. Hic fuit oppugnationis exitus. Iamque hiems appropinquabat, et tantis detrimentis acceptis Octavius desperata oppugnatione oppidi Dyrrachium sese ad Pompeium recepit.
3.9

At the departure of the **Liburnian ships** from Illyricum Marcus Octavius went to Salonae with the ships he had available. There he roused the Dalmatians and the other natives and deflected Issa from its allegiance to Caesar. Unable to influence the association of Roman citizens at Salonae either by promises or by heralding danger, he began to besiege the town. Salonae is fortified both by the nature of its site and by a hill, but the Roman citizens hastily built wooden siege towers and used these as their fortifications. Being incapable—because of their small numbers—of standing firm in resistance, of the adult male slaves; they also cut off all the women's
hair, making catapult ropes. Recognizing their decision Octavius surrounded the town with five camps and began to apply pressure with a blockade and siege works simultaneously. Although prepared to endure everything, the people of Salonae were in very great difficulties over provisioning. Therefore they sent a delegation to Caesar requesting his help. The remaining hardships they withstood on their own, insofar as they could. After a considerable interval, when the protracted siege had made Octavius' men rather careless, they took advantage of midday, when the enemy fell back. Deploying children and women on the walls so that nothing of their daily routine would be noticeably absent, they joined forces with the men they had very recently set free and burst into the closest of Octavius' camps. After storming this they continued straight on to the next, then the third and fourth, and finally the last. They ejected the enemy from every camp and killed a large number of them, then forced the rest and Octavius himself to take refuge on their ships. This put an end to the siege. For winter was approaching, and after suffering such substantial losses Octavius despaired of the siege of Salonae and went back to Pompey at Dyrrachium.
pp. 203-205

Latin and English: *Caesar. Civil War*, ed. and transl. C. Damon. LCL 39 (2016).

These two chapters describe a situation during the Civil War. More precisely, they describe the circumstances before the Battle of Dyrrachium fought in 48 BC between Pompey and the Optimates on one side, and Caesar on the other. In the beginning of 49 BC Caesar started a civil war against the forces of the senate led by Pompey. Pompey went first to Greece to gather forces to fight the war, and later he went to Dyrrachium in modern Albania, from which he commanded the navy. Marcus Octavius was one of Pompey's admirals who went to Salona with his fleet. Octavius convinced the inhabitants of the island of Issa to defect to him, however, he did not turn the Romans in Salona against Caesar. The event and the subsequent attempt to capture Salona is described in **B-1.2**. In **B-1.1** the fleet of the Liburni is described as a separate entity within Pompey's fleet, with the *Liburnica classis* distinct from the *naves* from Egypt etc., whereas in **B-1.2**, Caesar uses the word *liburna* to denote a liburnian as a type of ship.[205]

B-2 Sextus Propertius (second half of the 1st century BC)

Elegiae

scilicet incesti meretrix regina Canopi, / una Philippeo sanguine adusta nota, / ausa Iovi nostro latrantem opponere Anubim, / et Tiberim Nili cogere ferre minas, / Romanamque tubam crepitanti pellere sistro, / baridos et contis **rostra Liburna** sequi, / foedaque Tarpeio conopia tendere saxo, / iura dare et statuas inter et arma Mari! 3.11.29-46	To be sure, the harlot queen of licentious Canopus, the one disgrace branded on Philip's line, dared to pit barking Anubis against our Jupiter and to force the Tiber to endure the threats of the Nile, to drive out the Roman trumpet with the rattling sistrum and with the poles of her barge pursue **the beaks of our galleys**, to stretch effeminate mosquito-nets on the Tarpeian rock and give judgement amid the arms and statues of Marius. pp. 259-261

Latin and English: *Propertius. Elegies*, ed. and transl. G. P. Goold. LCL 18 (1990).

In this poem Propertius explains how Cynthia, his muse, governs his life. Propertius lists strong and important women in history. Beginning with the verse 39, he talks, quite 'chivalrously', of Cleopatra as a '*whore queen, a progeny of incestuous Canopus*'. He confronts, among other things, Egyptian 'barges' (*baridos*) with the Roman liburnians. The reference is to the battle of Actium between Octavian on one side, and Antony and Cleopatra on the other. In this single verse two things are important. First, liburnians symbolise the Roman fleet and its maritime power, implying that by the second half of the 1st century BC, liburnians were already an integral part of the Roman fleet and recognised as such, unless, of course, Propertius is not using a synecdoche. The second important point is that Propertius points to a *rostrum*, a beak, as an integral part of liburnians.

---

[205] Cf. Čače 2013: 36-38.

5. Written sources on lembs and Liburnians from the 4th c. BC to Late Antiquity

B-3 Quintus Horatius Flaccus (Horace) (65-8 BC)

Epodes

Ibis **Liburnis** inter alta navium, amice, propugnacula, paratus omne Caesaris periculum subire, Maecenas, tuo. quid nos, quibus te vita si superstite iucunda, si contra, gravis? 1.1-5	You, Maecenas, will sail on a **Liburnian galley** among ships with towering superstructures, prepared to undergo every danger that threatens Caesar. What about me, to whom life will be a delight if you survive, but otherwise a burden? p. 271

Latin and English: *Horace. Odes and Epodes*, ed. and transl. N. Rudd. LCL 33 (2004).

The *epode* describes the friendship between Horace and Maecenas. Maecenas is about to depart for the battle of Actium, fighting on the side of Octavian. He is facing an immediate danger and Horace writes about the fear of losing his friend. The context of the poem suggests it was written after the battle. The battle of Actium is usually described in terms of Antony's fleet being much stronger than Octavian's but due to the employment of the liburnians, swift and fast vessels, paired with Octavian's strategic prowess, Octavian won the battle. There is sufficient evidence to consider this account historically inaccurate.[206] However, this is a historical narrative that was promoted not only by Octavian himself, but also, as we can see here with the poet Horace, as well with many later historians. In line with this historical narrative, the context of this poem presents a liburnian as a smaller ship which moves boldly among Antony's galleons. In an alternate reading, i.e. Maecenas moving around Octavian's galley, Watson translates the sentence as '*you will sail in a light Liburnian craft amongst the towering vessels of Octavian's fleet*'. However, as Watson also points out it is more likely that Horace portrayed Maecenas boldly moving towards Antony's ships.[207]

B-4 Tiberius Catius Asconius Silius Italicus (*c.* 25/36 to *c.* 101/102)

Punica:

tranavit, volucris liquidas ceu scinderet auras, / hasta viri pectus rupitque immania membra: / quanta est vis agili per caerula summa **Liburnae**, / quae, pariter quotiens revocatae ad pectora tonsae / percussere fretum, ventis fugit ocior et se, / quam longa est, uno remorum praeterit ictu. 13.238-243	Like a bird cleaving the clear sky, the spear pierced the breast of Calenus and shattered his huge frame. With such force the light **Liburnian galley** skims over the surface of the deep; when the oars, drawn back to the rowers' chests, strike the water in unison, she flies swifter than the winds, and a single stroke of their blades carries her further than her own length. p. 223.

Latin and English: *Silius Italicus. Punica, Volume II: Books 9-17*, transl. J. D. Duff. LCL 278 (1934).

---

[206] See below.
[207] Watson 2003: 57-58.

Silius Italicus describes the death of Calenus: the spear that killed him is compared to movement of a liburnian. There follows a detailed poetical description of how swiftly and fast a liburnian moves by the force of rowers. It is described as faster than the winds, which suggest that it was not a sailing boat.

B-5 Marcus Annaeus Lucanus (Lucan) (39-65)

Bellum civile (Pharsalia)

[...] movit ab omni / Quisque suam statione ratem, paribusque lacertis / Caesaris hinc puppes, hinc Graio remige classis / Tollitur; inpulsae tonsis tremuere carinae, / Crebraque sublimes convellunt verbera puppes. / Cornua Romanae classis validaeque triremes / Quasque quater surgens extructi remigis ordo / Commovet et plures quae mergunt aequore pinus, / Multiplices cinxere rates. Hoc robur aperto Cornua Romanae classis validaeque triremes
/ Quasque quater surgens extructi remigis ordo / Commovet et plures quae mergunt aequore pinus, / Multiplices cinxere rates. Hoc robur aperto / Oppositum pelago: lunata classe recedant / Ordine contentae gemino crevisse **Liburnae**. / Celsior at cunctis Bruti praetoria puppis / Verberibus senis agitur molemque profundo / Invehit et summis longe petit aequora remis.
3.524-537

Then each man started his vessel from its anchorage, and the two fleets leaped forward with rival strength of arm—Caesar's ships on one side and the fleet rowed by Greeks on the other; the hulls tremble to the beat of the oars, and the rapid stroke tears the tall vessels through the water. The wings of the Roman fleet were closed in by ships of many kinds—stout triremes, and vessels driven by four tiers of rowers rising one above another, and others that dipped in the sea a still greater number of blades. These heavy ships were set as a barrier against the open sea; the **galleys**, content to rise aloft with but two banks of oars, were further back in crescent formation. Towering above them all, the flag-ship of Brutus, driven by six rows of oars and advancing its bulk over the deep, reaches for the water far below with its topmost tier.
p. 153

Latin and English: *Lucan. The Civil War (Pharsalia)*, transl. J. D. Duff. LCL 220 (1928).

The *Pharsalia* deals with the fight between Julius Caesar and the Senate, for which Pompey serves as military leader. This fight ended in Pompey's defeat at Pharsalus (48 BC). Morrison offers a detailed description of this text:

> It is possible that the Massaliots called them by that name [liburnians], but as Phocaean Greeks they are more likely to have called them pentecontors, which were originally of one file but in time adopted two, like the triacontors, and like the fast pirate galley, the hemioliai. [...] It is probably that liburnian is Lucan's name for the Massaliot pentecontors, a recognition that they were of the same design. He describes them as 'ordine contentae gemino crevasse liburnae' [...], 'content with the growth of a double oar system' (i.e. the modest achievement of two levels, eschewing the over-ambitious third level) is mildly humorous. It makes two points, growth, i.e. increase in height leading to doubling. Mere doubling of what could be (and once was) single would leave two possibilities: either a single level ship with oars double-manned or a two-level ship with one man to each oar. The word growth (crevasse) leave no doubt that the second possibility is one Lucan describes.[208]

---
[208] Morrison 1996: 131.

5. Written sources on lembs and Liburnians from the 4th c. BC to Late Antiquity

B-6 Gaius Plinius Secundus (Pliny the Elder) (23/24-79)

Naturalis historia

B-6.1

12. Balaenae et in nostra maria penetrant. in Gaditano oceano non ante brumam conspici eas tradunt, condi autem aestatis temporibus in quodam sinu placido et capaci, mire gaudentes ibi parere; hoc scire orcas, infestam iis beluam et cuius imago nulla repraesentatione exprimi possit alia quam carnis inmensae dentibus truculentae.
13. Inrumpunt ergo in secreta ac vitulos earum aut fetas vel etiamnum gravidas lancinant morsu incursuque **ceu Liburnicarum rostris fodiunt.**
illae ad flexum inmobiles, ad repugnandum inertes et pondere suo oneratae, tunc quidem et utero graves pariendi ve poenis invalidae, solum auxilium novere in altum profugere et se tuto defendere oceano.
9.12-13

Whales even penetrate into our seas. It is Whales attacked by grampuses. said that they are not seen in the Gulf of Cadiz before midwinter, but during the summer periods hide in a certain calm and spacious inlet, and take marvellous delight in breeding there; and that this is known to the killer whale, a creature that is the enemy of the other species and the appearance of which can be represented by no other description except that of an enormous mass of flesh with savage teeth. The killer whales therefore burst into their retreats and bite and mangle their calves or the females that have calved or are still in calf, and charge and **pierce them like warships ramming**. The whales being sluggish in bending and slow in retaliating, and burdened by their weight, and at this season also heavy with young or weakened by travail in giving birth, know only one refuge, to retreat to the deep sea and defend their safety by means of the ocean.
p. 171

B-6.2

Inter praecipua autem miracula est qui vocatur nautilos, ab aliis pompilos. Supinus in summa aequorum pervenit, ita se paulatim absubrigens ut emissa omni per fistulam aqua velut exoneratus sentina facile naviget. Postea prima duo bracchia retorquens membranam inter illa mirae tenuitatis extendit, qua velificante in aura ceteris subremigans bracchiis media se cauda ut gubernaculo regit. **Ita vadit alto Liburnicarum ludens imagine**, si quid pavoris interveniat, hausta se mergens aqua.
9.87-88

But among outstanding marvels is creature called the nautilus, and by others the pilot-fish. Lying on its back it comes to the surface of the sea, gradually raising itself up in such a way that by sending out all the water through a tube it so to speak unloads itself of bilge and sails easily. Afterwards it twists back its two foremost arms and spreads out between them a marvellously thin membrane, and with this serving as a sail in the breeze while it uses its other arms underneath it as oars, it steers itself with its tail between them as a rudder. **So it proceeds across the deep mimicking the likeness of a fast cutter**, if any alarm interrupts its voyage submerging itself by sucking in water.
p. 221

B-6.3
Simili anseres quoque et olores ratione commeant, sed horum volatus cernitur. **Liburnicarum more rostrato impetu feruntur**, facilius ita findentes aera quam si recta fronte inpellerent; a tergo sensim dilatante se cuneo porrigitur agmen largeque inpellenti praebetur aurae. colla inponunt praecedentibus, fessos duces ad terga recipiunt. (Ciconiae nidos eosdem repetunt. genetricum senectam invicem educant.) olorum morte narratur flebilis cantus, falso, ut arbitror aliquot experimentis. Idem mutua carne vescuntur inter se.
10.62

Geese and swans also migrate on a similar principle, but the flight of these is seen. **They travel in a pointed formation like fast galleys**, so cleaving the air more easily than if they drove at it with a straight front; while in the rear the flight stretches out in a gradually widening wedge, and presents a broad surface to the drive of a following breeze. They place their necks on the birds in front of them, and when the leaders are tired they receive them to the rear. (Storks return to the same nest. They nourish their parents' old age in their turn.) A story is told about the mournful song of swans at their death—a false story as I judge on the strength of a certain number of experiences. Swans are cannibals, and eat one another's flesh.
p. 333

B-6.4
Pinaster nihil est aliud quam pinus silvestris minor altitudine et a medio ramosa sicut pinus in vertice. Copiosiorem dat haec resinam quo dicemus modo. Gignitur et in planis. Easdem arbores alio nomine esse per oram Italiae quas tibulos vocant plerique arbitrantur, sed graciles succinctioresque et enodes **Liburnicarum** ad usus, paene sine resina.
16.17

The pinaster is nothing else but a wild pine tree of smaller height throwing out branches from the middle as the pine does at the top. This variety gives a larger quantity of resin, in the manner which we shall describe. It grows in flat countries also. Most people think that trees called *tibuli* that grow along the coasts of Italy are the same tree with another name, but the *tibulus* is a slender tree and more compact than the pinaster, and being free from knots is used for building **light gallies**; it is almost devoid of resin.
p. 415.

Latin and English: *Pliny. Natural History, Volume II: Books 3-7; Volume IV: Books 12-16,* transl. H. Rackham. LCL 352, 370 (1942-1945).

What is interesting about Pliny the Elder's account is that he does *not* mention liburnians although he was quite familiar with the Liburni as we see in *Nat. Hist.* 3.39, or 3.139-40, and used liburnians himself, as we can see from **B-9**. In the rest of the text he just uses the form of liburnians to describe killer whales, who attack other animals like liburnians with their rams (**B-6.1**), the nautilus, which moves like a liburnian raising the sails, using the oars and steering itself with a rudder (**B-6.3**), the geese and swans, who fly in a formation similar to liburnians (**B-6.4**). Finally, he notes that liburnians are built from a sort of a pine tree called *tibulus* (**B-6.4**)

5. Written sources on lembs and Liburnians from the 4th c. BC to Late Antiquity

B-7 Cornelius Tacitus (55 - c. 120)

Annales

B-7.1
Neque nescium habebat Anteium caritate Agrippinae invisum Neroni opesque eius praecipuas ad eliciendam cupidinem eamque causam multis exitio esse. Igitur interceptis Antei litteris, furatus etiam libellos, quibus dies genitalis eius et eventura secretis Pammenis occultabantur, simul repertis quae de ortu vitaque Ostorii Scapulae composita erant, scribit ad principem magna se et quae incolumitati eius conducerent adlaturum, si brevem exilii veniam inpetravisset: quippe Anteium et Ostorium inminere rebus et sua Caesarisque fata scrutari. Exim missae **liburnicae** advehiturque propere Sosianus. Ac vulgato eius indicio inter damnatos magis quam inter reos Anteius Ostoriusque habebantur, adeo ut testamentum Antei nemo obsignaret, nisi Tigellinus auctor extitisset, monito prius Anteio ne supremas tabulas moraretur. Atque ille hausto veneno, tarditatem eius perosus intercisis venis mortem adproperavit.
16.14.3

He [Antistius Sosianus] was further aware that Pammenes' affection for Agrippina had earned him the hatred of Nero; that his riches were admirably calculated to excite cupidity; and that this was a circumstance which proved fatal to many. He therefore intercepted a letter from Anteius, stole in addition the papers, concealed in Pammenes' archives, which contained his horoscope and career, and, lighting at the same time on the astrologer's calculations with regard to the birth and life of Ostorius Scapula, wrote to the emperor that, could he be granted a short respite from his banishment, he would bring him grave news conducive to his safety: for Anteius and Ostorius had designs upon the empire, and were peering into their destinies and that of the prince. **Fast galleys** were at once sent out, and Sosianus arrived in haste. The moment his information was divulged, Anteius and Ostorius were regarded, not as incriminated, but as condemned: so much so, that not a man would become signatory to the will of Anteius until Tigellinus came forward with his sanction, first warning the testator not to defer his final dispositions. Anteius swallowed poison; but, disgusted by its slowness, found a speedier death by cutting his arteries.
p. 357

Latin and English: *Tacitus. Annals: Books 13-16*, transl. J. Jackson. LCL 322 (1937).

De vita Iulii Agricolae

B-7.2
Eadem aestate cohors Usiporum per Germanias conscripta et in Britanniam transmissa magnum ac memorabile facinus ausa est. occiso centurione ac militibus, qui ad tradendam disciplinam immixti manipulis exemplum et rectores habebantur, tres **liburnicas** adactis per vim gubernatoribus ascendere; et uno remigante suspectis duobus eoque interfectis, nondum vulgato rumore ut miraculum praevehebantur. mox ubi aquam atque utilia raptum existent, cum plerisque Britannorum sua defensantium proelio congressi ac saepe victores, aliquando pulsi, eo ad extremum inopiae venere, ut infirmissimos suorum, mox sorte ductos vescerentur. atque ita circumvecti Britanniam, amissis per inscitiam regendi navibus, pro praedonibus habiti, primum a Suebis, mox a Frisiis intercepti sunt. ac fuere quos per commercia venumdatos et in nostram usque ripam mutatione ementium adductos indicium tanti casus inlustravit.
28.1

During the same summer a cohort of Usipi, enrolled in our German provinces and sent across to Britain, perpetrated a signal and memorable crime. After murdering their centurions and such soldiers as had been distributed among their companies to instill discipline, and who passed as models and instructors, they manned three **galleys**, violently coercing the helmsmen: making one of them join the rowers—for the other two fell under suspicion and were put to death—they caused great surprise as they sailed past before the news was abroad. Afterwards, disembarking for water and to forage for necessaries, they gave battle to various bodies of Britons defending their property, and after many victories and some defeats ultimately were reduced to such straits as to eat the weakest of their company, and after them the victims drawn by lot. In this fashion they sailed round Britain, and then lost the ships because of their ignorance of navigation. They were treated as pirates and put to death, some by the Suebi, others by the Frisii; some of them also were sold in the way of trade, and so reached by exchange of purchasers our bank of the river, and gained notoriety by the story of their remarkable adventures.
p. 77

Latin and English: *Tacitus. Agricola. Germania. Dialogue on Oratory*, transl. M. Hutton, W. Peterson, Rev. R. M. Ogilvie, E. H. Warmington, M. Winterbottom. LCL 35 (1914).

Historiae

B-7.3
vocatis principibus insulae consilium aperit et contra dicere ausos, Claudium Pyrrichum trierarchum **Liburnicarum ibi navium**, Quintium Certum equitem Romanum, interfici iubet [...].
2.16.2

Accordingly he [Decumus] summoned the leading men of the island and disclosed his purpose; when Claudius Pyrrichus, commander of **the Liburnian ships there**, and Quintius Certus, a Roman knight, dared to oppose him, he ordered them to be killed.
p. 187

5. Written sources on lembs and Liburnians from the 4th c. BC to Late Antiquity

B-7.4
Et erat insula amne medio, in quam gladiatores navibus molientes, Germani nando praelabebantur. Ac forte plures transgressos completis **Liburnicis** per promptissimos gladiatorum Macer adgreditur; sed neque ea constantia gladiatoribus ad proelia quae militibus, nec proinde **nutantes e navibus** quam stabili gradu e ripa volnera derigebant. Et cum variis trepidantium inclinationibus mixti remiges propugnatoresque turbarentur, desilire in vada ultro Germani, retentare puppes, scandere foros aut comminus mergere; quae cuncta in oculis utriusque exercitus quanto laetiora Vitellianis, tanto acrius Othoniani causam auctoremque cladis detestabantur.
2.35.1

In the middle of the river was an island, which the gladiators were trying to reach in boats, but the Germans swam across and anticipated them. When a considerable number of Germans had crossed, Macer filled some **light Liburnian vessels** and attacked them with the bravest of his gladiators. But gladiators have not the same steadfast courage in battle as regular soldiers, and now **in their unsteady boats** they could not shoot so accurately as the Germans, who had firm footing on the shore; and when the gladiators in their fright began to move about in confusion so that rowers and fighters were commingled and got in one another's way, the Germans actually jumped into the shallow water, held back the boats, and boarded them, or sank them with their hands. All this went on under the eyes of both armies, and the keener the delight it gave the Vitellians, the greater the indignation which Otho's followers felt toward Macer, who was the cause and author of their defeat.
p. 213

B-7.5
Tum progressus Lucilius auctorem se palam praebet. Classis Cornelium Fuscum praefectum sibi destinat, qui propere adcucurrit. Bassus honorata custodia **Liburnicis navibus** Atriam pervectus a praefecto alae Vibennio Rufino, praesidium illic agitante, vincitur, sed exsoluta statim vincula interventu Hormi Caesaris liberti: is quoque inter duces habebatur.
3.12.3

Then Lucilius appeared and showed himself openly as the ringleader. But the fleet chose Cornelius Fuscus as their prefect, who came to Ravenna with all speed. Bassus was taken to Adria[209] with an escort of **light vessels** under an honourable guard. He was put in chains by the prefect of cavalry, Vibennius Rufinus, who was on garrison duty there; but he was at once released through the intervention of Hormus, a freedman of Vespasian. Hormus also was counted among the leaders of the Flavian party.
p. 351

B-7.6
Haec singuli, haec universi, ut quemque dolor impulerat, vociferantes, initio a quinta legione orto, repositis Vitellii imaginibus vincla Caecinae iniciunt; Fabium Fabullum quintae legionis legatum et Cassium Longum praefectum castrorum duces deligunt; forte oblatos **trium**

With such cries, now separately, now in a body, as indignation moved each, the Fifth legion taking the lead, they replaced the statues of Vitellius and threw Caecina into chains. They chose as their commanders Fabius Fabullum, legate of the Fifth legion, and Cassius Longus, prefect of the camp.

---

[209] The translator changes Atriam to Adriam.

Liburnicarum milites ignaros et insontes trucidant [...].
3.14.1

Happening to meet the marines from **three light galleys** who had no knowledge or complicity in what had happened, they slew them.
p. 353.

B-7.7
Digresso Valente trepidos, qui Ariminum tenebant, Cornelius Fuscus admoto exercitu et **missis** per proxima litorum **Liburnicis** terra marique circumvenit: occupantur plana Umbriae et qua Picenus ager Hadria adluitur, omnisque Italia inter Vespasianum ac Vitellium Appennini iugis dividebatur. 2. Fabius Valens e sinu Pisano segnitia maris aut adversante vento portum Herculis Monoeci depellitur.
3.42.1

Valens' departure made the troops at Ariminum anxious and timid. Cornelius Fuscus brought up his land forces and sent light **men-of-war**[210] along the neighbouring coast and thereby cut the garrison off by land and sea. The Flavians now held the plains of Umbria and that part of Picenum that is washed by the Adriatic; in fact, all Italy was divided between Vespasian and Vitellius by the range of the Apennines. 2. Fabius Valens sailed from the harbour of Pisa, but was forced by calm or by head winds to put in at the port of Hercules Monoecus.
p. 399.

B-7.8
Ceterum ut mare tutius Valenti quam litora aut urbes, ita futuri ambiguus et magis quid vitaret quam cui fideret certus, adversa tempestate Stoechadas Massiliensium insulas adfertur. ibi eum missae a Paulino **Liburnicae** oppressere.
3.43.2

But while the sea seemed to Valens safer than shores or cities, he was still doubtful of the future and saw more clearly what to avoid than what to trust. An adverse storm drove him to the Stoechadae islands belonging to the Massilians, where he was captured by **some light galleys** which Paulinus sent after him.
p. 401

B-7.9
Caesa ibi cohors, regium auxilium olim; mox donati civitate Romana signa armaque in nostrum modum, desidiam licentiamque Graecorum retinebant. Classi quoque faces intulit, vacuo mari eludens, quia lectissimas **liburnicarum** omnemque militem Mucianus Byzantium adegerat.
Quin et barbari contempti<m> vagabantur fabricatis repente navibus: camaras vocant, artis lateribus latam alvom sine vinculo aeris aut ferri conexam; et tumido mari, prout fluctus attollitur, summa navium tabulis augent, donec in modum tecti claudantur. Sic inter undas volvuntur, pari utrimque prora et mutabili remigio, quando hinc vel illinc adpellere indiscretum et innoxium est.
3.47.3

There he [a certain Anicetus] massacred a cohort, which originally consisted of auxiliaries furnished by the king; later its members had been granted Roman citizenship and had adopted Roman standards and arms, but retained the indolence and licence of the Greeks. He also set fire to the fleet and escaped by sea, which was unpatrolled since Mucianus had concentrated the best **light galleys** and all the marines at Byzantium. Moreover, the barbarians had hastily built vessels and now roamed the sea at will, despising the power of Rome. Their boats they call camarae; they have a low freeboard but are broad of beam, and are fastened together without spikes of bronze or iron. When the sea is rough the sailors build up the bulwarks with planks to

---

[210] This is obviously the translator's misake: Cornelius Fuscus must have sent liburnians along the shores.

5. Written sources on lembs and Liburnians from the 4th c. BC to Late Antiquity

match the height of the waves, until they close in the hull like the roof of a house. Thus protected these vessels roll about amid the waves. They have a prow at both ends and their arrangement of oars may be shifted, so that they can be safely propelled in either direction at will.
p. 407

B-7.10
Advertit ea res Vespasiani animum, ut vexillarios e legionibus ducemque Virdium Geminum spectatae militiae deligeret. Ille incompositum et praedae cupidine vagum hostem adortus coegit in naves, effectisque raptim **Liburnicis** adsequitur Anicetum in ostio fluminis Chobi, tutum sub Sedochezorum regis auxilio, quem pecunia donisque ad societatem perpulerat. Ac primo rex minis armisque supplicem tueri: postquam merces proditionis aut bellum ostendebatur, fluxa, ut est barbaris, fide pactus Aniceti exitium perfugas tradidit, belloque servili finis impositus.
3.48.1

These events attracted Vespasian's attention, so that he sent detachments from his legions under the command of Virdius Geminus, whose military skill had been well tested. He attacked the enemy's troops when they were off their guard and were scattered in their greed for booty, and forced them to their boats; afterwards he quickly built some **light galleys** and caught up with Anicetus at the mouth of the river Chobus, where he had sought shelter under the protection of the king of the Sedochezi, whose alliance he had secured by bribes and gifts. At first the king sheltered his suppliant with the aid of threats and arms; but after the reward for treachery and the alternative of war were set before him, with the unstable loyalty of a barbarian he bargained away the life of Anicetus, gave up the refugees, and so an end was put to this servile war.
p. 409

B-7.11
Pauci gladiatorum resistentes neque inulti cecidere, ceteri ad naves ruebant, ubi cuncta pari formidine implicabantur, permixtis paganis, quos nullo discrimine Vitelliani trucidabant. Sex **Liburnicae** inter primum tumultum evasere, in quis praefectus classis Apollinaris; reliquae in litore captae, <a>ut nimio ruentium onere pressas mare hausit.
3.77.2

A few of the gladiators resisted and fell not without vengeance on their foes. The rest rushed to the ships; but there an equal panic caused utter confusion, for the Vitellians slew without distinction the townspeople who joined the soldiers in their flight. Six **Liburnian galleys** escaped at the first alarm with Apollinaris the prefect of the fleet on board; the rest of the ships were captured at the shore, or else were swamped by the excessive weight of those who rushed on board.
p. 463

B-7.12

| Civilem cupido incessit navalem aciem ostentandi: complet quod biremium quaeque simplici ordine agebantur; adiecta ingens luntrium vis: **tricenos quadragenosque \<ferunt\> armamenta Liburnicis solita**; et simul captae luntres sagulis versicoloribus haud indecore pro velis iuvabantur. spatium velut aequoris electum, quo Mosae fluminis os amnem Rhenum Oceano adfundit.
5.23.1 | Civilis was now seized with a desire to make a naval demonstration; he therefore manned all the biremes and all the ships that had but a single bank of oars; to this fleet he added a vast number of boats, [putting in each] thirty or forty men, **the ordinary complement of a Liburnian cruiser**; and at the same time the boats that he had captured were fitted with particoloured plaids for sails, which made a fine show and helped their movement. The place chosen for the display was a small sea, so to speak, formed at the point where the mouth of the Maas discharges the water of the Rhine into the ocean.
p. 215 |

Latin and English: *Tacitus. Histories: Books 1-3*, transl. C. H. Moore. LCL 111 (1925).

Tacitus. Histories: Books 4-5. Annals: Books 1-3, transl. C. H. Moore, J. Jackson. LCL 249 (1931).

Germania

B-7.13

| pars Sueborum et Isidi sacrificat: unde causa et origo peregrino sacro, parum comperi, nisi quod signum ipsum in modum **liburnae** figuratum docet advectam religionem.
9.1 | A section of the Suebi sacrifices also to Isis: the cause and origin of this foreign worship I have not succeeded in discovering, except that the emblem itself, which takes the shape of **a Liburnian galley**, shows that the ritual is imported.
p. 145. |

Latin and English: *Tacitus. Agricola. Germania. Dialogue on Oratory*, transl. M. Hutton, W. Peterson, Rev. R. M. Ogilvie, E. H. Warmington, M. Winterbottom. LCL 35 (1914).

Morrison commented upon Tacitus' mentions of liburnians at some length:

> *Liburnicae appear frequently in the* Histories, *first in Otho's fleet at the first battle of Cremona (2.35.4) [8.4], then at Ravenna (3.12 and 14, 'the soldier of three liburnians' [8.5 and 8.6] and 3.42.3 [8.8] as a squadron used by the Othonians on the east coast of Italy. Mucianus, in command of the Black sea fleet, concentrated all the best liburnians at Byzantion, leaving the rest of the area undefended (3.47.15 [8.9]), from which it may be concluded that the fleet consisted of liburnians. To meet Otho's naval ascendancy Vespasian built liburnians quickly (3.48.4 [8.10]). When Tarracina, occupied by the rebel Misenum fleet under Claudius Apollinaris fell to Vitellius's brother Lucius, Apollinaris escaped with six liburnians and the rest of the fleet (apparently all liburnians), were captured or put out of action (3.77.12 [8.11]). Liburnians also appear in Tacitus's later* Agricola *(28.5 [8.2] where he tells the story of three liburnians hijacked in Britain by a cohort of Usipi. From these passages the conclusion*

5. Written sources on lembs and Liburnians from the 4th c. BC to Late Antiquity

must be drawn that in the provinces the liburnian was the predominant warship of the early empire; that the three was present in strength in the home fleets but used for special purposes Fours were used, perhaps as a deliberate archaism, in AD 52 for Claudius's mock battle. Augusts in his mock battle used nothing larger than threes. Fours are not mentioned by Tacitus as in naval service, yet nineteen of them appear in the inscription of the early empire. It must be supposed that Antony's larger ships, fives to tens, which were sent to Forum Julii *in 31 BC were laid up together with most of Octavian's fives and sixes, only one or two fives and sixes being kept in commission for special service. It is not therefore surprising that ships of higher rating than fours are rarely represented in inscription after the first half of the first century AD.*[211]

Even if Morrison missed some of the sources and made some imprecisions (e.g. *Agricola* considered an earlier work than *Histories*), his evaluation is correct, and the missed sources do not contribute much to the overall picture he offered.

B-8 Gaius Plinius Caecilius Secundus (Pliny the Younger) (61/62 - before 117)

Epistulae

7. Magnum propiusque noscendum, ut eruditissimo viro, visum. Iubet **Liburnicam** aptari: mihi, si venire una vellem, facit copiam; respondi studere me malle, et forte ipse, quod scriberem, dederat.
8. Egrediebatur domo; accipit codicillos Rectinae Tasci imminent! periculo exterritae (nam villa eius subiacebat, nec ulla nisi navibus fuga): ut se tanto discrimini eriperet orabat.
9. Vertit ille consilium et quod studioso animo incohaverat obit maximo. Deducit quadriremes. ascendit ipse non Rectinae modo sed multis (erat enim frequens amoenitas orae) laturus auxilium.
10. Properat illuc unde alii fugiunt, rectumque cursum recta gubernacula in periculo tenet adeo solutus metu, ut omnes illius mali motus omnes figuras ut deprenderat oculis dictaret enotaretque.
11. Iam navibus cinis incidebat, quo propius accederent, calidior et densior; iam pumices etiam nigrique et ambusti et fracti igne lapides; iam vadum subitum ruinaque montis litora obstantia.

6.16.7–11

My uncle's scholarly acumen saw at once that it was important enough for a closer inspection, and he ordered **a fast boat** to be made ready, telling me I could come with him if I wished. I replied that I preferred to go on with my studies, and as it happened he had himself given me some writing to do.

As he was leaving the house he was handed a message from Rectina, wife of Tascius, whose house was at the foot of the mountain, so that escape was impossible except by boat. She was terrified by the danger threatening her and implored him to rescue her from her fate. He changed his plans, and what he had begun in a spirit of inquiry he completed as a hero. He gave orders for the warships to be launched and went on board himself with the intention of bringing help to many more people besides Rectina, for this lovely stretch of coast was thickly populated. He hurried to the place which everyone else was hastily leaving, steering his course straight for the danger zone. He was entirely fearless, describing each new movement and phase of the portent to be noted down exactly as he observed them.

Ashes were already falling, hotter and thicker as the ships drew near, followed by bits of pumice and blackened stones, charred and cracked by the flames: then suddenly they were in shallow

---

[211] Morrison 1996: 171-72.

> water, and the shore was blocked by the debris from the mountain.
> pp. 427-431.

Latin and English: *Pliny the Younger. Letters, Volume I: Books 1-7*, transl. B. Radice. LCL 55 (1969).

This famous letter offers a romanticised version of Pliny the Elder's death by his nephew during the eruption of Mount Vesuvius in AD 79. For us it is of interest that Pliny mentions a liburnian as a sort of exploratory, scientific ship that Pliny the Elder used to explore natural phenomena – in this case, he used a liburnian to get closer to the erupting volcano. In the rest of the text, the word *navis* is used as a general term for a ship, whereas specifically warships – here deployed in a rescue mission – are mentioned as *quadriremes*.

B-9 Decimus Iunius Iuvenalis (second half of 1st century)

Saturae

> si vocat officium, turba cedente vehetur / dives et ingenti curret super ora Liburna / atque obiter leget aut scribet vel dormiet intus / (namque facit somnum clausa lectica fenestra).
> 3.239-241

> If duty calls, the crowd gives way as the rich man is conveyed, racing along above their faces in his huge Liburnian galley, reading or writing on the way or sleeping inside (you know how a litter with its window closed brings on drowsiness).
> p. 187.

Latin and English: *Juvenal and Persius*, ed. and transl. S. Morton Braund. LCL 91 (2004).

In this poem a type of couch in which the rich people were carried through the busy streets of Rome (*letica*) is ironically compared with a liburnian. The comparison is probably based on how a liburnian '*cutting*' through the waves is similar to that of the litter '*cutting*' through the crowd.

5. Written sources on lembs and Liburnians from the 4th c. BC to Late Antiquity

B-10 Publius Annius Florus (1st/2nd century AD)

Epitoma de Tito Livio

XLIIII. EXPEDITIO IN CYPRUM
3.9 Aderat fatum insularum. igitur et Cypros recepta sine bello. Insulam veteribus divitiis abundantem et ob hoc Veneri sacram Ptolomaeus regebat. Et divitiarum tanta erat fama, nec falso, ut victor gentium populus et donare regna consuetus P. Clodio tribuno plebis duce socii vivi que regis confiscationem mandaverit. Et ille quidem ad rei famam veneno fata praecepit. Ceterum Porcius Cato Cyprias opes **Liburnis** per Tiberinum hostium invexit. Quae res latius aerarium P. R. quam ullus triumphus implevit.
1.44 (3.9.5)

XLIIII. The Expedition to Cyprus
3.9. The fate of the islands was sealed; and so Cyprus too was taken over without any fighting. This island, rich in ancient wealth and therefore dedicated to Venus, was under the rule of Ptolemy. But such was the fame of its riches (and not without cause) that a people which had conquered nations and was accustomed to make gifts of kingdoms ordered, on the proposal of Publius Clodius, the tribune of the people, that the property of a king, allied to themselves and still living, should be confiscated. Ptolemy, on hearing the news of this, anticipated fate by taking poison, and Porcius Cato brought the wealth of Cyprus in **Liburnian galleys** to the mouth of the Tiber. This replenished the treasury of the Roman people more effectively than any triumph.
p. 199

Latin: *Ivli Flori Epitomae de Tito Liuio bellorum omnium annorum DCC libri duo*, ed. O. Rossbach. Teubner, Leipzig 1896.
English: *Florus. Epitome of Roman History*, transl. E. S. Forster. LCL 231 (1929).
The fragment refers to Cato the Younger's mission in Cyprus.[212]

B-11 Gaius Suetonius Tranquillus (*c.* AD 100)

Divus Augustus

B-11.1
[...] repetit<a It>alia tempestate in traiectu bis conflictatus, primo inter promunturia Peloponensi atque Aetoliae, rursus circa montes Ceraunios utrubique parte **liburnicarum** demersa, simul eius, in qua vehebatur, fusis armamentis et gubernaculo diffracto [...].
17.3

[...] and on his way back to Italy he [Octavian August] twice encountered storms at sea, first between the headlands of the Peloponnesus and Aetolia, and again off the Ceraunian mountains. In both places a part of his **galleys** were sunk, while the rigging of the ship in which he was sailing was carried away and its rudder broken.
p. 173

---

[212] Plut. *Cat. Min.* 34-38, see more recently in Morrell 2017: 116-22.

Caligula

B-11.2
Fabricavit et deceris **Liburnicas** gemmatis puppibus, versicoloribus velis, magna thermarum et porticuum et tricliniorum laxitate magnaque etiam vitium et pomiferarum arborum varietate; quibus discumbens de die inter choros ac symphonias litora Campaniae peragraret.
in extructionibus praetoriorum atque villarum omni ratione posthabita nihil tam efficere concupiscebat quam quod posse effici negaretur.
37.2

He also built **Liburnian galleys** with ten banks of oars, with sterns set with gems, particoloured awnings, huge spacious baths, colonnades, and banquet-halls, and even a great variety of vines and fruit trees; that on board of them he might recline at table from an early hour, and coast along the shores of Campania amid dancers and musicians.
p. 475

Latin and English: *Suetonius. Lives of the Caesars, Volume I: Julius. Augustus. Tiberius. Gaius. Caligula*, transl. J. C. Rolfe. LCL 31 (1914).

Nero

B-11.3
Hoc consilio per conscios parum celato solutilem navem, cuius vel naufragio vel camarae ruina periret, commentus est atque ita reconciliatione simulata iucundissimis litteris Baias evocavit ad sollemnia Quinquatruum simul celebranda; datoque negotio trierarchis, qui **Liburnicam** qua advecta erat velut fortuito concursu confringerent, protraxit convivium repetentique Baulos in locum corrupti navigii machinosum illud optulit, hilare prosecutus atque in digressu papillas quoque exosculatus.
34.2

When this leaked out through some of those connected with the plot, he devised a collapsible boat, to destroy her by shipwreck or by the falling in of its cabin. Then he pretended a reconciliation and invited her in a most cordial letter to come to Baiae and celebrate the feast of Minerva with him. On her arrival, instructing his captains to wreck **the galley** in which she had come, by running into it as if by accident, he detained her at a banquet, and when she would return to Bauli, offered her his contrivance in place of the craft which had been damaged, escorting her to it in high spirits and even kissing her breasts as they parted.
p. 139

5. WRITTEN SOURCES ON LEMBS AND LIBURNIANS FROM THE 4TH C. BC TO LATE ANTIQUITY

Vita Plinii Secundi

B-11.4

Periit clade Campaniae; cum enim Misenensi classi praeesset et flagrante Vesuvio ad explorandas propius causas **Liburnica** pertendisset, nec aduersantibus ventis remeare posset, vi pulveris ac favillae oppressus est, vel ut quidam existimant a servo suo occisus, quem aestu deficiens, ut necem sibi maturaret, oraverit.

He lost his life in the disaster in Campania. He was commanding the fleet at Misenum, and setting out in **a Liburnian galley** during the eruption of Vesuvius to investigate the causes of the phenomenon from nearer at hand, he was unable to return because of head winds. He was suffocated by the shower of dust and ashes, although some think that he was killed by a slave, whom he begged to hasten his end when he was overcome by the intense heat.
p. 487

Latin and English: *Suetonius. Lives of the Caesars, Volume II: Claudius. Nero. Galba, Otho, and Vitellius. Vespasian. Titus, Domitian. Lives of Illustrious Men: Grammarians and Rhetoricians. Poets (Terence. Virgil. Horace. Tibullus. Persius. Lucan). Lives of Pliny the Elder and Passienus Crispus*, transl. J. C. Rolfe. LCL 38 (1914).

As for Suetonius's sources on liburnians, from the quotes it appears that his use of the term *liburnica* is less restricted to a certain specific type of ship. His description of Caligula's excessiveness (**B-11.2**) suggests a rather generic meaning of *liburnica*.[213] It is also noteworthy that Suetonius, while describing the end of Pliny the Elder, claims that he found his demise on a liburnian (**B-11.4**), contrary to what we read in Pliny the Younger's description, where he explicitly states that his uncle chose not to take a liburnian but rather sturdier warships (**B-8**).

B-12 Firmicus Maternus (first half of 4th century AD).

Matheseos libri VIII

B-12.1

In X. parte Piscium quicumque habuerit horoscopum, si Luna bene posita et Venus in parte horoscopi fuerit constituta, erit grandis potens notus omnibus, regibus coniunctus, multam peragrans terram, grande navigans pelagus, praeponetur **liburnis**. Navale illi conmittetur imperium, erit in pugna superior et victor, habebit amoris maximam gratiam,

Whoever has the ascendant in the tenth degree of Pisces, if the Moon is well-placed and Venus on the ascendant, will be great, powerful, well-known to all, and a friend of kings; he travels over many lands and sails great seas; **naval** commands will be entrusted to him. He will be a victor in battle, have great pleasure in love; he will be just and pious but

---

[213] However, see Caro (2017), who has recently argued that these gigantic ships were built using the same shipbuilding technology used in the production of the liburnians.

erit iustus religiosus, sed circa muliebres concubitus semper inpatiens, magna praestans et omnia inpetrans. Morietur autem aut ira regis aut odio. In XI. parte Piscium quicumque habuerit horoscopum, in desertis locis a bestiis †naturam perdet, aut in humidis locis, aut certe in prima aetate biothanatus sine filiis peribit.
8.30.4

always impatient with desire around women; demanding and obtaining many things. He will die from the anger of the king. Whoever has the ascendant in the eleventh degree of Pisces will be castrated by beasts in desert or humid places or will die a violent death in early youth, leaving no children.
p. 299

B-12.2
In X. parte Piscium est clara stella, in australi scilicet Pisce. Si itaque Venus cum horoscopo in ipsa stella fuerit inventa, et Luna bene sit collocata, quicumque natus fuerit in ea, erit magnus potens regi notus, multas terras peragrabit, grande navigabit pelagus. **Liburnicum illi et navale conmittetur imperium**. Erit victor in proeliis, ab omnibus amabitur, erit iustus religiosus, pronus sane ad muliebres concubitus, praestans multa, potiora impetrans. Ira regis vel odio morietur.
8.31.9

A bright star is in the tenth degree of Pisces, that is, in the East. If Venus is in that star on the ascendant and the Moon well located, whoever has that star on his ascendant will be great and powerful, a familiar of kings; he will travel many lands and sail the great sea. **He will lead navies**, be a victor in battles; will be beloved by all, be just and pious. But he will be involved in love affairs with many women. He will be outstanding in many areas and will always obtain his desires. He will die from the anger of the king.
p. 301

Latin: *Iulii Firmici Materni Matheseos libri VIII*, eds W. Kroll, F. Skutsch, K. Ziegler. Teubner, Leipzig 1968.

English: *Ancient Astrology - Theory and Practice, Matheseos Libri VIII*, transl. J. R. Bram. Noyes Press, Park Ridge 1979.

In the both passages above the translator omitted translating the phrase mentioning liburnians. In **B-12.1** *praeponitur liburnis* means, '*he will be made in charge of liburnians*', the translator obviously considered the phrase redundant and translated it as '*naval commands will be entrusted to him*'. The context of the both passages boils down to the same: the person born in that particular sign will be successful in naval military campaigns. Among other things, he will be put in charge of the liburnians which is paraphrased in the next phrase: *navale illi conmittetur imperium* – '*naval commands will be entrusted to him*'. In **B-12.2** the phrase *liburnicum illi et navale conmittetur imperium* is translated merely as '*he will lead navies*'.

5. Written sources on lembs and Liburnians from the 4th c. BC to Late Antiquity

B-13 *Anonymus de rebus bellicis* (after AD 337 - before AD 378).

De rebus bellicis

B-13.1
[7] Docebimus igitur uelocissimum **liburnae** genus decem nauibus ingenii magisterio praeualere, ita ut hae per eam sine auxilio cuiusquam turbae obruantur.
Praefatio 7

I shall, in fact, demonstrate how a particularly fast type of **warship** is able through a brilliant invention to outmatch ten other ships, sending them to the bottom without the aid of a large crew.
pp. 107-108

B-13.2
1. **liburnam** naualibus idoneam bellis, quam pro magnitudine sui uirorum exerceri manibus quodammodo imbecillitas humana prohibebat, quocumque utilitas uocat ad facilitatem cursus ingenii ope subnixa animalium uirtus impellit. 2. in cuius alueo uel capacitate bini boues machinis adiuncti adhaerentes rotas nauis lateribus uoluunt, quarum supra ambitum uel rotunditatem exstantes radii currentibus iisdem rotis in modum remorum aquam conatibus elidentes miro quodam artis effectu operantur, impetu parturiente discursum. 3. haec eadem tamen liburna pro mole sui pro que machinis in semet operantibus tanto uirium fremitu pugnam capescit ut omnes aduersarias liburnas comminus uenientes facili attritu comminuat.
17

1. **A Liburnian ship** suitable for naval warfare, so large that human weakness more or less precluded its being operated by men's hands, is propelled in any required direction by animal power harnessed by the aid of human ingenuity to provide easy locomotion. 2. In its hull or hold oxen are yoked to machines, two of each, and turn wheels attached to the ship's sides: the spokes project beyond the circumference or rim of the wheels, and, striking the water forcibly like oars as the wheels rotate, work with a wonderous and ingenious effect, their impetus producing locomotion. 3. The same warship, however, owing to its massiveness and the machines working inside it, joins battle with such furious strength that it easily crushes and destroys all opposing warships that come to close quarters with it.
pp. 119-120

B-13.3 Quod si nauali bello terras fugiens maria hostis obsideat, nouo celeritatis ingenio terrestri quodammodo ritu rotis et bubus subacta fluctibus **liburna** transcurrens restituet sine mora uictoriam.
18.7

But if the enemy flees the land and besets the seas with naval warfare, victory will be restored to you without delay by a **warship** swiftly traversing the waves by means of a new speed device, for it is propelled by wheels and oxen as though it were on land.
p. 121

Latin and English: *A Roman Reformer and Inventor, being a new text of the treatise* De rebus bellicis, transl. and intr. E. A. Thompson. Clarendon Press, Oxford 1952.

B-13.1-3 do not describe actual liburnians, rather, they are the inventor's vision of how liburnians were to be made more efficient.[214]

---

[214] Cf. Fleury 2015.

B-14 Eutropius (fl. c. AD 360)

Breviarium ab urbe condita

Quinto anno Punici belli, quod contra Afros gerebatur, primum Romani C. Duillio et Cn. Cornelio Asina consulibus in mari dimicaverunt, paratis navibus rostratis, quas **Liburnas** vocant.
2.20.2

In the fifth year of the Punic War, which was waged against the Africans, in the consulship of Gaius Duillius and Gnaeus Cornelius Asina, the Romans fought for the first time at sea, after building beaked warships, which they call **Liburnian**.
p. 12

Latin: *Eutropi Breviarium ab urbe condita*, ed. F. Ruehl. Leipzig: Teubner, 1887.

English: *The Breviarium ab Urbe Condita of Eutropius The Right Honourable Secretary of State for General Petitions Dedicated to Lord Valens Gothicus Maximus & Perpetual Emperor*, transl., comm. and intr. H. W. Bird. TTH 14 (1993).

Eutropius states that the Romans employed liburnians as early as the first Punic war. However, it is more likely that at the time of the composition of the text, the word *liburna* has already started denoting a warship in general, and liburnians mentioned by Eutropius are clear anachronism.

B-15 Claudius Mamertinus (fl. around AD 362)

Gratiarum actio de consulatu suo Iuliano imp.

See **A-18** above.

B-16 Prudentius (AD 348/349 - after AD 405)

Contra Symmachum

fluctibus Actiacis signum symphonia belli
Aegypto dederat, clangebat bucina contra.
institerant tenues cumbae fragilesque phaseli
inter turritas Memphitica rostra **Liburnas**:
nil potuit Serapis deus et latrator Anubis.
stirpis Iuleae ductore exercitus ardens
praevaluit, gelido quem miserat Algidus axe.
2.528-534

It was only a musical instrument that gave Egypt the signal for battle on the waters at Actium, a while on the other side the trumpet blared. Slight boats and frail yachts pressed their Egyptian rams amid towered **galleys**, but their god Serapis and their barking Anubis were powerless. The eager army led by a scion of the Julian stock and sent by Algidus from a cold clime outmatched them.
p. 49

5. Written Sources on lembs and Liburnians from the 4th c. BC to Late Antiquity

Latin and English: *Prudentius. Against Symmachus 2. Crowns of Martyrdom. Scenes From History. Epilogue*, transl. H. J. Thomson. LCL 398 (1953).

B-17 Paulinus Nolanus (AD 353-431)

Epistulae

8. [...] uigesimo demum et tertio die miserante iam deo finem errorum atque discriminum in Lucanis positura litoribus.
quibus propinquanti dominus aeternus, ut indefessa eius bonitas usque ad terminum susceptae sibi nauigationis operaretur mirabilia sua, tacito suae adspirationis instinctu conpulsos a littore duabus nauiculis piscatores obuiam misit, qui nauem hanc eminus conspicati primo aspectu territi refugerunt; plenam enim armatorum et **liburnae** aemulam sibi uisam ipsi postea retulerunt. deinde magnis et saepe repetitis senis nostri uocibus aliquando reuocati ratione se cum habita et agente sic domino, nihil sibi ab ea naui timendum, in quam uocarentur, intellexerunt.
48.8

8. [...] Finally God showed pity, and on the twenty-third day it was to put an end to its wanderings and dangers on the shores of Lucania. As it approached land the eternal Lord willed His unwearying goodness to perform its miracles to the very end of the voyage He had undertaken; so with the silent instigation of His inspiration, He induced some fishermen to put out from the shore in two skiffs and meet the ship. When they saw it from a distance they were terrified at first sight and fled; later they said that it seemed to them to be full of armed men and **like a fast pirate vessel**. But subsequently they were drawn back by the loud repeated shouts of that old man of ours. After taking stock of the situation, and under the guidance of the Lord, they realised that they had nothing to fear from the ship to which they had been summoned.
pp. 265-266

Latin: *S. Pontii Meropii Paulini Nolani Opera Vol. 1: Epistulae*, eds G. Hartel, W. von Hartel [Corpus Scriptorum Ecclesiasticorum Latinorum 29]. Österreichische Akademie der Wissenschaften, Vienna 1894.

English: *Letters of St. Paulinus of Nola, Vol. 2: Letters 23-51*, transl. P. G Walsh. The Newman Press, New York-Ramsey 1966.

B-18 Orosius (4th/5th century AD)

Historiarum aduersum paganos libri VIII

7 Inde Corcyram cepit; fugientes nauali proelio persecutus profligauit, multisque rebus cruentissime gestis ad Caesarem uenit. Antonius defectu et fame militum suorum permotus bellum maturare instituit ac repente instructis copiis ad Caesaris castra processit et uictus est.
8 Tertio post pugnam die Antonius castra ad Actium transtulit, nauali proelio decernere paratus. Ducentae triginta rostratae fuere Caesaris naues et triginta sine rostris, triremes uelocitate **Liburnicis** pares et octo legiones classi superpositae, absque cohortibus quinque praetoriis.
6.19.7-8

7. He [Agrippa] then captured Corfu, chased and crushed in a sea-battle those who fled, and, after performing many bloody deeds, returned to Caesar. Antony, troubled because his soldiers were deserting and starving, decided to force the issue. He suddenly marshalled his army, marched on Caesar's camp, and was defeated.
8. Two days after this battle, Antony moved his camp to Actium and made ready to fight at sea. There were 230 warships, and 30 ancillary ships – triremes equal in speed to **Liburnians** – in Caesar's fleet. Eight legions along with an additional five praetorian cohorts were embarked on the fleet.
p. 306-307.

Latin: *Orose: Histoires contre les païens*, 3 vols, ed. M.-P- Arnaud-Lindet. Les Belles Lettres, Paris 1990–91.

English: *Orosius Seven Books of History against the Pagans*, transl. and intr. A. T. Fear. TTH 54 (2010).

For us it is particularly interesting to compare Orosius' mentioning of liburnians with his (probable) contemporary Vegetius (below, **B-19**). For Orosius, as the context shows, *liburnica* was a name for a particular vessel to which triremes were compared. This was particularly in relation to one characteristic of the liburnians, their swiftness, which implies that triremes were not liburnians, properly speaking. This is quite a difference from Vegetius for whom *liburna* is a generic name for warships.

B-19 Flavius Vegetius Renatus (after AD 383 - before AD 450)

De re militari

B-19.1
Classis item duo genera sunt, unum **liburnarum**, aliud lusoriarum.
2.1

There are likewise two sorts of navy, one of **warships**: the other of river patrol-boats.
p. 30

5. Written sources on lembs and Liburnians from the 4th c. BC to Late Antiquity

B-19.2
XXXII. NOMINA IVDICVM QVI PRAEERANT CLASSI.
Liburnis autem, quae in Campania stabant, praefectus classis Misenatium praeerat, eas uero, quae Ionio mari locatae fuerant, praefectus classis Rauennatium retinebat; sub quibus erant deni tribuni per cohortes singulas constituti. Singulae autem **liburnae** singulos nauarchos, id est quasi nauicularios, habebant, qui exceptis ceteris nautarum officiis gubernatoribus atque remigibus et militibus exercendis cottidianam curam et iugem exhibebant industriam.
4.32

32. The titles of the officers commanding the fleet.
The prefect of the fleet at Misenum was in command of the **warships** stationed in Campania, whilst those located on the Ionian Sea were controlled by the prefect of the fleet at Ravenna. Under each of them were ten tribunes appointed one for each cohort. Each **warship** had a single navarch, that is, a kind of skipper, who was exempted from the other duties of sailors and put in a daily responsibility and unfailing efforts to training pilots, oarsmen and marines.
p. 141

B-19.3
XXXIII. VNDE APPELLENTVR **LIBVRNAE**.
Diuersae autem prouinciae quibusdam temporibus mari plurimum potuerunt, et ideo diuersa genera nauium fuerunt. Sed Augusto dimicante Actiaco proelio, cum Liburnorum auxiliis praecipue uictus fuisset Antonius, experimento tanti certaminis patuit **Liburnorum naues** ceteris aptiores. Ergo similitudine et nomine usurpato ad earundem instar classem Romani principes texuerunt. Liburnia namque Dalmatiae pars est Iadertinae subiacens ciuitati, cuius exemplo nunc naues bellicae fabricantur et appellantur **liburnae**.
4.33

33. How **warships** got their name
Different provinces at various times held considerable naval power and therefore the types of ships were diverse. But when Augustus was fighting at the battle of Actium and Antony was beaten mainly by the auxiliaries provided by the Liburni, it became clear from the experience of that great encounter that **the ships of the Liburni** were better-designed than the rest. Therefore usurping the likeness and the name, the Emperors built the fleet according to their pattern. Liburnia is a part of Dalmatia lying next to the city of Iadera; ships of war are built today on their model and are called *liburnae*.
pp. 141-142

B-19.4
XXXIIII. QVA DILIGENTIA FABRICENTVR **LIBVRNAE**.
Sed cum in domibus substruendis harenae uel lapidum qualitas requiratur, tanto magis in fabricandis nauibus diligenter cuncta quaerenda sunt, quia maius periculum est nauem uitiosam esse quam domum. Ex cupresso igitur et pino domestica siue siluestri larice et abiete praecipue **liburna** contexitur, utilius aereis clauis quam ferreis configenda; quamlibet enim grauior aliquanto uideatur expensa, tamen, quia amplius durat, lucrum probatur afferre; nam ferreos clauos tepore et umore celeriter robigo consumit, aerei autem etiam in fluctibus propriam substantiam seruant.
4.34

34. The care with which **warships** are built.
As when building houses the quality of the sand or stone of the foundations is important, so the more carefully should all materials be obtained when building ships, because it is more dangerous for a ship to be faulty than a house. So the **warship** is constructed principally from cypress, domestic or wild pine, larch, and fir. It is better to fasten it with bronze nails than iron; for although the cost seems somewhat heavier, it is proved to be worthwhile because it lasts longer, since iron nails are quickly corroded by rust in warm, moist conditions, whereas bronze preserve their own substance even below the water-line.
p. 142

B-19.5
XXXVII. DE MODO **LIBVRNARVM**.
Quod ad magnitudinem pertinet, minimae **liburnae** remorum habent singulos ordines, paulo maiores binos, idoneae mensurae ternos uel quaternos interdum quinos sortiuntur remigio gradus. Nec hoc cuiquam enorme uideatur, cum in Actiaco proelio longe maiora referantur concurrisse nauigia, ut senorum etiam uel ultra ordinum fuerint. Scaphae tamen maioribus **liburnis** exploratoriae sociantur, quae uicenos prope remiges in singulis partibus habeant, quas Britanni picatos uocant. Per has et superuentus fieri et commeatus aduersariorum nauium aliquando intercipi adsolet et speculandi studio aduentus earum uel consilium deprehendi. Ne tamen exploratoriae naues candore prodantur, colore Veneto, qui marinis est fluctibus similis, uela tinguntur et funes, cera etiam, qua ungere solent naues, inficitur. Nautae que uel milites uenetam uestem induunt, ut non solum per noctem sed etiam per diem facilius lateant explorantes.
4.37

37. The size of **warships**.
So far as size is concerned, the smallest **warships** have one rank of oars a side, those slightly bigger two ranks, those of appropriate dimensions three, four, sometimes five ranks for their oarage. This should not seem enormous to anyone, because at the battle of Actium far larger vessels are reported to have clashed, so that these were of six and even more ranks. But to the larger **warships** are attached scouting skiffs, having about twenty oarsmen on each side; these the Britons call *picati*. They are used on occasion to perform descents or to intercept convoys of enemy shipping or by studious surveillance to detect their approach or intentions. Lest scouting vessels be betrayed by white, the sails and rigging are dyed Venetian blue, which resembles the ocean waves; the wax used to pay ships' sides is also dyed. The sailors and marines put on Venetian blue uniforms too, so as to lie hidden with greater ease when scouting by day as by night.
pp. 143-144

Latin: Flavius Vegetius Renatus, *Epitoma rei militaris*, ed. A. Önnerfors. Teubner, Stuttgart 1995.

English: *Vegetius: Epitome of Military Science*. Transl. N. P. Milner, 2nd ed. TTH 16 (1996).

> V. was evidently no military practitioner: he draws on written sources and combines actual experiences with the thoughts of his predecessors. (H. Brandt, *NP*).

For our purposes, this text is of a particular interest. Vegetius claims (**B-19.3**) that the Roman warships, starting from Augustus, were built on the model of original warships built by Liburnians. He claims that at the battle at Actium, the Romans realised the ships' superiority in battle. This led to the word 'liburnian' being used for all warships of whatever size by the late Empire. From this perspective, (that the term 'liburnian' denotes *all* kinds, or at least predominant kind of warship) one can of course question how much his description corresponds to the 'original' liburnian of the 1st century BC. Nevertheless, it is evident that 'liburnian' did not refer to any specific size of ship in this instance, but rather the shape, which then can be built bigger or smaller (**B-19.5**).

6. Discussion

As already pointed out in the introduction, there is a prevailing view in a modern scholarship that lembs and liburnians were closely related. The two most authoritative Anglophone scholars, Casson and Morrison, explain it in the following way.

Casson writes:

> *The liburnian was a fast, two-banked galley adapted from a craft developed among the Liburnians, piratical-minded dwellers of the Dalmatian coast and its off shore islands. We know that in the third and second centuries B.C. these people were using lemboi, that some of these were pressed into Rome's service, and that at least one model of lembos had two levels of oarsmen. All this makes a strong case for there being a connection between the lembos and liburnian. There were, as noted above, many kinds of lemboi; no doubt the liburnian was the one that the Romans found particularly useful for their purposes. The earliest certain mention of liburnians is at the Battle of Naulochus in 36 B.C., but there is no reason why they could not have been in use long before.*[215]

Morrison accepts general positions of Casson, saying that: *These liburnians which played a large part in the Roman imperial fleets, the name eventually becoming like τριήρης; a general word for warship, are then a local kind of λέμβοι, essentially with oars at two levels.*[216] This is also restated later in the book: *As with lemboi, of which they [sc. liburnians] were a type...*[217]

Leaving aside the questionable claim whether the earliest *certain* mention of liburnians could be dated to 36 BC, as they were undoubtedly used by Caesar in 49 BC, and probably by Pompey in the 60s BC (see p. 177, below), Casson and Morrison argued that the Liburni were using lembs with an implication that lembs were authentically Liburnian vessels and that the liburnians are essentially a sub-type, or regional variation of lembs. These views remain accepted in some recent publications, within which we can also find factually unsupported claims, such as that the Romans first came into contact with lembs in the First Illyrian War in 229 BC, and that liburnians were developed from captured 'Illyrian' crafts.[218] However, such an idea seems to have no definite support in the available sources and needs to be discussed in some detail.

6.1. Lemb

Apart from the written evidence from the 3rd and early 2nd centuries BC, which comes mostly from the histories of Polybius and Livy, it is likely that some images of ships from the coins of the south Adriatic communities minted in this same period depict those Illyrian versions of lembs (see also 4.2.6. above).[219] There are no underwater finds which might be securely identified as this type of ship, although it was cautiously suggested that the wreck from Vela Svitnja bay near the island of Vis, dated to the 2nd or early 1st century BC, might have been

---
[215] Casson 1971: 141-42.
[216] Morrison 1996: 264.
[217] Morrison 1996: 317.
[218] Pitassi 2011: 89-90, 106-09.
[219] Casson 1971: 125-27; Kozličić 1980/81; 1993: 29-32; Morrison 1996: 263-64; Höckmann 1997: 194; 2000: 138; Medas 2004: 131-34, etc. The connection between Illyrian *lembos* and *liburnica* is made by almost all authorities who oversimplify complexity of identities amongst the communities from eastern Adriatic coast. For different opinions see Džino 2003 and Medas 2004.

a cargo version of this type of ship.[220] It does not even seem that south Adriatic communities invented the lemb. The lemb is mentioned a few times in the Aegean context a century before the first mention of south Adriatic (Illyrian) lembs found in accounts of the events from 231 BC leading up to First Illyrian War. Aristotle describes the shape of the birds as 'lemblike', Anaxandrides calls a flatterer trailing behind the flattered a 'lemb', while Demosthenes and Lycurgus understand lemb to be a small ship used around larger ones, probably a ship's boat. Furthermore, Pliny the Elder ascribes its invention to the Cyrenians, not Illyrians.[221] Thus, the initial purpose of this type of boat was clearly a towing boat, ship's boat, or transportation vehicle. Lemb was also used as a cargo vessel, for which the best evidence is the description of cargo from a lemb transporting oil from Samos and Miletus to Alexandria in 259/258 BC, mentioned in the *Cairo Zenon Papyri*. This role of lemb as a transport ship in the 3rd century BC is also confirmed in the evidence provided by the *Petrie papyrus* from 218 BC.[222]

Aside from these commercial activities, the lemb was used in naval warfare in different ways. Demetrios Poliocertes used the lembs as moving platforms for catapults and archers during the siege of Rhodes 305/304 BC. The use of lembs as carriers for light catapults in this period is also described by Philo of Byzantium (Philo Mechanicus) who lived in *c.* 280-220 BC. There is also evidence, coming from Polybius, that the Roman fleet utilised the lemb as a scout ship in the First Punic War.[223] The southeast Adriatic communities therefore did not invent the lemb, but rather adopted and improved this type of ship for different purposes: the swift transportation of troops, and the interception and robbing of sea-trade vehicles.[224] Some authors also suggest that the Aetolian League started to use lembs for the very same reason the Illyrians did contemporaneously in the late 3rd century BC.[225] As noted, this type of ship had advantages for piratical use because it was fast and spacious for carrying cargo, and it seems that the Aegean pirates also started to employ lembs around the same period. According to Livy, the pirate fleet which had base in Myonnesus, a promontory located in Asia Minor close to the island of Samos and in the vicinity of the modern-day islet of Çıfıtkalesi, also used lembs in 197 BC. Their fleet was composed of lembs, and the other type of fast galley, for piracy on account of its ability to carry cargo – the *keles*. This information must have been coming from Polybius, but unfortunately only a fragment mentioning this event was preserved in the Suda lexicon.[226] As such, we cannot be certain whether the pirates from Myonnesus adopted the Aegean lemb for their piratical needs, used the south Adriatic model, or improved the Macedonian prototype of the Illyrian lemb (see p. 175-76 below).

Posidonius mentions pirate lembs in the 2nd century BC, informing his readers that this type of ship was used for piracy, but does not describe a particular example of where they might have been used.[227] Another reference to lemb as a boat used by pirates outside the Adriatic is in

---

[220] Kirigin *et al.* 2006: 196-97, calculating the cargo of the shipwreck and comparing it with cargo of a lemb mentioned in the *P. Cairo Zen.* (see A-V). For the shipwreck see Cambi 1972; Vrsalović 2011 [1978]: 119-20.
[221] Arist. *De motu anim.* 710; Anaxandrides, Frag. 35 (ap. Ath. 6.242F); Dem. *C. Phorm.* 10.7; Zenoth. 6.4-7.5; Lucyrg. *Leoc.*17; Plin. *HN* 7.208.
[222] *P. Cairo Zen.* 59015; *P. Petr.* 2.20 iv. See p. 65-69 above.
[223] Rhodes: Diod. Sic. 20.85.3; platforms for catapults: Philo Mech. D21, D38; Murray 2012: 133, 135, 141-42, 164, 201; First Punic war: Polyb. 1.53.9, cf. App. *Pun.* 50.
[224] The sources for Illyrian *lembs* are abundant, see Džino 2003: 23-27 and Chapter 5 above, especially **A-IX** (Polybius) and **A-8** (Livy).
[225] Scholten 2000: 107, 150
[226] Livy, 37.27.4; Polyb. 21.12. See Casson 1971: 160-62 for *keles*, without mentioning its use for piracy, and **A-V** above.
[227] Posidonius, F28 §12 (=Strabo, 2.3.4).

the letter of Alciphron, who likely lived in the 2nd or 3rd century AD. Alciphron indicates that the lemb was used by the Attalian pirates in the eastern Mediterranean.[228] The letter appears to be fictious, but it is clear that the information about the pirates would have been believable for the intended audience of Alciphron's letter. Finally, there is a more cryptic connection between pirates and lembs in the Aegean made in Rufinus' epigram from *Anthologia Palatina*. Rufinus, who lived in the 3rd or 4th century AD, names two dockside prostitutes from the harbour of Samos 'Lembion' and 'Cercurion' after the small boats lemb and *kerkouros/cercurus*. The epigram compares the prostitutes with pirate ships who rob and sink other ships as, apparently, these ladies did with their customers by depraving them of money.[229] While this is not evidence for the lembs and *kerkouroi* being used in the 3rd or 4th century AD for piracy, the play on words with the names of the prostitutes and piracy must have been apparent for Rufinus' audience.[230]

An passage often quoted from Polybius shows how Illyrian lembs in 229 BC captured larger enemy ships, using infantry to board them without any hint that these same lembs had ramming capabilities. Instead, the Illyrians sacrificed some of their ships by lashing them together in batches of four to capture the ram of an opponent's ship which had rammed them. The crews of the lashed ships boarded the helpless ship and captured it.[231] This was probably a cleverly devised ad hoc strategy, for it would be too expensive to regularly sacrifice so many ships in a naval battle. It looks as though these south Adriatic ships were principally used for the transport of troops, for raids on unarmed opponents, and perhaps for policing the seas, but not for naval battles.[232] The Roman navy had not encountered even the slightest problem in defeating the navies of the Illyrian kingdom in 229, 219 and 168 BC. Most probably due to space restrictions, the rowers doubled as foot soldiers, and a number of around 30 rowers seems to be the standard for an Illyrian lemb, although written sources mention smaller and larger models with 16 and 50 rowers respectively.[233]

King Philip V of Macedon ordered 100 lembs to be built by Illyrian shipbuilders for the transportation of his troops in 216 BC.[234] His shipbuilders experimented further with the ship type by enlarging it into a bireme class warship in 214 BC, even fitting it with a ram. This is best seen in the battle of Chios against the combined fleets of the Rhodians and Pergamonians (201 BC), where the Macedonian lembs were fitted with rams.[235] The lembs were also recorded in different Hellenistic navies at that time, but it is not clear whether this relates specifically to Illyrian lembs, their modified Macedonian versions, or some other modifications of original Aegean lemb. Here, it is also important to note that the Adriatic Greeks, such as the Issaeans, of the time also used lembs, which are much more likely to be influenced by the south Adriatic

---

[228] Alciphr. 1.8.2.
[229] *Anth. Pal.* 5.44,
[230] Cf. Casson 1971: 163-66 for *kerkouros*, a supposed auxiliary ship and cargo carrier which Casson claims was not mentioned after first century BC; a claim which is disproven by Rufinus' epigram.
[231] Polyb. 2.10.3-5.
[232] Gabrielsen 2001 makes the valid point that the same types of the ships were used for raiding and policing the seas in the Hellenistic period. Thus, it is likely that Illyrian lembs were used for this purpose too.
[233] Different numbers of rowers are given by the sources: 50 rowers: Polyb. 2.3.1; 30 rowers: Polyb. 21.43.13, 16 rowers: Livy, 34.35.5.
[234] Polyb. 5.101.1-3, cf. Kleu 2015: 46-49.
[235] Livy, 24.40.2, cf. Kleu 2015: 58 n.259 (*lembs* as biremes); Polyb. 16.4.8-12 (ramming capabilities). It seems that *lemb* with ram was called *pristis*, see Casson 1971: 126-27; Morrison 1996: 263.

model.²³⁶ It seems that modifying the lemb for naval warfare was ultimately not successful, and the lemb was not mentioned in a military context in the ancient sources after the mid 2nd century BC. The only exception was the mention of lembs in connection with the legate Otacilius from 89(?) BC, and Caesar's brief reference to small lembs (*lenunculi*) used for the evacuation of his troops commanded by G. Scribonius Curio (Younger) in Numidia.²³⁷ While Otacilius might have had used lembs for fighting, Caesar's troops in North Africa used them as an auxiliary ship of rather small size, as they quickly sunk when panicking troops started to board them.

This form of the vessel was, in later periods, mentioned only very rarely. There is only one mention of a lemb as a cargo ship in Strabo's description of the African circumnavigation of Eudoxos (*c.* 118 or 116 BC), an account originally coming from Poseidonius.²³⁸ In the Roman era, lembs are mentioned, for example, in a list of ship types given by Aulus Gellius in the 2nd century AD,²³⁹ and as later sources, listed in section 5.2 above, demonstrate, the term continues in later periods to describe a small vessel, river boat or fishermen's boat in any part of the Mediterranean. Epigraphy is not helpful in this regard, but at least it is possible to ascribe the *V corpora* of *lenunculari* in the port of Ostia as a professional association of ferrymen in this large port.²⁴⁰

6.2. Liburnian

Liburnica, *liburna*, or simply liburnian, is a well-known type of warship widely used in the Roman imperial fleets as the smallest class of warships from *c.* the 1st century BC. Two sources ascribe the origins of the ship to the Liburni: Appian and Vegetius.²⁴¹ These ships are never mentioned in the original context before their inclusion in the Roman fleets. The Λιβυρνικὰ σκάφη in Stephanus of Byzantium's reference to the Liburni is more likely a later addition, although Stephanus cites Hecataeus from the 6th century BC as a source for some information before mentioning the boat of the Liburni.²⁴² It is very difficult to assess what kind of changes were introduced when the Romans adopted the liburnian, when it is, at this point in time, impossible to establish how the original looked. The only possible hint comes from the fragment of Philoxenus of Alexandria, who lived in the 1st century BC and compared liburnian (*liburnos*) with *gaulos* – the Phoenician beamy and rounded type of merchant ship. The fragment is problematic for reading on account of its fragmentary nature, and its original context is unclear.²⁴³ The *gaulos* is relatively well known to the Greek authors from *c.* the 5th to the 3rd centuries BC, and is even mentioned as late as Plutarch in the late 1st and early 2nd centuries AD.²⁴⁴ It is impossible to deduct whether Philoxenus refers to his present, or (perhaps more likely) transfers the information from an earlier source. The implication of this comparison is that the liburnians were originally of a beamy and rounded shape. This could certainly be

---

[236] Livy, 33.19.10; 34.26.11 (*lembs* in the Seleucid and Rhodian navies); 31.45.10; 32.21.27 (Issaean *lembs*).
[237] Sisenna, *Hist.* Fr. 38 (= Cornell 2013, Fr. 47).
[238] *BNJ*, 87 F28 §12-13 (=Strabo, 2.3.4).
[239] Gell. *NA* 10.25.5.
[240] CIL 14.4144; AE 1987, 176a and 195-96.
[241] App. *Ill.* 3.7; Veg. *Mil.* 4.33. Appian (*B Civ.* 2.39) states that the Romans call all fast ships *liburnica*, because the ships of the Liburni were the first fast ships they came in conflict with.
[242] Steph. Byz. 415.10 s.v. Λιβυρνοί (BNJ 1 F93). This reminds to the use of the term 'Illyrians' in Stephanus, cf. Dzino 2014: 47-49.
[243] Philox., Frag. 79b, see also **B-II** above.
[244] Casson 1971: 66, 159 n.7.

applied to the *serilia*, which were beamy and rounded cargo ships, and it is even possible to draw distant parallels with the representation of the ship from the Nesactium *situla* (above, 4.2.3). However, it is more difficult to imagine the Liburnian, a ship renowned for speed, as beamy and rounded. If we accept, for the sake of the argument, this fragment of Philoxenus as trustworthy, it is possible to speculate that the liburnian from the 1st century BC evolved from a ship used in a similar contexts as the lemb – specifically, a spacious ship useful for trade and piracy. In all certainty, the Roman liburnian was a product of several centuries of development, influenced by global changes in Mediterranean shipbuilding design, and the changing needs of the local Liburnian communities who developed the original prototype. Hypothetically, what the early Roman sources called *liburnica* might have been a recent type of indigenous ship.[245]

The term *liburnica* became the standard name for Roman warships in the later imperial period. Thus, it starts to be difficult to distinguish whether the term is used in the sources generically or in reference to a specific type of warship. This makes a significant problem when assessing sources from the 2nd century AD onwards, as they might have used the term as a general term for warships or light warships. The recent assumption that the liburnian was already introduced into the Roman fleets during Pompey's war against the pirates in the 60s BC is acceptable, but needs more substantial evidence.[246] One of the earliest mentions of the liburnian relates to Cato Minor's governorship at Cyprus in 58 BC, but it does not come from a contemporary source.[247] In a contemporary context, the liburnian is not mentioned before the civil war between Caesar and Pompey. Describing the events from 49 BC, Caesar states that Pompeian commander M. Octavius arrived with his ships to besiege Salona, the future capital of Roman Dalmatia, after the departure of *liburnarum ex Illyrico*. As Čače pointed out, the word *liburnarum* relates to the type of ship, not the Liburnian squadron, which would make this the earliest contemporary mention of liburnians in written sources.[248] In this conflict, the contingent of Liburnian allies is specifically mentioned by Caesar as a part of the Pompeian combined Liburnian-Achaean squadron commanded by Scribonius Libo and Marcus Octavius. However, Caesar does not specify which kind of warships were used by the Liburni. Plutarch mentions 500 warships in the Pompeian fleet, together with undetermined large number of liburnians and swift vessels, distinguishing the liburnians from the warships.[249] It is again unclear whether Plutarch refers to light ships in general, as would align with the perspective of his times, or whether his original source referred to a liburnian as a particular type of the warship.

In a contemporary context, this type of ship is not mentioned again until the Augustan era.[250] They participated in the battle against Sextus Pompeius in 36 BC, when Octavian made a liburnian his flagship.[251] Liburnians were also prominent in the decisive naval battle off Actium, between Octavian and Marc Antony in 31 BC. However, the myth that the liburnians decided the outcome of this battle comes from the contemporary propaganda, which presented

---

[245] Dzino, Boršić 2020.
[246] Bérchez Castaño 2010: 70; Reddé 1986: 105-06, cf. Čače 1993: 17-23.
[247] Flor. 1.44 (3.9.5). Eutr. 2.20.2 uses the term 'liburnian' in relation to the First Punic war, but taking into account the date of the composition, we cannot assume that liburnian was really used at that time. Rather, Eutropius anachronistically used the term which in his times depicted generic warship.
[248] Caes. *BCiv.* 3.9; Čače 2013: 36-38.
[249] Caes. *BCiv* 3.5; Plut. *Pomp.* 64.
[250] Prop. 3.11.44.
[251] App. *B Civ.* 5.111.1; Cass. Dio, 50.31.3.

Octavian as an underdog to Antony, developing into a quite persistent tradition.[252] As can be deduced from the written sources, the liburnians were used in this battle as auxiliary ships to warships of larger classes, and only occasionally were they involved in actual fighting, so it is impossible to maintain the claim that they decided the battle.[253] There are numerous mentions of liburnians in written, papyrological and epigraphic sources from the imperial period, whether as depictions of liburnian proper, or as generic depictions of the Roman warships in general.

Scholars disagree which existing visual representations from reliefs, paintings and coins depict a Roman liburnian, and therefore, disagree on its distinguishing features.[254] Appian describes the Roman liburnian as a fast ship of biremes/δίκροτα class with a distinguishable pointy prow.[255] As said at the beginning of the book (above p. 2), these terms are generally taken to depict a ship with two rows of oars, with one rower per oar. However, if the already mentioned hypothesis that the prefix 2 in the names of ancient warship-classes depicts a total number of rowers at the rowing bench (cross-side), not the number of banks of rowers on one side of the ship is correct,[256] this search for liburnian is founded on the wrong premises, as scholars looked for the ship with two banks of oars, instead of the ship that has one row of oars on each side.[257] Thus, the question of the *liburnica*'s visual representation in antiquity remains an unresolved issue.

6.3. Etymology

Although etymology has been brought as evidence for linking south Adriatic lembs and liburnians, with a more careful analysis it seems to be of little help. For instance, in the case of 'liburnian', the name of the vessel is obviously derived from an ethnonym, one which refers to the communities described as the Liburni in ancient sources, and who inhabited Liburnia: the territory and adjacent islands between the rivers Arsia (modern Raša) and Titius (modern Krka) (see p. 11). The Latin names of the vessel, *liburnica* and *liburna*, were promiscuously used – *liburnica* being an adjective, describing the (often missing) noun *navis* – literally, 'Liburnian ship' – whereas *liburna* is a noun relating to the type of the ship. This is very clear from the epigraphy, for in imperial times only *liburna* is mentioned as word that describes a type of warship used in the Roman imperial fleets.[258] Although the Greeks must have been familiar with the Liburni from fairly early times, as we can see from the *periploi* of Pseudo-Scylax and Pseudo-Scymnus, and probably even Hecataeus, there is no extant and available Greek source mentioning liburnian (ships) that would predate Latin sources. The only exception would be if Philoxenus of Alexandria used some older sources, but this is not possible to claim with certainty, as has been outlined above. Thus, the Greek terms λιβύρνα/ λίβυρνος/λιβυρνίς should be considered as derivatives from the Latin *liburn(ic)a*.

---

[252] Hor. *Epod.* 1.5; Prop. 3.11.44; Plut. *Ant.* 62; Veg. *Mil.* 4.33, see Tarn 1931: 193 n.8 and Panciera 1956: 136-38.
[253] Morrison 1996: 157-70; Murray 2012: 235-44.
[254] Panciera 1956: 154-56; Reddé 1986: 106-10; Orna-Ornstein 1995; Morrison 1996: 248-53; Höckmann 1997; Pitassi 2011: 106-09, 138-44 – all making more or less valid and often mutually different points.
[255] App. *Ill.* 3.7.
[256] Tilley 2007.
[257] We should not forget Pliny's likening of the liburnian from his times (ie. the Roman liburnian) to the *nautilus*, which might indicate that the Roman version of the ship could have indeed been a bireme, Pliny, HN 9.88, cf. Panciera 1956: 140.
[258] See epigraphic evidence listed in Morrison 1996: 172-74, with the additions of AE 1939, 231; 1961, 120; 1967, 429; 1990, 205, etc.

6. Discussion

The Greek term λέμβος can be found in texts that are older than the oldest Latin mentions of *lembus*, making it almost certainly a loan word from Greek. However, the etymology of λέμβος is inconclusive. According to *Etymological Dictionary of Greek*, λέμβος is [a] *foreign word without etymology; perhaps Illyrian*.[259] There are also other older etymological lexica that offer a similar interpretation of the word λέμβος which is uncritically adopted by Casson as well.[260] Thus, the scholars who argue that liburnians were a type of lemb rely, among other things, upon the alleged Illyrian etymological origin of the word 'lemb', which brings the two types of vessels into close linguistic proximity.

There are two problems with this position. First, there seems to be no hard linguistic evidence that would support an 'Illyrian' etymological origin of the word, demonstrated by authorial reservations such as 'perhaps' in Beekes, van Beek, and 'may be' in Whatmough. Even Krahe, who thought of linking the two names more firmly, put a question-mark in the title of his article and inserted several other carefully restricting claims in his text, which testify that his attempt to etymologically link the two names should remain open. However, there is an even bigger problem with this etymological linking, and that is the extremely scarce linguistic evidence for an 'Illyrian' language, which does not allow even a basic reconstruction of the language(s) spoken by the indigenous populations of the western Balkan peninsula.[261] Even if we, for the sake of the argument, accept the hypothesis that the word lemb is of 'Illyrian' origin, it is difficult to link it to the Liburni, since the Iron Age Liburni likely did not speak an 'Illyrian' language, but rather a language akin to northern Adriatic groups, such as the Histri and Veneti (see also p. 11).[262]

---

[259] Beekes, Beek 2010: 847.
[260] Casson 1971: 126 n.104, citing Conway *et al.* 1933: 64 and Krahe 1952. Also in similar tune Pokorny (1959-1969: 660-61): 'illyr. *lembus* (*$lengʷho$-s*) 'leichtes Fahrzeug', daraus gr. λέμβος, lat. *lembus* ds.; oberital. FlN *Lambrus* (: ἐλαφρός)'.
[261] See recently Dzino 2012: 70, 82, 87.
[262] Also called the 'north Adriatic group of languages' - Katičić 1976: 179; Polomé 1982.

Table 1: Lemb in Greek and Roman written sources (L. Boršić)

Author	The time of writing	Text nr.	Time referred to in the text	Origin of lembs	Deployment of lembs	Description of lembs
Demosthenes	c. 350 BC?	A-I.1	c. 350 BC	Massalia?	Athens-Syracuse	A smaller vessel accompanying a larger ship, that could serve also as a salvage vessel, could carry more than thirty people.
		A-I.2		Greece?	Athens-Bosporus	
Lycurgus	c. 330 BC	A-II	338 BC	Athens?	Athens-Rhodes	A smaller boat made to reach a larger one.
Aristotle	c. 350 BC?	A-III	-	-	-	Birds' breastbones are compared to lembs: strong and acute.
Anaxandrides	376-349 BC	A-IV.1	-	-	-	'Lemb' is used as a term the size of which cannot be determined.
		A-IV 2	-	-	-	Derogatory, referring to a person that follows someone else, presumably more important.
Zeno of Kaunos	c. 258 BC	A-V	c. 258	Milet, Samos?	Alexandria, Egypt	Freight boats
Plautus	c. 250 – c. 184 BC	A-1.1	-	-	-	Pirate fast-sailer
		A-1.2	-	-	-	Pirate ship
		A-1.3	-	-	-	Pirate ship
Theophilos?	c. 218 BC	A-VI	c. 218	-	The Nile	Freight boat (22.5 tons, Casson 1993)
Theocritus	c. 3rd c. BC	A-VII	-	-	-	Fishermen's boat
Philo of Byzantium	c. 280 – 220 BC	A-VIII.1-3	-	-	-	Can be tied up together; war machines (catapults?) can be put on them; can be equipped with some protective sheds with holes so that stone-throwers can catapult the stones.
Polybius	3rd/2nd c. BC	A-IX.1	261 BC	-	-	Lembs are used as symbols of the weakest kind of ships that can be used in a war.
		A-IX.2	249 BC	Roman	Lilybaeum	Lembs are used to sail in front of the fleet.
		A-IX.3	233-232 BC	Illyria	Medion (Acarnania)	100 lembs can cary 5000 people.
		A-IX.4				
		A-IX.5	230 BC		Epirus	-

6. Discussion

Author	The time of writing	Text nr.	Time referred to in the text	Origin of lembs	Deployment of lembs	Description of lembs
		A-IX.6	229 BC		Corcyra	-
		A-IX.7			Paxi (Corcyra)	Four lembs were lashed together to trap the Achaean ships.
		A-IX.8	229 BC?		Nutria?	-
		A-IX.9	228 BC		-	-
		A-IX.10	220 BC		Cycladas	-
		A-IX.11	219 BC		Pharos	-
		A-IX.12	218 BC	the Celts? used by Hannibal	The Rhone	Lembs bigger than 'canoes'; lembs can be used on rivers; there are usually many lembs in an expedition; used to carry the elephants across the river.
		A-IX.13				
		A-IX.14				
		A-IX.15	220-216 BC	Illyria (the Labeateans, Scerdilaïdas)	Illyria-Lissus-Pylos-Cyclades-Naupactus	-
		A-IX.16				
		A-IX.17	218 BC		-	-
		A-IX.18	219 BC	Illyria (Demetrius of Pharos)	Gulf of Ambracia	-
		A-IX.19	218 BC?	Illyria (the Labeateans, Scerdilaïdas)	Cephallenia	-
		A-IX.20	217 BC		Leucas	-
		A-IX.21	217 BC?		Cape Malea	-
		A-IX.22	216 BC?	Macedonians	-	The Macedonian king Philip ordered 100 lembs to be built by Illyrian shipwrights. Polybius however does not claim that lemb was originally an Illyrian ship. Lembs are fast boats.
		A-IX.23	201 BC?		Samos	-
		A-IX.24	201 BC?		Chios	-
		A-IX.25				

Author	The time of writing	Text nr.	Time referred to in the text	Origin of lembs	Deployment of lembs	Description of lembs
		A-IX.26				
		A-IX.27				
		A-IX.28	198-197 BC		Malian Gulf	–
		A-IX.29	168 BC?		The Aegean	–
		A-IX.30	168 BC?	Egyptians	Egypt–Rhodes	–
L. Accius	170 – c. 90 BC	A-2	–	–	–	Small fishermen's boat?
S. Turpilius	2nd c. BC	A-3	–	–	–	–
Posidonius	135 – c. 51 BC	A-X.1-2	c. 130 BC	Eudoxus	circumnavigation of Africa	Tow-boats are compared to 'pirate lembs'; a lemb is size of a penteconter; a sailing boat.
L. Cornelius Sisenna	c. 118 – 67 BC	A-4	c. 90 BC	Roman?	–	Lembs were already part of Roman fleet around 90 BC.
M. Terentius Varro	116-27 BC	A-5	–	–	–	The term 'lemb' used for the zodiac.
Caesar	100-44 BC	A-6	49 BC	Roman?	Tunisia?	Small boat?
Vergi	70-19 BC	A-7	–	–	–	Skiff which can be driven by one oarsman, used in metaphor of forcing boat upstream.
Diodorus Siculus	c. 60-36 BC	A-XI	305-304 BC	Demetrios Poliocretes (the Macedonian Greek)	Rhodes	Fortified with planks, with catapults (cf. Philo of Byzantium).

6. Discussion

Author	The time of writing	Text nr.	Time referred to in the text	Origin of lembs	Deployment of lembs	Description of lembs
Livy	59 BC – 17 AD	A-8.1	214 BC	Macedonian	Apollonia	Lembs had or could have two rows of oars (biremes, 8.1), but not more than sixteen oars (8.9). They were light and speedy (8.11, 8.17), could be armed (8.14), had a recognizable shape by pointing their prows head-on (8.17). As for their freight capacity, ten lembs could carry twenty horses and two hundred prisoners (8.17). Lembs could also come in rather large numbers: the booty acquired after the fall of Gentius consisted of two hundred and twenty lembs (8.23).
		A-8.2	-	Macedonian	-	
		A-8.3	-	Issaean	-	
		A-8.4	-	Issaean	-	
		A-8.5	-	Macedonian	-	
		A-8.6	-	Macedonian	-	
		A-8.7	-	Cyprian	-	
		A-8.8	-	Pergamonian	-	
		A-8.9	-	Spartan	-	
		A-8.10	-	Spartan	-	
		A-8.11	-	Pirates from Asia Minor	Chios, shores of Asia Minor	
		A-8.12	-	Ardiaeian (Illyrian)	-	
		A-8.13	-	Gaulic	-	
		A-8.14	198 BC?	Thessalian	Thessalonica to Aenia	
		A-8.15	-	Dyrrhachian (Illyrian)	-	
		A-8.16	-	Ardiaeian (Illyrian)	-	
		A-8.17	-	Macedonian	-	
		A-8.18	-	Ardiaeian (Illyrian)	-	
		A-8.19	-	Ardiaeian (Illyrian)	-	
		A-8.20	-	Cretan	Samothrace	
		A-8.21	-	Macedonian	-	
		A-8.22	-	Macedonian	-	

Author	The time of writing	Text nr.	Time referred to in the text	Origin of lembs	Deployment of lembs	Description of lembs
		A-8.23		Ardiaeian (Illyrian)	-	-
Q. Curtius Rufus	1st c. AD	A-9.1	331 BC	Persian	Chios	Pirate boats, large in number.
		A-9.2	332 BC	Methymne	Chios	Pirate boats?
Pliny the Elder	23/24–79 AD	A-10	-	-	-	Pliny attributes the origin of the lemb to the Cyrenians without any reference as to the origin of this claim.
Plutarch	c. 45–125 AD	A-XII	168 BC?	Crete?	Samothrace	-
Frontinus	Second half of 1st c. AD	A-11	191 BC	Roman	Ambracia	-
Appian	90–160 AD	A-XIII.1	218–201 BC	Roman?	Africa (Carthage)	Many in number, i.e. smaller ships.
		A-XIII.2	c. 250–231 BC	Illyria	Issa	Explicitly called 'Illyrian lembs'.
		A-XIII.3	c. 44 – 36 BC	Roman?	Sicily	-
P. Hamb. 4 248	145 AD	A-XIV	145 AD	Siknopaiu Nesos?	Siknopaiu Nesos?	Probably a smaller freight boat to transport goods on the Nile.
Harpocration	2nd c. AD	A-XV	Demosthenes (see I)	-	-	Lemb is a sort/form of ship.
Dionysius of Byzantium	2nd c. AD	A-XVI	-	-	Bosporus	A high pointed vessel named after lemb due to its shape.
Aelius Aristides	2nd c. AD	A-XVII.1	2nd c. AD	-	-	Probably a small boat that can be overturned by a single person.
		A-XVII.2		-	Egypt (Elephantine)	A sailing boat.
M. Cornelius Fronto	2nd c. AD	A-12	-	-	-	-
Pausanias of Halicarnassus		A-XVIII	-	-	-	-

6. Discussion

Author	The time of writing	Text nr.	Time referred to in the text	Origin of lembs	Deployment of lembs	Description of lembs
Aulus Gellius	fl. 2nd c. AD	A-13.1	-	-	-	-
		A-13.2	-	-	-	Smaller boat which one oarsman can force upstream, a metaphor from Vergil.
Achilles Tatius	fl. late 2nd c. AD	A-XIX.1	2nd c. AD	-	Sarepta	Smaller boat with a few people aboard, swift.
		A-XIX.2		-	-	-
Alciphron	c. 2nd or early 3rd c. AD	A-XX.1	c. 2nd or early 3rd c. AD	-	-	Smaller boat.
		A-XX.2		Attalian pirates	Corcyria	Bigger Attalian pirate boat with many oars and many rowers.
P. Mich.	249 AD	A-XXI	249 AD	Oxyrhynchus?	-	-
Diogenes Laertius	fl. 3rd c. AD	A-XXII	2nd c. BC	-	-	The origin of the name of Heraclides Lembus (Greek: Ἡρακλείδης Λέμβος), a Greek statesman, historian and philosophical writer.
Heliodorus	3rd c. AD	A-XXIII	3rd c. AD	-	Aegina	-
Solinus	mid 3rd c. AD	A-14	-	-	India?	Canoe-like boat?
Rufinus	3rd or 4th c. AD	A-XXIV	3rd or 4th c. AD	Samos	-	Small boat, could be connected with piracy.
Themistius	c. 317-385 AD	A-XXV	c. 350 AD	Isaurian	-	Fragile craft, without tillers or rest of the gear.
Ammianus Marcellinus	c. 330-400 AD	A-15.1	353 AD?	-	Melas River (Manavgat)	-
		A-15.2	-	-	-	Symbol of modest and expedient military activity.
		A-15.3	332 AD	-	-	Native Sarmatian boats.
		A-15.4	359 AD	-	The Tigris	Smaller boat?
		A-15.5	c. 360 AD	-	The Danube?	-
		A-15.6	363 AD?	Roman	Mesopotamia?	Smaller boat.
		A-15.7	-	Roman	Central Italy	Ornamented pleasure yachts.
		A-15.8	-	-	-	Smaller boat, skiff.
		A-15.9	-	-	-	-
Avienus	mid 4th c. AD	A-16	-	-	Oestreminis (Portugal)	-

185

Author	The time of writing	Text nr.	Time referred to in the text	Origin of lembs	Deployment of lembs	Description of lembs
Ausonius	c. 310 - c. 394	A-17	-	-	Moselle	Skiff made of bark with oars.
St Basil the Great	329/330-381/382 AD	A-XXVI	mid 4th c. AD	-	-	Smaller boat, similar to ἀκάτιον.
Libanius	314-393 AD	A-XXVII.1	314-393 AD	-	-	Smaller boat that can be attached to a larger vessel.
		A-XXVII.2		-	-	Smaller boat.
		A-XXVII.3		Demosthenes (see I)	Bosporus to Athens	-
Claudius Mamertinus	fl. 362 AD	A-18	362 AD	-	-	Lembs and liburnians mentioned together as symbols of Julian's fleet (could have been for poetic reasons, but this is not clear).
St Jerome	347-420 AD	A-19.1	-	-	-	Used in metaphor from Vergil and Aulus Gellius.
		A-19.1	c. 350 AD	-	Cyprus	Smaller boat, skiff, that can be stolen by an old person alone.
Aurelius Prudentius Clemens	348/349 - after 405 AD	A-20	-	-	-	Small boat.
Historia Augusta	turn of the 4th to 5th c. AD	A-21	c. 270 AD	-	The Danube	The only mention of 'lembarius', 'boatman on lembs'.
Claudius Claudianus	c. 400 AD	A-22	-	-	-	-
Sidonius Apollinaris	c. 430-489 AD	A-23.2	-	-	-	Lemb is a swift boat, big enough to hold a couch and a load of fish together with a boatman and a number of oarsmen; used for fishing.
		A-23.3	-	-	-	Lemb is a stitched boat.
Fulgentius	c. AD 500	A-24	-	.	-	Lemb is a very fast small boat,
Zosimus	c. 500 AD	A-XXVIII	mid 4th c. BC	-	The Rhine Delta, Britain	Smaller boat.
Hesychius	5th or 6th c. AD	A-XXVIX	-	-	-	Synonymous with epholkion.

6. Discussion

Author	The time of writing	Text nr.	Time referred to in the text	Origin of lembs	Deployment of lembs	Description of lembs
Procopius	c. 507 - after 555 AD	A-XXX.1	535-555	-	Carthage	-
		A-XXX.2		-	Palermo	Lembs have masts from where bowmen can shoot their arrows.
		A-XXX.3		-	Rome (Tiber)	Lembs are used to make an improvised mill on the Tiber.
		A-XXX.4		-	Central Italy	Lembs are protected with a fence of high planks.
		A-XXX.5		-	North Italy	Lembs are put on wagons and carried on land.
		A-XXX.6		-	Naples	-
		A-XXX.7		-	Central Italy	Small boat filled with inflammable materials and put on the top of the tower.
		A-XXX.8		-	Lake Mariout (Egypt)	Lembs are also called diaremata, mentioned only here.
		A-XXX.9		-	Syrtes	Small boats.
Agathias (Scholasticus)	c. 532 - after 580 AD	A-XXXI	541-562	-	Lazica	-
Isidorus Hispalensis	c. 560-636	A-25	-	-	-	A small boat, also known as cumba, caupulus, and lintris or carabus, used on the Po and in marshes.

187

Table 2: Liburnian in Greek and Roman written sources (L. Boršić)

Author	Time of writing of the text	Text nr.	Time referred to in the text	Origin of liburnian	Deployment of liburnians	Description of liburnians
Hecataeus	c. 560-480 BC? 6th c. AD	B-I	-	-	-	Boat (σκάφη) associated with the Liburnians? Likely comes from 6th century AD interpretation of Stephanus from Byzantium.
Philoxenus Alexandrinus	1st c. BC	B-II	-	-	-	Gaulos is a sort of libyrnos when used in context of piracy.
Caesar	c. 45 BC	B-1.1 B-1.2	49 BC	Issa?	Dyrrachium, Issa?	-
Propertius	after 31 BC	B-2	31 BC	Roman	Actium	Ships with beaks, symbol of Roman fleet?
Horace	c. or after 31 BC	B-3	31 BC	Roman (Octavian's)	Actium	Smaller, swift ship?
Silius Italicus	Second half of the 1st c. AD	B-4	83-103	-	-	Swift, fast, oared ship.
Lucan	Second half of the 1st c. AD	B-5	48 BC	Roman	Pharsalus	A 'modest' ship with two rows of oars (depending on how ironically Lucan described two rows).
Plutarch	45-125 AD	B-III.1	31 BC	Roman	Caesar's fleet, Battle of Actium	Liburnians had a deck; liburnians had a bronze ram; not considered as warships properly, but rather in the same group as scout-ships.
		B-III.2 B-III.3	49-48 BC		Pompey's fleet	
Pliny the Elder	77-79 AD	B-6.1	-	-	-	Liburnians attack with their rams (like killer whales).
		B-6.2	-	-	-	Liburnians use sails, oars, and rudders to move (compared with nautilus).
		B-6.3	-	-	-	Liburnians sail in a formation like geese and swans.
		B-6.4	-	-	-	Liburnians are made from tibulus, a sort of a pine tree.

6. Discussion

Author	Time of writing of the text	Text nr.	Time referred to in the text	Origin of liburnian	Deployment of liburnians	Description of liburnians
Tacitus	c. 100 AD	B-7.1	1st c. AD	Roman	-	-
		B-7.2			Britain	Oared ship.
		B-7.3			Aleria (Corsica)	-
		B-7.4			Cremona	Unsteady, shaky boats.
		B-7.5			Ravenna	-
		B-7.6			Ravenna	-
		B-7.7			Rimini	-
		B-7.8			Îles d'Hyères	-
		B-7.9			Byzantion	-
		B-7.10			The Chobus River (Pontus)	Liburnians can be quickly built.
		B-7.11			Miseno?	-
		B-7.12			Rhine-Meuse-Scheldt Delta ?	The ordinary complement of a Liburnian cruiser contained thirty or forty men.
		B-7.13			-	The emblem of the goddess Isis takes the shape of a liburnian, which, so Tacitus says, shows that it must have been imported to the Suebi.
Pliny the Younger	After 79 AD	B-8	79 AD	Roman	The Bay of Naples	-
Iuvenal	Second half of 1st c. AD	B-9	-	-	-	A couch on which rich Romans were carried through Rome is compared to a liburnian.
Florus	1st/2nd c. AD	B-10	58 BC	Roman	Cyprus to Rome	-
Suetonius	c. 100 AD	B-11.1	after 31 BC	Roman	Peloponnesus, Aetolia	-
		B-11.2	37-41 AD		Campania	Caligula built a gigantic liburnica with ten banks of oars.
		B-11.3	59 AD		Baia	-
		B-11.4	79 AD		The Bay of Naples	-

Author	Time of writing of the text	Text nr.	Time referred to in the text	Origin of liburnian	Deployment of liburnians	Description of liburnians
Appian	First half of 2nd c. AD	B-IV.1	-	-	-	The origin of the name 'liburnian': the Romans call swift ships liburnians because the Romans first came in conflict with Liburnian pirates who used swift ships.
		B-IV.2	36 BC	Roman	Tarentum	-
		B-IV.3	36 BC	Roman	Messana?	-
		B-IV.4	-	Liburnian	-	Light, fast-sailing pinnacles of the Liburnians were Roman light, swift biremes.
		B-IV.5	32?	Liburnian	-	-
Firmicus Maternus	First half of 4th c. AD	B-12	first half of 4th c. AD	-	-	Generic term for ship, navy.
(Pseudo)Lucian	First half of 4th c. AD	B-V	-	Liburnian	-	Swift, double-banked ship.
Pseudo-Callisthenes	c. 338 AD	B-VI.1	c. 335 BC	Macedonian	from Macedonia via the River Thermodon to Thrace	-
		B-VI.2	332 BC		Egypt to Tripolis	-
Anonymus de rebus bellicis	c. 337-378 AD	B-13.1	-	-	-	Particularly fast type of warship.
		B-13.2	-	-	-	Suitable for naval warfare, built too big to be operated by men's hands.
		B-13.3	-	-	-	-
Eutropius	fl. c. 360 AD	B-14	259 BC	Roman	Punic Wars	The Romans employed liburnians as early as the first Punic War. Likely an anachronism.
Claudius Mamertinus	fl. 362 AD	B-15	362 AD	-	-	Lembs and liburnians mentioned together as symbols of Julian's fleet (could have been for poetic reasons, but this is not clear).
Eunapius	c. 347-414 AD	B-VII	-	-	-	Type of vessel: a small κάραβος, it could be thirty-oared (triakontērēs).
Prudentius	c. 348-405 AD	B-16	31 BC	Roman	Actium	-

6. Discussion

Author	Time of writing of the text	Text nr.	Time referred to in the text	Origin of liburnian	Deployment of liburnians	Description of liburnians
Paulinus of Nola	353-431	B-17	-	-	Lucania	Full of armed men and 'like a pirate ship'.
Orosius	4th-5th c. AD	B-18				Distinguishes triremes from liburnians; liburnians are characterised by swiftness.
Vegetius	After 383 – before 450 AD	B-19.1	-	Roman	-	Liburna is a universal name of warship.
		B-19.2	-		-	The fleet of liburnians is commanded from Campania and Ravenna by ten tribunes; each Liburnian had a single navarch.
		B-19.3	-		-	In the battle at Actium, the Romans saw that the Liburnian ships were better designed and adopted their shape for building subsequent ships.
		B-19.4	-		-	Liburnians were built from cypress, pine, larch, and fir; it is fastened with bronze nails.
		B-19.5	-		-	Liburnians have from one to five ranks of oars depending on their size; they have attached scouting ships with twenty oarsmen on each side; the sails of liburnians are dyed in 'Venetian blue', as well as the uniforms of the sailors and marines.
Zosimus	c. 500 AD	B-VIII	c. 360 AD	-	-	The origin of liburnians named after a certain town in Italy [sic].

6.4. Overview of usage of the terms lembos and liburnica in ancient sources from the 4th century BC until Late Antiquity

Several conclusions can be made from these two tables:

Table 1 shows a variety of descriptions of lembs. Lembs can denote vessels ranging from canoes, small fishermen's boats, and skiffs, to galleys carrying catapults, or large pirate ships with numerous oars and rowers. We can see a structural difference in our sources as well: lembs are typically distinguished from ships with rams, whereas having a beak or a ram was evidently a typical trait of liburnians. Table 2 also shows a change in the meaning of the term liburnian through time: from the light biremes of the 1st century BC, to a generic term for any (war)ship, even as big as a quinquereme. In later periods, this was the universal term for a warship, with some possible exceptions. For example, a 6th-century Oxyrhynchus papyrus, where the term liburnian is last recorded in a contemporary context, makes it very clear that liburnian has already become antiquated by the beginning of the 6th century, if not earlier.[263] The authors who mention both vessels (Greek: Plutarch, Appian, and Zosimus; Latin: Caesar and Pliny the Elder) never mention lembs and liburnians in the same context, nor do they connect them in any way. The only ancient author who connects these ships is the 4th-century Latin panegyrist Claudius Mamertinus, who identifies them as separate ships – see pp. 193-94. It is quite interesting that Pliny the Elder, in his list of the origin of ships, does not mention liburnians at all. Moreover, none of our sources claim that liburnians are a type of lembs, nor do they connect them in any way.

The idea that liburnians are a type of lemb, can be traced back to some later texts, mostly the *glossaria*. The *Liber Glossarum*, or *Glossarium Ansileubi*, probably dating from the 8th century AD, under the entry *lembus* writes: *lemniculus barca scapha cumba liburna eos*. This can be explained by widening the radius of the meaning of the terms 'lemb' and 'liburnian'. Since it is likely that in Late Antiquity both terms denoted a certain shape or ship form, it may be that, for some late authors, 'lemb' becomes synonymous with 'ship'. Besides relying on late sources, how else can we explain the connection between lembs and liburnians found in much of the modern discussion on ancient ships? First, there is perception that the Liburni belonged to an overarching ancient 'Illyrian' identity ('nation', 'race', 'people'). This assumption has been advanced in post-medieval scholarly and popular discourses, and structured into the scholarly paradigms of the 19th and earlier 20th century.[264] As observed a few times in this book, modern scholarship rejects the idea of an overarching 'Illyrian' identity, and finds no evidence that the Iron Age Liburnian communities shared the same identity-discourse with south Adriatic communities.[265] The etymological interpretations of an 'Illyrian' origin of the word λέμβος and *confusio nominum* might have been influenced by the prejudicial linking of the name of the ship with the Liburnian 'tribe', of which even the overenthusiastic pan-Illyrist Krahe was also well-aware.[266] To this should be added the fact that in the 3rd to 4th centuries AD the term 'liburnica' indicated any type of warship, as has been mentioned, and the term lembos was used in more generic sense, as it references a wider variety of forms, shapes, functions and

---

[263] *P. Oxy.* 16.2032,52 and 54: λιβερνίου πλοίου (540-41), cf. Pryor, Jeffreys 2006: 123-31.
[264] Dzino 2014b.
[265] Dzino 2014a and above p. 17.
[266] *Da beide keine sprachliche Begründung für ihre Auffassung geben, dürfte diese auf der sachlichen Beobachtung beruhen, daß auch sonst Schiffsbenennungen der klassischen Sprachen, vor allem solche für kleinere und schnell bewegliche Fahrzeuge, dem illyrischen Bereich entnommen wurden, zum mindesten für ihm entnommen gelten.* Krahe 1952: 79.

origins. The connection between the two ships may have developed in Late Antique sources, when the terms 'lemb' and 'liburnian' became disassociated from their original contexts, even antiquated, especially the liburnian, which was replaced by a new type of warship, the *dromon*, likely evolved from the Roman liburnian.[267] However, as pointed out, there is no evidence that these two vessels originally had any connection whatsoever.

6.5. Lemb and liburnian: the same ship?

Therefore, the assumption that the south Adriatic (Illyrian) *lembos* and Liburnian *liburnian* were versions of the same ship type, or the products of a common shipbuilding tradition centred upon geographic proximity, is very problematic. The southern and northern Adriatic in the first millennium BC represented two different (but not entirely separate) regional social networks, and the assumption that they must share common naval traditions does not necessarily reflect the actual situation. Archaeology presents the Liburnian material culture as a reflection of a distinct social network spreading over the northern Adriatic and its hinterland, with southern Adriatic communities belonging to a different social network stretching into the hinterland through modern Albania, Montenegro, southeast Bosnia, and western Serbia.[268] As pointed out earlier (p. 8-9, 57-58) the different 'maritime cultural landscapes' of the southern and northern Adriatic must have had an impact on local shipbuilding traditions. Some images of the ships from the Daunian stelae not discussed here, indicate different naval traditions appearing in the southern Adriatic from the 6th century BC onwards, which were more influenced by the Aegean and Greek models.[269] In contrast, the northern Adriatic evidence reveals the existence of an integrated network of several local 'techno-communities', who continued to produce sewn-plank ships in the Roman times.[270]

There are many problems with the identification of the Illyrian lemb and the liburnian as the same type of ship. The lemb was a wide term that covered a range of small ships used for different civilian and military purposes. At least three sub-types of lembs were used in warfare: an original lemb used for civilian purposes adapted into platforms for the catapults, the south Adriatic lemb for piracy, trade and transport of the troops, and an upgraded Macedonian type of fighting lemb.[271] This type of boat was never connected by the sources with the Liburni or liburnian, but with a different indigenous group from the southern Adriatic called the 'Illyrians'. Its mention in the Adriatic and in military contexts is limited to the 3rd and early 2nd centuries BC, a century before the first mention of the liburnian. Finally, it is important to note that the term *lembos* continues to be used in the imperial period parallel with *liburnica* – as is apparent in Appian's *Illyrike*, where both ships are mentioned separately.[272] The evidence from Ammianus Marcellinus also confirms that in the later 4th century AD the term *lembos* still referred to specific type of ship, something also evidenced in the *Panegyric to emperor Julian*

---

[267] Pryor, Jeffreys 2006: 127.
[268] Dzino 2012, with literature.
[269] Medas 2016: 157-59.
[270] 'Techno-communities' are the groups of people that carry out particular activities by employing particular technologies. When a technology is transferred, the recipient 'techno-community' adopts it by inventing new functional types whose performance characteristics are more suitable for participating in activities of their own social/political/ethnic group, Schiffer 2009: 825-26; 2011: 175-76; Skibo, Schiffer 2008: 125-33. See Dzino, Boršić 2020 for protohistoric Adriatic shipbuilding 'techno-communities' with examples of lembs and liburnians.
[271] Cf. Torr 1895: 103, 116; Casson 1995: 122; Morrison 1996: 263, who recognise two sub-types of *lemb*: the earlier Greek small ship and ship's boats, and the latter Illyrian *lemb*.
[272] App. *Ill*. 3.7 and *Ill*. 7.18.

by Claudius Mamertinus, who is the only author who actually mentions both types of ships together in the same sentence.[273]

The images of the ships from the southern Adriatic coinage, which could be considered with reasonable certainty to represent an Illyrian lemb, show no similarities with other Iron Age depictions of the ships from the Adriatic area, such as those from Nesactium, Novilara, Cattolica, and Glasinac. The ships from Novilara, and probably also those from Cattolica and Glasinac, have zoomorphic bows facing outwards, and the ships from Glasinac have different stern extensions. The ship from Nesactium has a recognisable rounded hull and stern, that has no similarities with the ship-images from the coins, but, as pointed out earlier (p. 177) it could hypothetically be related to the possible comparison made by Philoxenus of Alexandria between the liburnian and Phoenician *gaulos*. The reconstruction of the ships shown on the southern Adriatic coinage indicates that the Illyrian lemb might have been smaller than the Roman-era liburnian, although it is worth observing that the images used for reconstruction are not necessarily accurate, as there is no agreement on which visual representations of the ship from the Roman period in fact represent a liburnian.[274]

6.6. Conclusion

This discussion on lembs and liburnians has enlightened some aspects related to these ships, but many other matters will still remain unsolved, such as the shape and characteristics of the original liburnian developed by the Liburni, and its differences with the liburnians used in the Roman imperial fleets. However, one can observe that the dominating opinion that the south Adriatic (Illyrian) lemb and original prototype of liburnian were products of a specific and distinct Adriatic shipbuilding tradition is very problematic. The evidence for the Adriatic ships dating prior to the 3rd century BC cannot be reconciled with later literary descriptions of the lemb and liburnian. Even if we did not have images from southern Adriatic coins and the Prozor belt-buckle, it would be impossible to maintain that the shipbuilding technologies in the eastern Adriatic did not change between the 8th/7th century BC and the 3rd century BC. The impact of more advanced naval technologies coming from the Etruscans and Greeks must have caused some degree of local adaptation within protohistoric northern and southern Adriatic shipbuilding practices after *c.* 500 BC, changing many aspects of early Iron Age Adriatic naval traditions. Social transformations taking place in this same period, including a subsequent rise in the need for imported Mediterranean artefacts as signifiers of social status and social practices, must have pushed the elites of these communities to invest resources into redeveloping and improving their ships for new purposes: trade and raiding.[275]

The development of these ships, and the adaptation and adoption of advanced naval technologies also impacted on communities in the southern and northern Adriatic. In the case of south Adriatic communities, the adoption and use of an efficient small, fast, and spacious boat from the Aegean, among other things, impacted on the local economy by increasingly drawing its new elites towards sea-raiding and its related profits. Likewise, the development

---

[273] Amm. Marc. 25.8.3; Cl. Mam. 8.3: *... de Iuliani lembis liburnisque...*
[274] Kozličić (19881: 170) calculated the length of the warship lemb on the waterline to be 15 m, with a waterline breadth of 4.5 m, while Morrison (1996: 317) sees the ships from Trajan's column, assumed to be liburnians, as being 18 m in length on the waterline, with a 3-m waterline breadth.
[275] Technological innovation is most frequently driven by the needs of the elites, Schiffer 2011: 43-53.

of the liburnian could have also played an important role in the transformation of Liburnian communities, as it made them desirable naval allies for the Romans, and was thus an additional factor in their rapid inclusion in Roman imperial networks. Finally, the fact that naval warfare became obsolete in the Mediterranean after the Battle of Actium led the Roman imperial shipbuilders to prioritise smaller and faster ships, so it is no wonder that the liburnian became a common sight in imperial fleets.

On the other hand, not everything was changing with regard to Adriatic shipbuilding. There is firm evidence coming from more recent underwater finds of shipwrecks that the sewn shipbuilding tradition persisted in the Histrian and Liburnian regions, even into the Roman period. These ships, primarily used for trade, were called *serilia* in the Roman literature and have nothing to do with the liburnians that were developed locally and adopted into the Roman navy. They do, however, provide evidence for the existence of distinct shipbuilding traditions in the prehistoric and ancient north Adriatic, which were a reflection of the 'maritime cultural landscape' shared by the communities in the area. The shipwrecks of sewn ships are not discovered in the southern Adriatic, and ancient sources do not connect these kinds of ships with this region, which is strong evidence that this area had different shipbuilding traditions from that of the north Adriatic. The distinct regional variety of the ships discovered in the Histrian and Liburnian waters also confirms the existence of regional networks between the shipbuilders, and even more specific local adjustment to the 'maritime cultural landscape' of the north-eastern Adriatic.

Another important point made in this book is that the assumption that liburnian and the Illyrian lemb represent the same type or sub-type of warship is unfounded, for there is no sound supporting evidence. The design of these ships was the outcome of differing naval traditions and products of different 'maritime cultural landscapes'. These designs also developed in the southern and northern Adriatic contemporary to social transformations, caused primarily by increased contacts with the Mediterranean world after *c.* 500 BC. The identification of Illyrian lembs and liburnians as more or less the same ship originates from the stereotypical assumption in older scholarship that the prehistoric indigenous population of the eastern Adriatic shared the same culture and, roughly, the same identities. By this logic, it was then easy to conclude that the people who share the same identity and culture develop identical or very similar ships. However, recent research shows that such an assumption is wrong, and that the eastern Adriatic and its hinterland was not inhabited by 'Illyrians' in prehistory and protohistory, but rather by a plurality of different communities connected by different networks, who in no way shared the same identity and/or culture. Thus, the frequently repeated claim in earlier literature that the lemb and liburnian are similar, or even identical, because the 'Liburni are an Illyrian tribe' must be discarded once and for all.

Two different terms, lemb and liburnian, were used in the sources depicting these as two different kinds of ship, rather than being interchangeable terms depicting the same ship type. The southern Adriatic shipbuilders adopted and developed an existing type of small and fast Aegean ship to be used to attack unarmed ships, to swiftly transport troops, and also to conduct naval warfare. It was named a *lembos* by the ancient authors, the most appropriate term they had at their disposal. There is much less information available on the origins of the liburnian. While it is indeed likely to see its original design developing in Liburnia, no reliable information exist about this original design before the ship was included in the Roman fleets, apart from the fact that it was a relatively small, but extremely fast and manoeuvrable vessel. Even if one could determine beyond any doubt which image of the ships from Roman times that depicts a liburnian, it is impossible within the present corpus of evidence to know what kinds of changes and adaptations were made by the Roman shipbuilders.

Finally, at the end, it is important to state that all discussions of Illyrian lembs and liburnians before their inclusion in Roman imperial fleets are based on very fragmentary evidence. We have tried to base our conclusions on a wide range of available written, archaeological, and iconographic evidence, as well as use existing modern literature on the subject, all the while attempting to make local scholarship more visible to the wider scholarly audience. It is understandable that our assessment of the evidence invites future discussions and re-interpretations, and we can only hope that sooner or later new material evidence will lead us in the right direction, confirming or challenging present conclusions.

Bibliography

Ancient authors not listed in Chapter 5

Aeschylus, Persians. Seven against Thebes. Suppliants. Prometheus Bound, ed. and transl. A. H. Sommerstein LCL. 145 (2009).
Apollonius Rhodius. Argonautica, ed. and transl. W.H. Race. LCL 1 (2009).
Appian. Roman History, Volume III, ed. and transl. B. McGing. LCL 4 (2019).
Dio Cassius. Roman History, Volume II: Books 12-35, transl. E. Cary, H.B. Foster. LCL 37 (1914).
Dio Cassius. Roman History, Volume III: Books 36-40, transl. E. Cary, H.B. Foster. LCL 53 (1914).
Dio Cassius. Roman History, Volume V: Books 46-50, transl. E. Cary, H.B. Foster. LCL 82. (1917).
Herodotus. The Persian Wars, Volume I: Books 1-2, transl. A.D. Godley. LCL 117 (1920).
Livy, Julius Obsequens. History of Rome, Volume XIV: Summaries. Fragments. Julius Obsequens. General Index, transl. A.C. Schlesinger. LCL 404 (1959).
Pseudo-Scylax's Periplous: The Circumnavigation of the Inhabited World, text, transl. and comm. G. Shipley. Bristol Phoenix Press, Bristol 2011.
Scymni Chii Periegesis et Dionysii Descriptio Graeciae, ed. A. Meineke. F. Nicolai, Berlin 1896.
Strabo. Geography, Volume I: Books 1-2, transl. H.L. Jones. LCL 49 (1917).
Strabo. Geography, Volume III: Books 6-7, transl. H.L. Jones. LCL 182 (1924).
Velleius Paterculus. Compendium of Roman History, transl. F.W. Shipley. LCL 152 (1924).

Modern sources

Anastasi, A. 2003. Lembos. Elementi per una ricerca. In: *Atti del II Convegno Nazionale di Archeologia Subacquea, Castiglioncello 2001*. Bibliotheca Archaeologica 12 : 253-258. Edipuglia: Bari.
Arnaud, P. 2006. La navigation en Adriatique d'après les données chiffrées des géographes anciens. In: S. Čače, A. Kurilić, F. Tassaux (eds) *Les routes de l'Adriatique antique: Géographie et économie*: 39-53. Ausonius éditions 17. Bordeaux-Zadar: Institut Ausonius - Sveučilište u Zadru.
Artegiani, A., Bregant, D., Paschini, E., Pinardi, N., Raicich, F., Russo, A. 1996. The Adriatic Sea general circulation. Part I: air-sea interactions and water mass structure. *Journal of Physical Oceanography* 27: 1492-1514.
Babić, S. 2002. Princely Graves of the Central Balkans – A Critical History of Research. *European Journal of Archaeology* 5(1): 70-88.
Babić, S. 2004. *Поглаварство и полис - Старије гвоздено доба Централног Балкана и грчки свет*. Belgrade: Balkanološki Institut.
Balen-Letunić, D. 1996. Figuralno ukrašene pojasne kopče tipa Prozor. *VAMZ* 28-29: 23-38.
Bandelli, G. 1981. La guerra istrica del 221.a.C. e la spedizione alpina del 220.a.C. *Athenaeum* 59(1-2): 3-29.
Bandelli, G. 1983. Aquileia, la Dalmazia e l'illirico. *Antichità Altoadriatiche* 26: 62-66.
Barnett, Ch. 2016. Rethinking Identities in pre-Roman Liburnia. *Miscellanea Hadriatica et Mediterranea* 3: 63-97.
Barnett, Ch. 2019. *Cultural Integration, Social Change and Identities in Late Iron Age and Roman Liburnia*. Unpublished PhD thesis. Macquarie University, Sydney.

Basch, L. 1983. Bow and Stern Appendages in the Ancient Mediterranean. *Mariner's Mirror* 69(4): 395-412.

Basch, L. 1987. *Le musée imaginaire de la marine antique*. Athens: Hellenic Institute for the Preservation of Nautical Traditions.

Batović, A., Batović, Š. 2013. *Helenistički grobovi iz Nadina u okviru V. (zadnje) faze liburnske kulture: građa za povijest Liburna, IV.-I. st. prije Krista*. Zadar: Arheološki muzej Zadar.

Batović, Š. 1977. Caractéristiques des agglomérations fortifiées dans la region des Liburniens. *GodCBI* 15/13: 201-225.

Batović, Š. 1987. Istarska kultura željeznog doba. *RFFZd* 26: 5-74.

Batović, Š. 1987. Liburnska grupa. *PJZ*, vol. 5: 339-390.

Beek, A. L. 2015. Where have all the pirates gone? In: F. Carrer, V. Gheller (eds) *Invisible Cultures: Historical and Archaeological Perspectives*: 270-284. Newcastle-upon-Tyne: Cambridge Scholars Publishing.

Beekes, R., van Beek, L. 2010. *Etymological Dictionary of Greek* 1. Boston-Leiden: Brill.

Bekić, L. (ed.) 2009. *Jurišićev zbornik: Zbornik radova u znak sjećanja na Marija Jurišića*. Zagreb-Zadar: Hrvatski restauratorski zavod–Međunarodni centar za podvodnu arheologiju u Zadru.

Beltrame, C. 2000. *Sutiles naves* of Roman age. New evidence and technological comparisons with pre-Roman sewn boats. In: J. Litwin (ed.) *Down the River to the Sea. VIIIth International Symposium on Boat and Ship Archaeology*: 91-96. Gdansk: Polish Maritime Museum.

Beltrame, C. 2012. *Archeologia marittima del Mediterraneo, Navi, merci e porti dell'antichita all'eta moderna*. Rome: Carocci Editore.

Beltrame, C., Gaddi, D. 2013. Fragments of Boats from the Canale Anfora of Aquileia, Italy, and Comparison of Sewn-Plank Ships in the Roman Era. *IJNA* 42(2): 296-304.

Benac, A. 1987. O etničkim zajednicama starijeg željeznog doba u Jugoslaviji. *PJZ*, vol. 5: 737-802.

Benac, A., Čović, B. 1957. *Glasinac II: Željezno doba*. Sarajevo: Zemaljski muzej.

Bérchez Castaño, E. 2010. La liburna en el Contexto de la Flota Romana. *Liburna* 3: 69-87.

Bilić Dujmušić, S. 2004. Excavations at Cape Ploca, near Sibenik, Croatia. In: L. Braccesi, M. Luni (eds) *I Greci in Adriatico* 2: 123-140. Hesperìa: studi sulla grecità di occidente 18. Rome: 'L'Erma' di Bretschneider.

Bilić-Dujmušić, S. 2017. The Battle of Pharos. *VAHD* 100(1): 327-336.

Bilić-Dujmušić, S., Milivojević, F. 2014. Tko je pobijedio Demetrija? In: L. Mirošević, V. Graovac Matassi (eds) *Dalmacija u prostoru i vremenu. Što Dalmacija jest, a što nije, zbornik radova sa znanstvenog skupa održanog 14.-16. lipnja 2012*: 35-42. Zadar: Sveučilište u Zadru.

Blečić, M. 2007a. Reflections of Picen's impact in the Kvarner Bay. In: M. Guštin, P. Ettel, M. Buora (eds) *Piceni ed Europa: Atti del convegno*: 109-122. Archeologia di frontiera 6. Udine-Koper-Jena: Civici musei di Udine, Museo Archeologico - Univerza na Primorskem -Znanstveno-raziskovalno središče Koper, Inštitut za dediščino Sredozemlja - Universität Jena, philosophische Fakultät, Bereich für Ur-und Frühgeschichte.

Blečić, M. 2007b. Status, Symbols, Sacrifices, Offerings. The Diverse Meanings of Illyrian Helmets. *VAMZ* 40: 73-116.

Blečić Kavur, M. 2009. Universal and Original. Transformation of style in the North-Adriatic Region. In: G. Tiefengraber, B. Kavur, A. Gaspari (eds) *Keltske študije II. Studies in Celtic Archaeology. Papers in honour of Mitja Guštin*: 197-208. Protohistoire Européenne 11. Montagnac: Éditions Monique Mergoil.

Blečić Kavur, M. 2015. *A coherence of perspective – Osor in cultural contacts during the Late Iron Age*. Koper–Mali Lošinj: University of Primorska - Lošinjski muzej.

Blečić Kavur, M. 2019/20. Complexity and dynamics of the cultural processes of the Late Bronze Age on the eastern Adriatic coast: The example of more recent finds from Nin. *Diadora* 33-34: 7-18.

Boetto, G., Rousse, C. 2012. Traditions régionales d'architecture navale en Adriatique à l'époque romaine. *HistAntiq* 21: 427-441.

Boetto, G., Koncani Uhač, I., Uhač, M. 2014. Navires de l'âge du Bronze à l'époque romaine en Istrie. In: P. Pomey (ed.) *Ports et Navires dans l'Antiquité et à l'époque byzantine*: 22-25. Dossiers d'Archéologie 364.

Boetto, G., Koncani Uhač, I., Uhač, M. 2017. Sewn ship finds from Istria (Croatia): the shipwrecks of Zambratija and Pula. In: J. Litwin (ed) *Baltic and Beyond, Change and Continuity in shipbuilding, Proceedings of the 14th International Symposium on Boat and Ship Archaeology, Gdansk (Poland), 21st-25th September 2015*: 189-198. Gdansk: Polish Maritime Museum.

Boetto, G., Radić Rossi, I. 2017. Ancient Ships from the Bay of Caska (Island of Pag, Croatia). In: J. Litwin (ed.) *Baltic and Beyond, Change and Continuity in shipbuilding, Proceedings of the 14th International Symposium on Boat and Ship Archaeology, Gdansk (Poland), 21st-25th September 2015*: 279-288. Gdansk: Polish Maritime Museum.

Bonino, M. 1975. The Picene ships of the 7th century BC engraved at Novilara (Pesaro, Italy). *IJNA* 4(1): 11-20.

Borri, F. 2017. Captains and Pirates: Ninth Century Dalmatia and its Rulers. In: S. Gasparri, S. Gelichi (eds) *The Age of Affirmation. Venice, the Adriatic and the Hinterland between the 9th and 10th Centuries*: 11-38. Turnhout: Brepols.

Bracessi, L. 1984. *La legenda di Antenore: da Troia a Padova*. Il mito e la storia 1. Padova: Signum.

Braccesi, L. 2000. Per una interpretazione della stele di Novilara con naumachia. *Hesperìa: Studi sulla grecità d'Occidente* 10: 237-244.

Braccesi, L. (ed.) 2004. *La pirateria nell'Adriatico antico*. Hesperìa: studi sulla grecità di occidente 19. Rome: 'L'Erma' di Bretschneider.

Braccesi, L. 2010. Sofocle, Tritolemo e . . . l'Adriatico (per una rilettura provocatoria). In: E. Giovi (ed.) *Dal Mediterraneo all'Europa: Conversazioni adriatiche*. Hesperia: Studi sulla grecità d'Occidente 25. Rome: 'L'Erma' di Bretschneider.

Brajković, T. 2008. Antika. In: E. Podrug, T. Brajković, Ž. Krnčević, *Arheološki tragovi kultova i religija na Šibenskom području*: 55-89. Šibenik: Muzej grada Šibenika.

Braudel, F. 1972. *The Mediterranean and the Mediterranean World in the Age of Philip II*. Vol. 1. New York: Collins.

Brstilo Rešetar, M., Gotić, K. (eds) 2019. *Varvaria - Breberium - Bribir. Historical layers revealed*. Exhibition catalogue. Zagreb: Hrvatski povijesni muzej.

Brunšmid, J. 1898. *Die Inschriften und Münzen der griechischen Städte Dalmatiens*. Wien: A Hölder.

Brusić, Z. 1969. Rezultat podvodnih istraživanja u Ninu. *Mornarički glasnik* 22: 219-222.

Brusić, Z. 1970. Problemi plovidbe Jadranom u predhistoriji i antici. *Pomorski zbornik* 8: 549-567.

Brusić, Z. 1972. Podmorska arheološka istraživanja kod Nina. *Radovi Instituta JAZU u Zadru* 19: 245-251.

Brusić, Z. 1987. Zaton kod Zadra. Antička luka Aenone. *Arheološki pregled* 28: 121-122.

Brusić, Z. 1995. Serilia liburnica. *Radovi Zavoda za povijesne znanosti HAZU u Zadru* 37: 39-59.

Brusić, Z. 2002. Nekropole liburnskih naselja Nina i Kose kod Ljupča. *HistAntiq* 8: 213-242.

Brusić, Z. 2010. A selection of Liburnian jewellery. *Prilozi Instituta za arheologiju u Zagrebu* 27: 241-248.

Brusić, Z., Domijan, M. 1985. Liburnian Boats: Their Construction and Form. In: S. McGrail, E. Kentley (eds) *Sewn Plank Boats: Archaeological and Ethnographic papers based on those presented to a conference at Greenwich in November, 1984*: 67-86. British Archaeological Reports International Series 276. National Maritime Museum Archaeological Series 10. London-Oxford: National Maritime Museum, Greenwich - British Archaeological Reports.

Buršić Matijašić, K. 2008. *Gradinska naselja. Gradine Istre u vremenu i prostoru*. Zagreb: Leykam international.

Buršić Matijašić, K. 2012. Istra na prapovijesnim pomorskim putevima. *HistAntiq* 21: 203-214.

Cabanes, P. 1988. *Les Illyriens de Bardylis à Genthios: IV - IIe siècles avant J.- C*. Paris: Sedes.

Cabanes, P. 2008. Greek Colonisation in the Adriatic. In: G. Tsetskhladze (ed.) *Greek Colonisation: An Account of Greek Colonies and other Settlements Overseas* 2: 155-186. Mnemosyne Supplement 193. Leiden-Boston: Brill.

Cambi, N. 1972. Vis, uvala Vela Svitnja - brodolom antičkog broda. *Arheološki pregled* 14: 80-82.

Cambi, N. 2012. O nekim toponimima i opisu ratovanja na istočnom Jadranu u Lukanovom građanskom ratu. *Radovi HAZU. Razred za društvene znanosti* 49/512: 1-28.

Cambi, N. 2013. Roman military *tropaea* from Dalmatia. In: M. Sanader, D. Tončinić, I. Radman-Livaja (eds) *Proceedings of the XVIIth Roman military equipment conference: Weapons and Military Equipment in a Funerary Context*: 9-22. Zagreb: FF Press-Arheološki muzej u Zagrebu.

Cambi, N., Čače, S., Kirigin, B. (eds) 2002. *Greek Influence along the East Adriatic Coast*. Knjiga Mediterana 26. Split: Književni krug.

Cambi, N., Kirigin, B., Marin, E. 1980. Excavations at Issa, Island of Vis, Yugoslavia 1976, 1979 - A Preliminary Report. *Rivista di archeologia* 4: 83-86.

Caro, D. 2017. *Deceres liburnicae*. Le colossali navi imperiali della flotta di Ravenna. In: A. Panaino, P. Ognibene (eds) *'Salso mar' "Ἁλμυρὸς Πόντος. Atti del Seminario di Studi storico-navali tenutosi tra il 4 ed il 6 maio 2015 presso il Dipartimento di Beni Culturali dell'Università di Bologna in occasione del XIX Raduno Nazionale dei Marinai d'Italia (1-10 maggio 2015)*: 101-111. Milan: Mimesis.

Casson, L. 1951. Speed under Sail of Ancient Ships. *Transactions and Proceedings of the American Philological Association* 82: 136-148.

Casson, L. 1971. *Ships and Seamanship in the Ancient World*. Princeton: Princeton University Press.

Casson, L. 1993. A Petrie Papyrus and the Battle of Raphia. *The Bulletin of the American Society of Papyrologists* 30(3-4): 87-92.

Casson, L. 1995. Merchant Ships. In: R. Gardiner (ed.) *The Age of the Galley*: 117-126. London: Conway Maritime.

Castiglioni, M.-P. 2008. The Cult of Diomedes in the Adriatic: complementary contributions from literary sources and archaeology. In: J. Carvalho (ed.) *Bridging the Gaps: Sources, Methodology and Approaches to Religion in History*: 9-28. Pisa: PLUS-Pisa University Press.

Cavallaro, M. A. 2004. *Da Teuta e Epulo, Interpretazione delle guerre illyriche e histriche tra 229 e 177 a. C*. Bonn: Rudolf Habelt.

Ceka, H. 1972. *Questions de numismatique Illyrienne: avec un catalogue des monnaies d'Apollonie et de Durrhachium*. Tirana: University of Tirana.

Ceka, N. 1985. Aperçu sur le développement de la vie urbaine chez les Illyriens du Sud. *Iliria* 15(2): 137-162.

Cerva, M. 1996. Rome e la 'sottomissione' della Liburnia. *Atti e Memorie della Società Istriana di Archeologia e Storia Patria* n.s. 44: 7-18.

Cestnik, V. 2009. *Iron Age necropolis Kaštel near Buje. Analysis of burial practice in the Iron Age Istria*. Monografije i katalozi 18. Pula: Arheološki muzej Istre.

Chapman, J., Shiel, R., Batović, Š. 1996. *The Changing Face of Dalmatia. Archaeological and Ecological Studies in a Mediterranean Landscape*. Reports of the Research Committee of the Society of Antiquaries of London 54. London: Leicester University Press.

Cobau, M. 1994. *Le navi di Novilara*. Pesaro: Provincia di Pesaro e Urbino.

Conway, R.S., Whatmough, J., Jackson Johnson, S.E. 1933. *The Pre-ltalic Dialects of Italy 2*. Cambridge, Mass: Harvard University Press.

Cornell, T. J. (ed.) 2013. *The Fragments of the Roman Historians 3*. Oxford: Oxford University Press.

Čače, S. 1985. *Liburnija od 4. do 1. stoljeća prije nove ere*. Unpublished PhD thesis, University of Split.

Čače, S. 1989. Rimski pohod 221.g. i pitanje političkog uređenja Histrije. *RFFZd* 28: 5-17.

Čače, S. 1991. Rim, Liburnija i istočni Jadran u 2. st. pr. n. e. *Diadora* 13: 55-76.

Čače, S. 1993. Prilozi povijesti Liburnije u 1. stoljeću pr. Kr. *Radovi Zavoda za povijesne znanosti HAZU u Zadru* 35: 1-35.

Čače, S. 1994. Prilozi raspravi o osnivanju grčkih naseobina na Jadranu u 4. stoljeću pr. K. *RFFZd* 33: 33-54.

Čače, S. 1998. Manijski zaljev, Jadastini i Salona. *VAHD* 90-91: 57-87.

Čače, S. 2002. Corcira e la tradizione greca dell'espansione dei Liburni nell'Adriatico orientale. In: N. Cambi, S. Čače, B. Kirigin, B. (eds) *Greek Influence along the East Adriatic Coast*: 83-100. Knjiga Mediterana 26. Split: Književni krug.

Čače, S. 2005. Liburnski pirati: mit i stvarnost. *Bakarski zbornik* 10: 169-181.

Čače, S. 2006. South Liburnia in the beginning of the Principate. In: S. Čače, A. Kurilić, F. Tassaux (eds) *Les routes de l'Adriatique antique: Géographie et économie*: 65-79. Ausonius éditions 17. Bordeaux-Zadar: Institut Ausonius - Sveučilište u Zadru.

Čače, S. 2013. Notes on the relations between the Liburnian communities. *Asseria* 11: 11-50.

Čače, S., Kuntić-Makvić, B. 2010. Pregled povijesti jadranskih Grka. In: J. Poklečki Stošić (ed.) *Antički Grci na tlu Hrvatske*: 63-71. Exhibition catalogue. Zagreb: Galerija Klovićevi dvori.

Čače, S., Kurilić, A., Tassaux, F. (eds) 2006. *Les routes de l'Adriatique antique: Géographie et économie*. Ausonius éditions 17. Bordeaux-Zadar: Institut Ausonius - Sveučilište u Zadru.

Čače, S., Šešelj, L. 2005. Finds from the Diomedes sanctuary on the Cape Ploča: new contributions to the discussion about the Hellenistic period on the east Adriatic. In: M. Šegvić, I. Mirnik (eds) 2005. *Illyrica Antiqua: Ob honorem Duje Rendić-Miočević*: 163-186. Zagreb: FF Press.

Čargo, B. 2010. *Research into the south-west Issa necropolis up to 1970*. Split: Arheološki muzej u Splitu.

Čargo, B., Miše, M. 2010. Pottery production in Issa. *VAPD* 103: 7-40.

Čelhar, M. 2008. The underwater interdisciplinary project in Caska Bay, Pag island. In: I. Radić Rossi, A. Gaspari, A. Pydyn (eds) *Proceedings of the 13th Annual Meeting of the European Association of Archaeologists (Zadar - Zagreb Croatia, 18th-23rd September 2007), Session: Underwater Archaeology*: 176-186. Zagreb: Hrvatsko arheološko društvo.

Čelhar, M., Parica, M., Ilkić, M., Vujević, D. 2017. A Bronze Age underwater site near the islet of Ričul in northern Dalmatia (Croatia). *Skyllis* 17(1): 21-34.

Čović, B. 1967. O izvorima za istoriju Autarijata. *GodCBI* 5/3: 103-122.

Čović, B. 1976. *Od Butmira do Ilira*. Sarajevo: Veselin Masleša.

Čović, B. 1984. Umjetnost kasnog bronzanog i ranog željeznog doba na istočno jadranskoj obali i u njenom zaleđu. In: A. Benac (ed.) *Duhovna kultura Ilira*: 7-40. Posebna izdanja 67/11. Sarajevo: ANU BiH.

Čović, B. 1987. Glasinačka kultura. *PJZ*, Vol. 5: 575-643.

De Boer, J.G. 1993. Etruscan Sea-Going Vessels from the 10th to 5th century BC. *Talanta* 24-25: 11-22.

Dell, H.J. 1967. The origin and nature of Illyrian piracy. *Historia* 16(3): 344-358.

Dellaporta, K. 2011. Early Seafaring in the Ionian Sea. *Skyllis* 11(2): 19-24.

De Souza, P. 1999. *Piracy in the Graeco-Roman World*. Cambridge: Cambridge University Press.

Dimitrijević, M. 2016. Прилог познавања хеленистичке фортификације на југоисточном Јадрану. *Vojnoistorijski glasnik* 1: 185-206.

Dimitrijević, M. 2018. Socioeconomic relations and identities in the Southeastern Adriatic Iron Age. *GodCBI* 47: 7-26.

Downey, G. 1958. Themistius' First Oration. *Greek, Roman and Byzantine Studies* 1(1): 49-69.

Dragićević, I. 2016. Daorsi coins and a contribution to the understanding of the circulation of coinage in Daorsi territory. *VAHD* 109: 107-128.

Duplančić Leder, Ujević, T., Čala, M. 2004. Coastline lengths and areas of islands in the Croatian part of the Adriatic Sea determined from the topographical maps at scale 1:25000. *Geoadria* 9(1): 5-32.

Dyczek, P. 2009. *Rhizon/Risinium: Od iliryjskiej osady do rzymskiego municipium*. Xenia Posnaniesia 35. Poznań: Wydawnictwo Naukowe.

Dyczek, P. 2017a. *Terra incognita*: Results of Polish excavations in Albania and Montenegro, *Studia Europaea Gnesnensia* 16: 351-369.

Dyczek, P. 2017b. Rhizon/Risinium de novo in lucem proditus. *New Archaeological Discoveries in the Albanian Regions* 1: 375-392.

Džino, D. 2003. The Influence of Dalmatian shipbuilders on the ancient warships and naval warfare: the *lembos* and *liburnica*. *Diadora* 21: 19-36.

Dzino, D. 2006. Welcome to the Mediterranean semi-periphery: The place of Illyricum in book 7 of Strabo. *Živa antika* 56: 113-128.

Džino, D. 2007. Daorsi i 'Daorsi': o metanarativima i paralelnim narativima prošlosti. *Život. Časopis za književnost i kulturu* 55(3): 63-74.

Dzino, D. 2010. *Illyricum and Roman politics 229 BC - AD 68*. Cambridge: Cambridge University Press.

Dzino, D. 2012. Contesting identities of pre-Roman Illyricum. *Ancient West and East* 11: 69-96.

Dzino, D. 2014a. 'Illyrians' in ancient ethnographic discourse. *Dialogues d'Histoire Ancienne* 40(2): 45-65.

Dzino, D. 2014b. Constructing Illyrians: Prehistoric inhabitants of the Balkan Peninsula in the early modern and modern perceptions. *Balkanistica* 27: 1-39.

Džino, D. 2016. Appian's *Illyrike:* the final stage of the Roman construction of Illyricum. *Istraživanja* 27: 69-83.

Dzino, D. 2017. '*Liburni gens Asiatica*': Anatomy Of Classical Stereotype. *Arheološki radovi i rasprave* 18: 63-77.

Dzino, D., Boršić, L. 2020. Transfer of Technology in the Late Republican Adriatic. The Case Study of the liburnica. In: V.D. Mihajlović, M.A. Janković (eds) *Pervading Empire relationality and diversity in the Roman provinces*: 183-200. Potsdamer Altertumswissenschaftliche Beiträge 73. Stuttgart: Franz Steiner Verlag.

Džino, D., Domić Kunić, A. 2013. *Rimski ratovi u Iliriku. Povijesni antinarativ*. Biblioteka Lucius 9. Zagreb: Školska knjiga.

Dzino, D., Domić Kunić, A. 2018. A view from the frontier-zone: Roman conquest of Illyricum. In: M. Milićević Bradač, D. Demicheli (eds) *A Century of Brave: Roman conquest and indigenous resistance in Illyricum during the time of Augustus and his heirs*: 77-87. Zagreb: FF Press.

Edgar, C.C. 1923. Selected papyri from the archives of Zenon (Nos. 73-76). *Annales du Service des Antiquités de l'Égypte* 23: 73-98.

Edgar, C.C. 1925. *Zenon Papyri Vol. 1. Catalogue Général des Antiquités Égyptiennes du Musée du Caire Nos. 59901-59139*. Cairo: L'Institut français d'archéologie orientale.

Elez, P. 2015. Historical-geographical and geopolitical constants of the Adriatic and Adriatic region in the context of Braudel's vision of the Mediterranean. *Miscellanea Hadriatica et Mediterranea* 2: 85-108.

Ferone, C. 2004. Appiano, 'Illyr.' 3 e la pirateria illirica nel IV sec. a. C. *Hermes* 132: 326-337.

Fiala, F. 1895. Die Ergebnisse der Untersuchung prähistorischer Grablügel auf dem Glasinac im Jahre 1893. *Wissenschaftliche Mittheilungen aus Bosnien und der Herzegovina* 3: 3-38.

Flatman, J. 2012. Places of Special Meaning: Westerdahl's Comet, 'Agency', and the Concept of the 'Maritime Cultural Landscape'. In: B. Ford (ed.) *The Archaeology of Maritime Landscape*: 311-329. New York: Springer.

Fleury, Ph. 2015. La liburne automotrice du *De rebus bellicis*. In: Ph. Fleury, C. Jacquemard, S. Madeleine (eds) *La technologie gréco-romaine: Transmission, restitution et mediation*: 77-96. Caen: Presses Universitaires de Caen.

Forenbaher, S. 2018. *Special Place, Interesting Times. The island of Palagruža and transitional periods in Adriatic prehistory*. Oxford: Archaeopress.

Fraschetti, A. 1983. La Pietas di Cesare e la colonia di Pola. *Annali del Seminario di Studi del Mondo Classico. Archeologia e storia antica* 5: 77-102.

Frey, O.-H. 2011. The World of Situla Art. In: L. Bonfante (ed.) *The Barbarians of Ancient Europe: Realities and Interactions*: 282-312. Cambridge: Cambridge University Press.

Fuscagni, S., Marcaccini C. 2002. *Illiri, hostes communes omnium*: l'immagine di una conquista. In: L. Moscati Castelnuovo (ed.) *Identità e Prassi Storica nel Mediterraneo Greco*: 103-114. Milan: Edizioni ET.

Fuscagni, S., Marcaccini, C. 2004. La pirateria in Adriatico. Riflessioni e divagazioni. In: L. Braccesi (ed.) *La pirateria nell'Adriatico antico*: 139-144. Hesperìa: studi sulla grecità di occidente 19. Rome: 'L'Erma' di Bretschneider.

Gabrielsen, V. 2001. Economic activity, maritime trade and piracy in the Hellenistic Aegean. *Revue des Études Anciennes* 103: 219-240.

Gabrielsen, V. 2003. Piracy and Slave-Trade. In: A. Erskine (ed.) *Companion to the Hellenistic World*: 398-404: Malden MA–Oxford: Blackwell.

Gabrielsen, V. 2013. Warfare, Statehood, and Piracy in the Greek World. In: N. Jaspert, S Kolditz (eds) *Seeraub im Mittelmeerraum. Piraterie, Korsarentum und maritime Gewalt von der Antike bis zur Neuzeit*: 133-153. Mittelmeerstudien 3. Leiden-Boston: Brill.

Gabrovec, S., Mihovilić K. 1987. 'Istarska grupa'. *PJZ*, Vol. 5: 293-338.

Galaty, M.L. 2002. Modelling the Formation and Evolution of an Illyrian Tribal System: Ethnographic and Archaeological Analogs. In: W.A. Parkinson (ed.) *The Archaeology of Tribal Societies*: 109-122: Archaeological Series 15. Ann Arbor MI: International Monographs in Prehistory.

Gardiner, R. (ed.) 1995. *The Age of the Galley*. London: Conway Maritime.

Glogović, D. 1989. *Prilozi poznavanju željeznog doba na Sjevernom Jadranu: Hrvatsko primorje i Kvarnerski otoci*. Zagreb: JAZU: Zavod za arheologiju.

Glogović, D. 2014. *The Fifth Phase of the Iron Age of Liburnia and the Cemetery of the Hillfort of Dragišić*. British Archaeological Reports International Series 2689. Oxford: Archaeopress.

Gluščević, S. 1986a. Zaton kod Zadra. *Arheološki pregled* 27: 131-132.

Gluščević, S. 1986b. Neki oblici staklenog materijala iz antičke luke u Zatonu kod Zadra. *Arheološki vestnik* 37: 255-278.

Gluščević, S. 1987. Vađenje antičkog broda iz Zatona. *Obavijesti Hrvatskog arheološkog društva* 19(3): 43-44.

Gluščević, S. 2002. Hidroarheološko istraživanje i nalaz trećeg liburnskog broda u antičkoj luci u Zatonu kod Zadra. *Obavijesti Hrvatskog arheološkog društva* 34(3): 76-86.

Gluščević, S. 2004. Hydroarchaeological excavation and the discovery of the third 'sewn' Liburnian ship – seriliae – in the Roman port of Zaton near Zadar. *Archaeologia Maritima Mediterranea* 1: 41-52.

Govedarica, B. 2002. Zwischen Hallstatt und Griechenland: die Fürstengräber in den frühen Eisenzeit des Mittelbalkans. *GodCBI* 32/30: 317-328.

Hinshiranan, N. 2001. Kabang: the living boat. *Techniques & Culture: Revue semestrielle d'anthropologie des techniques* 35-36: 1-9.

Höckmann, O. 1997. The Liburnian: some observations and insights. *IJNA* 26(3): 192-216.

Höckmann, O. 2000. Stern rams in antiquity. *IJNA* 29(1): 136-142.

Horden, P., Purcell, N. 2000. *The Corrupting Sea: A Study of Mediterranean History*. Malden MA–Oxford: Blackwell.

Islami, S. 1972. Le monnayage de Skodra, Lissus et Genthios. *Iliria* 2: 379-408.

Ivanoff, J. 1999. *The Moken Boat: Symbolic Technology*. Banglamung: White Lotus Press.

Jaspert, N., Kolditz S. (eds) 2013a. *Seeraub im Mittelmeerraum. Piraterie, Korsarentum und maritime Gewalt von der Antike bis zur Neuzeit*. Mittelmeerstudien 3. Leiden-Boston: Brill.

Jaspert, N., Kolditz, S. 2013b. Seeraub im Mittelmeerraum: Bemerkungen und Perspektiven. In: N. Jaspert, S. Kolditz (eds) *Seeraub im Mittelmeerraum. Piraterie, Korsarentum und maritime Gewalt von der Antike bis zur Neuzeit*: 11-37. Mittelmeerstudien 3. Leiden-Boston: Brill.

Jašarević, A. 2014. Socio-ekonomska i simbolička uloga importovanih metalnih posuda s Glasinca. *GodCBI* 43: 51-99.

Jeličić Radonić, J. 2005. The foundation of the Greek city of Pharos on the island of Hvar. In: M. Šegvić, I. Mirnik (eds) *Illyrica Antiqua: Ob honorem Duje Rendić-Miočević*: 315-328. Zagreb: FF Press.

Jeličić Radonić, J., Katić, M. 2015. *Faros: Osnivanje antičkog grada* 1. Književni krug, Split.

Jurišić, M. 1983. Prilog poznavanju ilirskog brodovlja na Jadranu do 2. st. p.n.e. *Prinosi odjela za arheologiju* 1: 5-16.

Kaster, R.A. 1992. *Studies on the Text of Suetonius, de grammaticis et rhetoribus*. Atlanta GA: Scholars Press.

Katičić, R. 1976. *Ancient Languages of the Balkans*. The Hague–Paris: Mouton.

Katičić, R. 1995. *Illyricum Mythologicum*. Zagreb: Antibarbarus.

Kilian, K. 1973. Zu geschnürten Schienen der Hallstattzeit aus der Ilijak–Nekropole in Bosnien. *Germania* 51: 528-535.

Kirigin, B. 1985. Zapažanja o helenističkoj nekropoli Isse. In: *Sahranjivanje pokojnika sa aspekta ekonomskih i društvenih kretanja u praistoriji i antici*: 91-104. Materijali 20. Belgrade: Savez arheoloških društava Jugoslavije–Arheološko društvo Bosne i Hercegovine.

Kirigin, B. 2004. The Beginning of *Promunturum Diomedis*: Preliminary Pottery report. In: L. Braccesi, M. Luni (eds) *I Greci in Adriatico* 2: 141-150. Hesperia: Studi sulla grecità d'Occidente 18. Rome: 'L'Erma' di Bretschneider.

Kirigin, B. 2006a. The Greek background. In: D. Davison, V. Gaffney, E. Marin (eds) *Dalmatia: Research in the Roman Province 1970-2001. Papers in honour of. J. J. Wilkes*: 17-26. British Archaeological Reports International Series 1576. Oxford: Archaeopress.

Kirigin, B. 2006b. *Pharos the Parian settlement in Dalmatia: a study of a Greek colony in the Adriatic*. British Archaeological Reports International Series 1561. Oxford: Archaeopress.

Kirigin, B. 2009. Ancient Greeks in Croatia. In: J.J. Norwich (ed.) *Croatia: Aspects of art, architecture and cultural heritage*: 20-31. London: Frances Lincoln.

Kirigin, B. 2010. Otok Korčula (Kórkyra hē Mélaina, Crna Korčula). In: J. Poklečki Stošić (ed.) *Antički Grci na tlu Hrvatske*: 113-117. Exhibition catalogue. Zagreb: Galerija Klovićevi dvori.

Kirigin, B. 2018a. Book review: J. Jeličić Radonić and M. Katić, *Pharos - the foundation of the ancient city* (Jeličić Radonić, Katić 2015). *Journal of Greek Archaeology* 3: 477-482.

Kirigin, B. 2018b. Pharos, Greek Amphorae and Wine Production. In: D. Katsonopoulou (ed.) *Paros IV: Paros and its colonies. Proceedings of the Fourth International Conference on the archaeology of Paros and the Cyclades. Paroikia, Paros, 11-14 June, 2015*: 397-420. Athens: The Institute for Archaeology of Paros and the Cyclades.

Kirigin, B., Barbarić, V. 2019. The beginning of Pharos - the present archaeological evidence. *GodCBI* 48: 219-230.

Kirigin, B., Čače, S. 1998. Archaeological Evidence for the Cult of Diomedes in the Adriatic. *Hesperìa: Studi sulla grecità d'Occidente* 9: 63-110.

Kirigin, B., Marin, E. 1985. Issa '80: preliminarni izvještaj sa zaštitnih arheoloških iskopavanja helenističke nekropole Martvilo u Visu: Novi i neobjelodanjeni natpisi iz Visa. *VAHD* 78: 45-72.

Kirigin, B., Katunarić, T., Šešelj, L. 2006. Preliminary notes on some economic and social aspects of amphorae and fine ware pottery from Central Dalmatia, 4th-1st c. BC. In: F. Lenzi (ed.) *Rimini e l'Adriatico nell'età delle guerre puniche. Atti del Convegno Internazionale di Studi Rimini, Musei Comunali, 25-27 marzo 2004*: 191-226. Bologna: Ante Quem.

Kirigin, B., Johnston, A., Vučetić, M., Lušić, Z. 2009. Palagruža – The Island of Diomedes – and notes on ancient Greek navigation in the Adriatic. In: S. Forenbaher (ed.) *A Connecting Sea: Maritime Interaction in Adriatic Prehistory*: 137-155. British Archaeological Reports International Series 2037. Oxford: Archaeopress.

Kleu, M. 2015. *Die Seepolitik Philipps V. von Makedonien*. Kleine Schriftenreihe zur Militär- und Marinegeschichte 24. Bochum: Verlag Dr Dieter Winkler.

Knapp, A.B., Demesticha, S. 2017. *Mediterranean Connections. Maritime Transport Containers and Seaborne Trade in the Bronze and Early Iron Ages*. London: Routledge.

Koncani Uhač, I. 2009. Podvodna arheološka istraživanja u uvali Zambratija. *HistAntiq* 17: 263-268.

Koncani Uhač, I., Uhač, M. 2012. Prapovijesni brod iz uvale Zambratija – prva kampanja istraživanja. *HistAntiq* 21: 533-538.

Koncani Uhač, I., Boetto, G., Uhač, M. (eds) 2017. *Zambratija: Prehistoric sewn boat*. Katalog 85. Pula: Arheološki muzej Istre.

Kozličić, M. 1981. Prikazi brodova na novcu Daorsa. *Glasnik Zemaljskog muzeja u Sarajevu* n.s. 35-36: 163-188.

Kozličić, M. 1983. O problemu japodske prisutnosti u primorju istočnog Jadrana. In: *Arheološka problematika zapadne Bosne*: 109-18. Zbornik Arheološkog društva Bosne i Hercegovine 1. Sarajevo: Zbornik Arheološkog društva Bosne i Hercegovine.

Kozličić, M. 1993. *Croatian Shipping*. Knjiga Mediterana 10. Split-Zagreb: Književni krug–AGM.

Kozličić, M. 2012. Adriatic Sea routes from the Antiquity to the early Modern Age. *HistAntiq* 21: 13-20.

Kozličić, M., Bratanić, M. 2006. Ancient Sailing Routes in Adriatic. In: S. Čače, A Kurilić, F Tassaux (eds) *Les routes de l'Adriatique antique: Géographie et économie*: 107-124. Ausonius éditions 17. Bordeaux-Zadar: Institut Ausonius - Sveučilište u Zadru.

Krahe, H. 1952. Griech. λέμβος, lat. Lembus – eine illyrische Schiffsbezeichnung? *Gymnasium* 59: 79.

Kukoč, S. 1998. Grčki simboli u Ilirskom svijetu. *Opuscula Archaeologica* 22: 7-26.

Kukoč, S. 2009. *Japodi – fragmenta simbolica*. Biblioteka znanstvenih djela 164. Split: Književni krug.

Kukoč, S., Čelhar, M. 2019. Nadin (*Nedinum*): spatial concept of the Liburnian necropolis. *VAHD* 112: 9-31.

Kuntić Makvić, B. 1997. *De bello Histrico*. In: B. Čečuk (ed.) *Arheološka istraživanja u Istri*: 169-175. Izdanja hrvatskog arheološkog društva 18. Zagreb: Hrvatsko arheološko društvo.

Kuntić-Makvić, B., Marohnić, J. 2010. Natpisi. In: J. Poklečki Stošić (ed.) *Antički Grci na tlu Hrvatske*: 73-90. Exhibition catalogue. Zagreb: Galerija Klovićevi dvori.

Kurilić, A. 2008. *Ususret Liburnima: Studije o društvenoj povijesti ranorimske Liburnije*. Zadar: Odjel za povijest sveučilišta u Zadru.

Kurilić, A. 2011. Otok Pag od prapovijesti do kraja antičkog razdoblja. In: V. Skračić (ed.) *Toponimija otoka Paga*: 51-91. Zadar: Sveučilište u Zadru.

Kuzmanović, Z., Vranić, I. 2013. On the reflexive nature of archaeologies of Western Balkan Iron Age. A Case study of Illyrian argument. *Anthropologie* 51: 249-259.

Lewis, D.M. 2019. Piracy and Slave Trading in Action in Classical and Hellenistic Greece. *Mare Nostrum* 10(2): 79-108.

Lieu, S.N.C. 1989. From Caesar to Augustus: A Speech of Thanks to the Emperor Julian given by Claudius Mamertinus (Latin Panegyric XI/3). Introduction. In: S.N.C. Lieu (ed.) *The Emperor Julian Panegyric and Polemic: Claudius Mamertinus, John Chrysostom, Ephrem the Syrian* (2nd edn): 3-12. TTH 2. Liverpool: Liverpool University Press.

Lightfoot, E., Šlaus, M., O'Connell, T.C. 2012. Changing Cultures, Changing Cuisines: Cultural Transitions and Dietary Change in Iron Age, Roman, and Early Medieval Croatia. *American Journal of Physical Anthropology* 148: 543-556.

Lindhagen, A. 2009. The transport amphoras Lamboglia 2 and Dressel 6° revisited. *Journal of Roman Archaeology* 22: 83-108.

Lindhagen, A. 2016. Narona in Dalmatia—the Rise and Fall of a "Gateway Settlement". In: K. Höghammar, B. Alroth & A. Lindhagen (eds) *Ancient Ports The Geography of Connections. Proceedings of an International Conference at the Department of Archaeology and Ancient History, Uppsala University, 23-25 September 2010*: 225-251. BOREAS. Uppsala Studies in Ancient Mediterranean and Near Eastern Civilizations 34. Uppsala: Uppsala Universitet).

Liphschitz, N, Gluščević, S. 2015. The Zaton Boat 2 (Croatia): A Dendochronological Investigation. *Skyllis* 15(2): 158-160.

Litwin, J. (ed.) 2017. *Baltic and Beyond, Change and Continuity in shipbuilding, Proceedings of the 14th International Symposium on Boat and Ship Archaeology, Gdansk (Poland), 21st-25th September 2015*. Gdansk: Polish Maritime Museum.

Ljubić, Š. 1889. *Popis arkeologičkoga odjela Narodnog Zemaljskog muzeja u Zagrebu, Odsjek I*, vol. 1. Zagreb: Tiskarski i litografijski zavod C. Albrechta.

Lo Schiavo, F. 1970. Il gruppo liburnico-japodico, per una definizione nell'ambito della protostoria balcanica. *Atti della Accademia Nazionale dei Lincei* 14 (series 8): 363-524.

Magaš, D. 2013. *Geografija Hrvatske*. Zadar: Sveučilište u Zadru–Meridijani.

Majnarić-Pandžić, N. 1998. Brončano i željezno doba. In: S. Dimitrijević, T. Težak-Gregl, N. Majnarić-Pandžić *Prapovijest*: 159-358. Povijest umjetnosti u Hrvatskoj 1. Zagreb: Naklada Naprijed.

Marijan, B. 1999. Željezno doba na južnojadranskom području (istočna Hercegovina, južna Dalmacija). *VAHD* 93: 7-221.

Marohnić, J. 2010. The birds of Diomedes. *VAPD* 103: 41-61.

Maršić, D. 1997. Problemi izučavanja antičkog Epetija. *Diadora* 18-19: 47-76.

Matijašić, R. 1991. L'Istria tra Epulone e Augusto: archeologia e storia della romanizzazione dell'Istria (II sec.a.G-1 sec.d.C). *Antichità altoadriatiche* 37: 235-251.

Matijašić, R. 1996. *Antička Pula: s okolicom*. Pula: Zavičajna naklada 'Žakan Juri'.

Matijašić, R. 2009. *Povijest hrvatskih zemalja u antici do cara Dioklecijana*. Zagreb: Leykam International.

McCabe, A. 2012. Demosthenes and the Origins of the Maritime Lien. *Journal of Maritime Law and Commerce* 43(4): 581-591

Medas, S. 1997. La navigazione adriatica nella prima età del ferro. In: *Atti del Convegno 'Adriatico. Mare di molte genti, incontro di civiltà'*: 91-133: Cesena: Società di studi romagnoli.

Medas, S. 2004. Λέμβοι e liburnae. In: L. Braccesi (ed.) *La pirateria nell'Adriatico antico*: 129-138. Hesperìa: studi sulla grecità di occidente 19. Rome: 'L'Erma' di Bretschneider.

Medas, S. 2016. La navigazione tardo-arcaica in Adriatico. l'iconografia navale e la peculiarità della tradizione nautica. *Cuadernos de Prehistoria y Arqueología Universidad Autónoma de Madrid* 42: 143-166.

Medas, S. 2019. Le Stelle Daunia di Cattolica: L'icographie navale. In: M. Stoppioni (ed.) *Il signore dell'Adriatico. La stele daunia del Museo di Cattolica*: 49-76. Cattolica: Museo della Regina.

Mihajlović, V.D. 2014. Tracing ethnicity backwards : the case of the 'Central Balkan Tribes'. In: C.N. Popa, S. Stoddart (eds) *Fingerprinting the Iron Age: Approaches to identity in the European Iron Age. Integrating South-Eastern Europe into the debate*: 97-107. Oxford: Oxbow Books.

Mihajlović, V.D. 2019. *Skordisci između antičkih i modernih tumačenja*. Novi Sad: Centar za istorijska istraživanja, Filozofski fakultet.

Mihovilić, K. 1992. Die Situla mit Schiffskampfszene aus Nesactium. *Arheološki vestnik* 43: 67-78.

Mihovilić, K. 1995. Školjić (Funtana) i tragovi prapovijesnih obalnih i otočnih lokaliteta Istre. *Histria archaeologica* 26: 28-57.

Mihovilić, K. 1996. *Nesactium: The Discovery of the Grave Vault in 1981*. Monografije i katalozi 6. Pula: Arheološki muzej Istre.

Mihovilić, K. 2004. La situla del Nesazio con naumachia. In: L. Bracessi (ed.) *La pirateria nell'Adriatico antico*: 93-107. Hesperìa: studi sulla grecità di occidente 19. Rome: 'L'Erma' di Bretschneider.

Mihovilić, K. 2014. *The Histri in Istria: the Iron Age in Istria* (2nd edn). Exhibition catalogue. Monografije i katalozi 23. Pula: Arheološki muzej Istre.

Milićević Bradač, M. 2009. Horacije, Jadran i jugo. In: L. Bekić (ed.) *Jurišićev zbornik: Zbornik radova u znak sjećanja na Marija Jurišića*: 284-290. Zagreb-Zadar: Hrvatski restauratorski zavod–Međunarodni centar za podvodnu arheologiju u Zadru.

Milićević Bradač, M. 2010. Grčka kolonizacija na Sredozemlju. In: J. Poklečki Stošić (ed.) *Antički Grci na tlu Hrvatske*: 41-51. Exhibition catalogue. Zagreb: Galerija Klovićevi dvori.

Milošević, A., Krnčević, Ž. 2017. Two Bribir reliefs with Lewd Contents. In: A. Milošević (ed.) *Colloquium on Bribir II, Bribirska glavica 5.-6. V. 2017*: 33-35. Book of conference abstracts. Split: Muzej Hrvatskih arheoloških spomenika.

Mirošević, L., Graovac Matassi, V. (eds) 2014. *Dalmacija u prostoru i vremenu. Što Dalmacija jest, a što nije, zbornik radova sa znanstvenog skupa održanog 14.-16. lipnja 2012*. Zadar: Sveučilište u Zadru.

Miše, M. 2015. *Gnathia and Related Hellenistic ware on the East Adriatic coast*. Oxford: Archaeopress.

Miše, M. 2017. The Hellenistic ware from the indigenous necropolis at Gradina in Dragišić near Šibenik, Croatia. In: D. Demicheli (ed.) *Illyrica Antiqua II – In honorem Duje Rendić-Miočević, Proceedings of the international conference*: 83-104. Zagreb: FF Press.

Miše, M. 2019. Drinking wine in Liburnia: Hellenistic ware in the indigenous necropolis Dragišić in Dalmatia, Croatia. In: A. Peignard-Giros (ed.) *Daily Life in a Cosmopolitan World: Pottery and culture during the Hellenistic period. Proceedings of the 2nd Conference of IARPotHP, Lyon, November 2015, 5th-8th*: 175-185. Vienna: Phoibos Verlag.

Montebelli, C.R. 2007. *Archeologia navale. Cronaca di un rinvenimento adriatico: Le stele di Novilara*. Pesaro: Museo della marineria Washington Patrignani.

Morrell, K. 2017. *Pompey, Cato and the Governance of the Roman Empire*. Oxford: Oxford University Press.

Morrison, J.S. 1995. Hellenistic Oared Warships 399-31 BC. In: R. Gardiner (ed.) *The Age of the Galley*: 66-77. London: Conway Maritime.

Morrison, J.S. 1996. *Greek and Roman Oared Warships*. Oxbow Monographs 62. Oxford: Oxbow Books.

Morton, J.N. 2017. *Shifting Landscapes, Policies, and Morals: A Topographically Driven Analysis of the Roman Wars in Greece from 200 BC to 168 BC*. Unpublished PhD thesis, University of Pennsylvania, Philadelphia.

Murray, W.M. 2012. *The Age of Titans: The Rise and Fall of the Great Hellenistic Navies*. Oxford: Oxford University Press.

Murray, W.M., Fereiro, L.D., Vardalas, J., Royal, G. 2017. Cutwaters Before Rams: an experimental investigation into the origins and development of the waterline ram. *IJNA* 46(1): 72-82.

Nava, M.L. 2019. La stele daunia rinvenuta a Cattolica nel contesto della produzione della puglia protostorica. In: M. Stoppioni (ed.) *Il signore dell'Adriatico. La stele daunia del Museo di Cattolica*: 27-48. Cattolica: Museo della Regina.

Nawotka, K. 2017. *The Alexander Romance by Ps.-Callisthenes*. Leiden: Brill.

Nikolanci, M. 1958. Iliri, Grci i Rimljani na Jadranu. *Mornarički glasnik* 8: 50-66.

Novak, G. 1962. *Jadransko more u sukobima i borbama kroz stoljeća*. Belgrade: Vojno delo.

Orna-Ornstein, J. 1995. Ships on Roman coins. *Oxford Journal of Archaeology* 14(2): 179-200.

Page, D. 1978. *The Epigrams of Rufinus*. Cambridge Classical Texts and Commentaries 21. Cambridge: Cambridge University Press.

Panciera, S. 1956. Liburna. *Epigraphica* 18: 130-156.

Papazoglu, F. 1969. *Srednjobalkanska plemena u predrimsko doba*. Sarajevo: ANUBiH.

Papazoglu, F. 1988. Les royaumes d'Illyrie et de Dardanie. In: M. Garašanin (ed.) *Les Illyriens et les Albanais, Serie de conférences tenues du 21 mai au 4 juin 1986*: 173-200. Belgrade: SANU.

Parzinger, H. 1991. Archäologisches zur Frage der Illyrier. *Bericht der Römisch-Germanischen Kommission* 72: 205-261.

Peroni, R. 1976. La 'Koine' adriatica e il suo process do formazione. In: M. Suić (ed.) *Jadranska obala u protohistoriji: Kulturni i etnički problemi*: 95-116. Zagreb: Sveučilišna naklada Liber.

Pitassi, M. P. 2011. *The Roman Navy. Ships, Men & Warfare*. Barnsley: Seaworth Publishing.

Pitassi, M. 2016. *Roman Warships*. Woodbridge: The Boydell Press.

Poklečki Stošić, J. (ed.) 2010. *Antički Grci na tlu Hrvatske*. Exhibition catalogue. Zagreb: Galerija Klovićevi dvori.

Pokorny, J. 1959-1969. *Indogermanisches etymologishes Wörterbuch*. Bern: Francke.

Polomé, E. 1982. Balkan Languages (Illyrian, Thracian and Daco-Moesian). In: J. Boardman, I. Edwards, N. Hammond, E. Sollberger (eds) *The Cambridge Ancient History* 3.1: 866-898. Cambridge: Cambridge University Press.

Polzer, M.E. 2010. The 16th-Century B.C. Shipwreck at Pabuç Burnu, Turkey. Evidence for Transition from Lacing to Mortise-and-Tenon Joinery in Late Archaic Greek Shipbuilding. In: P. Pomey (ed.) *Transferts technologiques en architecture navale méditerranéenne de l'Antiquité aux temps modernes: identité technique et identité culturelle. Actes de la Table Ronde d'Istanbul 19-22 mai 2007*: 27-44.Varia Anatolica 20. Paris: De Boccard.

Pomey, P. (ed.) 2010a. *Transferts technologiques en architecture navale méditerranéenne de l'Antiquité aux temps modernes: identité technique et identité culturelle. Actes de la Table Ronde d'Istanbul 19-22 mai 2007*. Varia Anatolica 20. Paris: De Boccard.

Pomey, P. 2010b. Introduction. In: P. Pomey (ed.) *Transferts technologiques en architecture navale méditerranéenne de l'Antiquité aux temps modernes: identité technique et identité culturelle. Actes de la Table Ronde d'Istanbul 19-22 mai 2007*: 15-26. Varia Anatolica 20. Paris: De Boccard.

Pomey, P., Kahanov, Y., Rieth, E. 2012. Transition from Shell to Skeleton in Ancient Mediterranean Ship-Construction: analysis, problems, and future research. *IJNA* 41(2): 235-314.

Pomey, P., Boetto, G. 2019. Ancient Mediterranean Sewn-Boat Tradition. *IJNA* 48(1): 5-51.

Poparić, B. 1899. *O pomorskoj sili Hrvata u dobe narodnih vladara*. Zagreb: Matica Hrvatska.

Poulain, P.-M., Raicich, F. 2001. Forcings. In: B. Cushman-Roisin, M. Gacic, P.-M. Poulain, A. Artegiani (eds) *Physical Oceanography of the Adriatic Sea: Past, Present and Future*: 45-65. New York: Springer Science.

Prendi, F. 1975. Un aperçu sur la civilisation de la première period du Fer en Albanie. *Iliria* 3: 109-138.

Prusac Lindhagen, M. 2009. Illyriske pirater – mellom myter, minner og fortolkninger. *Viking: Norsk arkeologisk årbok* 72: 73-90.

Pryor, J.H., Jeffreys, E.J. 2006. *The Age of the ΔΡΟΜΩΝ: The Byzantine Navy ca 500-1204*. The Medieval Mediterranean: Peoples, economies and cultures, 400-1500 26. Leiden–Boston: Brill

Pungetti, G. 2012. Islands, culture, landscape and seascape. *Journal of Marine and Island Cultures* 1: 51-54.

Raban, A. 1984. The Thera Ships: Another Interpretation. *American Journal of Archaeology* 88: 11-19.

Radić, D., Borzić, I. 2017. The island of Korčula: Illyrians and Greeks. *VAHD* 110(1): 303-325.

Radić Rossi, I. 2007. Caska – podmorje. *HAG* 4: 371-373.

Radić Rossi, I. 2009. Caska – podmorje. *HAG* 6: 467-469.

Radić Rossi, I. 2010a. Plovidba Jadranom u grčko doba. In: J. Poklečki Stošić (ed.) *Antički Grci na tlu Hrvatske*: 91-101. Exhibition catalogue. Zagreb: Galerija Klovićevi dvori.

Radić Rossi, I. 2010b. Caska – podmorje. *HAG* 7: 483-487.

Radić Rossi, I. 2011. *Problematika prapovijesnih i antičkih nalazišta u hrvatskom podmorju*. Unpublished PhD thesis, University of Zadar, Zadar.

Radić Rossi, I., Boetto, G. 2010. Arheologija broda i plovidbe: šivani brod u uvali Caski na Pagu – Istraživačka kampanja 2009. *HistAntiq* 19: 299-304.

Radić Rossi, I., Boetto, G. 2011. Šivani brod u uvali Caska na Pagu – Istraživačka kampanja 2010. *HistAntiq* 20: 505-513.

Radić Rossi, I., Boetto, G. (forthcoming). Ancient ships from Cissa (Caska, Island of Pag, Croatia) in their cultural and historical context. *IJNA*.

Radić Rossi, I., Brusić, Z. 2014. Tisućljetno pomorstvo Nina na razmeđi Liburnije i Dalmacije. In: L. Mirošević, V. Graovac Matassi (eds) *Dalmacija u prostoru i vremenu. Što Dalmacija jest, a što nije, zbornik radova sa znanstvenog skupa održanog 14.-16. lipnja 2012*: 21-33. Zadar: Sveučilište u Zadru.

Reddé, M. 1986. *Mare Nostrum. Les infrastructures, le dispositif et l'histoire de la marine militaire sous l'empire romain*. Bibliothèque des Écoles françaises d'Athènes et de Rome 260. Rome: Écoles françaises d'Athènes et de Rome.

Rendić Miočević, D. 1989. *Iliri i antički svijet. Iriloške studije: povijest – arheologija – numizmatika – onomastika*. Biblioteka znanstvenih djela 33. Split: Književni krug.

Royal, J. 2012. The Illyrian Coastal Exploration Program, first interim report (2007-9): the Roman and Late-Roman finds. *American Journal of Archaeology* 116(3): 405-460.

Royal, J. 2013. Erforschung der Antike in den Meeren des alten Illyrien. In: M. Reinfeld (ed.) *Archäologie im Mittelmeer. Auf der suche nach versunkenen Schiffswracks und vergessenen Häfen*: 90-98. Darmstadt: Verlag Philipp von Zabern.

Royal, J. 2015. Maritime Evidence for Overseas Trade along the Illyrian Coast: the Eastern Mediterranean Connections. In: S. Demesticha (ed.) *Per Terram, Per Mare: Seaborne Trade and the Distribution of Roman Amphorae in the Mediterranean, Proceedings of the Per Terram Per Mare Conference, 11-15 April 2013, Nicosia, Cyprus*: 199-218. Oskarshamn: Åströms förlag.

Sampson, G. 2016. *Rome Spreads its Wings: Territorial Expansion between the Punic Wars*. Barnsley: Pen and Sword.

Schiffer, M.B. 2009. Expanding Ethnoarchaeology: Historical Evidence and Model-Building in the Study of Technological Change. In: J.P. Oleson (ed.) *The Oxford Handbook of Engineering and Technology in the Classical World*: 821-835. Oxford: Oxford University Press.

Schiffer, M.B. 2011. *Studying Technological Change: A Behavioural Approach*. Utah NA: University of Utah Press.

Scholten, J. 2000. *The Politics of Plunder: Aitolians and their Koinon in the Early Hellenistic Era, 279-217 B.C*. Berkeley CA: University of California Press.

Siewert, P. 2004. Politische Organisationsformen im vorrömischen Südillyrien. In: G. Urso (ed.) *Dall'Adriatico al Danubio: L'Illirico nell'età greca e romana*: 53-62. I Convegni della Fondazione Niccolò Canussio 3. Pisa: Edizioni ETS.

Sijpesteijn, P.J. 1996. A Labour Contract to Build a Boat. *Zeitschrift für Papyrologie und Epigraphik* 111: 159-162.

Skibo, J.M., Schiffer, M.B. 2009. *People and Things: A Behavioural Approach to material Culture*. New York: Springer.

Skok, P. 1971. *Etimologijski rječnik hrvatskoga ili srpskoga jezika*. Zagreb: JAZU.

Srdoč, D., Obelić, B., Horvatinić, N., Krajcar, I. 1984. Rudjer Bošković Institute Radiocarbon measurements VIII. *Radiocarbon* 26(3): 449-460.

Starac, A. 1999. *Rimsko vladanje u Histriji i Liburniji – Društveno i pravno uređenje prema literarnoj, natpisnoj i arheološkoj građi I: Istra*. Monografije i katalozi 10/1. Pula: Arheološki muzej Istre.

Stipčević, A. 1973. Jesu li ilirski brodovi imali pulene u obliku zmija. *Radovi Instituta JAZU u Zadru* 20: 413-418.

Stoppioni, M. (ed.) 2019. *Il signore dell'Adriatico. La stele daunia del Museo di Cattolica*. Cattolica: Museo della Regina.

Stylianou, J. 1998. *A Historical Commentary on Diodorus Siculus book 15*. Oxford: Oxford University Press.

Suić, M. 1975. Lukanov Jader (IV, 405) – rijeka Jadro ili grad Zadar. *Diadora* 8: 5-27.

Šašel Kos, M. 2005. *Appian and Illyricum*. Situla 43. Ljubljana: Narodni muzej.

Šegvić, M., Mirnik, I. (eds) 2005. *Illyrica Antiqua: Ob honorem Duje Rendić-Miočević*. Zagreb: FF Press.

Tarn, W.W. 1931. The Battle at Actium. *Journal of Roman Studies* 21: 173-199.

Tiboni, F. 2009. The Ships on the Novilara Stele, Italy: Questions of Interpretation and Dating. *IJNA* 38(2): 400-423.

Tiboni, F. 2017. Dalla pisside della Pania alla situla di Nesazio. Il tema navale nell'arte delle situle. *Archaeologia Maritima Mediterranea* 14: 79-94.

Tiboni, F. 2018. Per una rilettura delle navi di Glasinac. *Archaeologia Maritima Mediterranea* 15: 57-68.

Tilley, A. 2007. Rowing Ancient Warships: Evidence from a Newly Published Ship-Model. *IJNA* 36(2): 293-299.

Tonc, A. 2012. Silver pendants with Anthropomorphic Representations on the Territory of the Eastern Adriatic Protohistoric Societies. In: H. Meller, R. Maraszek (eds) *Masken der Vorzeit in Europa (II). Internationale Tagung vom 19. bis 21. November 2010 in Halle (Saale)*: 63-70. Tagungen des Landesmuseums für Vorgeschichte Halle 7. Halle: Landesamt für Denkmalpflege und Archäologie, Sachsen-Anhalt.

Torr, C. 1895. *Ancient Ships*. Cambridge: Cambridge University Press.

Tröster, M. 2009. Roman Hegemony and Non-State Violence: A Fresh Look at Pompey's Campaign against the Pirates. *Greece & Rome* 56(1): 14-33.

Ugarković, M. 2019. *Geometrija smrti: isejski pogrebni obredi, identiteti i kulturna interakcija. Antička nekropola na Vlaškoj njivi, na otoku Visu*. 2 Vols. Katalozi i monografije 6. Split–Zagreb: Arheološki muzej u Splitu–Institut za arheologiju u Zagrebu.

Vasić, R. 2003. To the North of Trebenishte. In: C.M. Stibbe (ed.) *Trebenishte: the fortunes of an unusual excavation*: 111-133. Studia Archaeologica 121. Rome: 'L'Erma' di Bretschneider.

Vasić, R. 2010. Белешке о Гласинцу - хронологија кнежевских гробова. *Starinar* 59: 109-117.

Vrsalović, D. 2011 [1978]. *Arheološka istraživanja u podmorju istočnog Jadrana: prilog poznavanju trgovačkih plovnih putova i privrednih prilika na Jadranu u antici*. Split: Književni krug–Arheološki muzej u Splitu.

Wallace, Sh. L. 1969. *Taxation in Egypt: from Augustus to Diocletian*. New York: Greenwood Press.

Watson, L.C. 2003. *A commentary on Horace's Epodes*. Oxford: Oxford University Press.

Wendt, C. 2016. Piraterie als definitorisches Moment von Seeherrschaft. In: E. Baltrusch, H. Kopp, C. Wendt (eds) *Seemacht, Seeherrschaft und die Antike*: 79-92. Historia Einzelschriften 244. Stuttgart: Franz Steiner Verlag.

Westerdahl, C. 2012. The Binary Relationship of Sea and Land. In: B. Ford (ed.) *The Archaeology of Maritime Landscape*: 291-310. New York: Springer.

Wilkes, J.J. 1969. *Dalmatia*. London: Routledge.

Wilkes, J.J. 1992. *The Illyrians*. Cambridge–Oxford: Blackwell.

Zaninović, M. 1976. Liburna. In: *Modeli naših brodova*: 158-161. Split: Čuvar Jadrana.

Zaninović, M. 1988. Liburnia militaris. *Opuscula Archaeologica* 13: 43-67.

Zaninović, M. 2004. Antički Grci na hrvatskoj obali. *Arheološki radovi i rasprave* 14: 1-57.

Zaro, G, Čelhar, M. 2018. Landscape as legacy in northern Dalmatia. In: L. Mirošević, G. Zaro, M. Katić, D. Birt (eds) *Landscape in Southeastern Europe*: 49-68. Studies on South East Europe 21. Vienna: Lit Verlag.